12.50

ADVANCES IN
MASS
SPECTROMETRY
Volume 16

ADVANCES IN

MASS

SPECTROMETRY

Volume 16

Plenary and Keynote Lectures of the 16th International Mass Spectrometry Conference held in Edinburgh, UK 31 August–5 September 2003

EDITED BY
ALISON E. ASHCROFT
The School of Biochemistry and Molecular Biology
University of Leeds
Leeds LS2 9JT, UK

GARETH BRENTON
Chemistry Department
University of Wales Swansea
Singleton Park, Swansea SA2 8PP, UK

JOHN J. MONAGHAN
School of Chemistry
University of Edinburgh
Edinburgh EH9 3JJ, UK

ELSEVIER
2004
Amsterdam – Boston – Heidelberg – London – New York – Oxford
Paris – San Diego –San Francisco – Singapore – Sydney –Tokyo

ELSEVIER B.V.
Sara Burgerhartstraat 25
P.O. Box 211, 1000 AE Amsterdam
The Netherlands

ELSEVIER Inc.
525 B Street, Suite 1900
San Diego, CA 92101-4495
USA

ELSEVIER Ltd
The Boulevard, Langford Lane
Kidlington, Oxford OX5 1GB
UK

ELSEVIER Ltd
84 Theobalds Road
London WC1X 8RR
UK

First edition 2004

Library of Congress Cataloging in Publication Data
A catalog record is available from the Library of Congress.

British Library Cataloguing in Publication Data
A catalogue record is available from the British Library.

ISBN: 0-444-51528-3

⊗ The paper used in this publication meets the requirements of ANSI/NISO Z39.48-1992 (Permanence of Paper).

Printed in the Netherlands

Contents

Preface

The 16th International Mass Spectrometry Conference was held in Edinburgh, UK, 31 August–5 September 2003. The Conference had been held 30 years previously in Edinburgh, in 1973. Much had changed over these 30 years in mass spectrometry and in the city of Edinburgh. Edinburgh now had a custom-built Conference Centre that the IMSC filled to bursting point. Mass spectrometry had moved from the chemical to the biological application area, and even into space. What had not changed in the IMSC since 1973 were the continuing themes of excellent science, a strong mass spectrometry community and a good social environment.

The aim of the Organising Committee was to organise a Conference that retained the best of the fundamentals of mass spectrometry and instrumentation, but that pointed the way, through Plenary Lectures, Scientific Sessions (oral and poster), Workshops and Seminars, and the Exhibition, to the future applications of the technique.

Over 1400 scientists, from 43 countries, attended the Conference and presented more than 900 papers.

The Proceedings of the 16th International Mass Spectrometry Conference include abstracts for all the accepted papers on the CD-ROM enclosed with this book. Full papers for inclusion in the book were requested from all the Plenary and Keynote Lecturers, and from the winner of the Curt Brunnée Award.

We would like to thank all the authors for their contributions and all those who worked so hard to make the 16th International Mass Spectrometry Conference a success.

A.E. Ashcroft,
A.G. Brenton
and
J.J. Monaghan

The Editors

Opening of the 16th International Mass Spectrometry Conference

by

Nico M.M. Nibbering
President of the International Mass Spectrometry Society

Good morning, Ladies and Gentlemen,

It is a great pleasure for me to welcome you all at this 16th International Mass Spectrometry Conference, which is the first conference being fully organized under the rules of the International Mass Spectrometry Society since the ratification of its Constitution at the previous International Mass Spectrometry Conference in Barcelona three years ago.

I would like to take the opportunity to thank first the Society's Vice-President for the Conference, John Monaghan, and his National Committee for the enormous amount of work in organizing this triennial meeting.

Similarly I would like to thank the members of the Society's Executive Committee who have assisted the last three years in running the Society in a variety of aspects and who have chaired subcommittees for the Awards Program.

That is, Emilio Gelpi as Vice-President of the Society who chaired the Thomson Medal Awards Committee, John Traeger as Secretary, Jan Vink as Treasurer, Yuzo Nakagawa who chaired the Curt Brunnee Award Committee, Marcos Eberlin who chaired the Journal of Mass Spectrometry Awards Committee, and Alison Ashcroft for taking care of having the Thomson Medals made and Graham Cooks as Past President of the Society for his advice when requested.

Then I would like to acknowledge the companies Thermo Finnigan for sponsoring the Curt Brunnee Award and Wiley for sponsoring the Journal of Mass Spectrometry Awards.

Yesterday, the Governing Committee of the Society, representing 34 different nations, has elected in its meeting a new Executive Committee and the next President of the Society will be John Monaghan of Scotland, UK, the Vice-President

John Traeger of Australia, the Secretary Alison Ashcroft of the UK, and the Treasurer Jan Vink of The Netherlands, while Jurgen Grotemeyer of Germany, Itsuo Katakuse of Japan, and Marcos Eberlin of Brazil will serve as representatives of regions A, B, and C of the Society, respectively.

It was already decided in Barcelona that the 17th International Mass Spectrometry Conference will take place in 2006 in Prague, Czech Republic, for which Zdenek Herman has been elected by the Governing Committee as the Society's Vice-President of the Conference, while Vladimír Havlíček will serve as Vice-Chairman.

Yesterday the Governing Committee chose also the venue of the 18th International Mass Spectrometry Conference, which is Bremen, Germany under Jurgen Grotemeyer.

I would like to remind you that it is exactly 30 years ago that the 6th International Mass Spectrometry Conference took place here in Edinburgh where I have seen yesterday already that many veterans in the field have come back. Appreciate also that in Cumbria in the North-West of England, near to the Scottish border, John Dalton, well-known to us as mass spectrometrists, grew up and developed almost 200 years to the day of this Conference the first part of his theory of atoms. To mark this anniversary a small John Dalton Exhibition has been arranged here at the Conference Centre.

I wish you all a scientifically very rewarding and socially a very enjoyable conference here in Edinburgh, the capital of Scotland, but this week also the Capital of Mass Spectrometry!

Welcome to the 16th IMSC in Edinburgh, UK from the Chairman of the Organising Committee, Professor John Monaghan

On behalf of the National Organising Committee, I would like to welcome all the delegates and accompanying persons to the Festival City of Edinburgh for the16th International Mass Spectrometry Conference. The Conference has been organised by the British Mass Spectrometry Society (BMSS) under the auspices of the International MS Society (IMSS).

It is 30 years since the Conference was last held in Edinburgh. Much has changed over these 30 years in mass spectrometry and in the city of Edinburgh. Edinburgh has a custom-built Conference Centre, which we will fill to bursting point. Mass spectrometry has moved from the chemical to the biological application area, and even into space. I am sure, however, that those of you who were in Edinburgh in 1973 will recognise the continuing themes of excellent science, a strong MS community, and a good social environment. Our aim has been to organise a Conference that retains the best of the fundamentals of mass spectrometry but that points the way, through Plenary Lectures, Scientific Sessions, Workshops, and the Exhibition, to the future applications of the technique. We hope also that you will find the Social events to be held in two of the city's finest buildings—old and new—enjoyable.

Since we last met at the magnificent Barcelona Conference in 2000, the obvious highlight for mass spectrometry has been the sharing of the 2002 Nobel Prize for Chemistry by two of our community, John Fenn and Koichi Tanaka. It is a particular pleasure to welcome both of them to speak in Edinburgh. It is, however, a matter of regret that John Beynon, the doyen of British mass spectrometry and a former colleague of mine at ICI Blackley, is too ill to attend.

This Conference could not have taken place without the support, financial and otherwise, of many organisations and individuals. The Executive Committee of the IMSS, particularly Nico Nibbering and Emilio Gelpi, have provided assistance when requested. The BMSS has provided us with the financial backing required to undertake an event of this magnitude. Without the contribution made by Exhibitors and other financial sponsors and supporters no Conference of this size could run successfully. Scientists from throughout the world have helped to arrange Scientific Sessions and I thank them for their assistance. I would finally like to thank the members of the Organising Committee. In particular, I would highlight the work carried out by those Committee members who took responsibility for specific tasks. These members have done more than seems humanly possible and no words of mine can recognise their contribution.

Finally, please enjoy the Conference but also take time to enjoy all that the City of Edinburgh and the surrounding parts of Scotland have to offer.

John J. Monaghan

Chairman 16th International
Mass Spectrometry Conference
Past-Chairman British Mass Spectrometry Society

The Organizing Committee of the 16th IMSC

Prof. J.J. Monaghan (Chair)
Dr A.E. Ashcroft (Vice-Chair)
A.K. Upton (Secretary)
Dr J. Oxford (Treasurer)

Committee Members

Prof. A.G. Brenton
Dr C. Eckers
Dr H. Evans
Prof. S.J. Gaskell
Dr N. Haskins
Dr G.J. Langley
Dr M.A. McDowall
Dr F. Mellon
Prof. F.S. Pullen
Prof. J.H. Scrivens

Sponsors and Supporters of the 16th IMSC

Major Sponsors

ABRF
Applied Biosystems
PerkinElmer Life and Analytical Sciences
Waters Micromass MS Technologies

Sponsors

Bruker Daltonics
Jeol (UK) Ltd
Quintiles Ltd
Shimadzu Biotech
ThermoElectron Corporation
Varian Inc

Other Sponsors and Supporters

AstraZeneca
British Mass Spectrometry Society
Genomic Solutions Inc
GlaxoSmithKline
Lothian and Edinburgh Enterprise
Pfizer
Syngenta
15th IMSC

John Fenn
Virginia, Commonwealth University

Koichi Tanaka
Shimadzu, Corporation

The Bicentenary of the Atomic Theory
John Dalton (1766–1844)

John Dalton first described the Atomic Theory in a lecture to the Manchester Literary and Philosophical Society on October 21 1803. In 1830, in a paper read to the Society, he said, "A series of Essays read before this Society and afterwards published in the 5th Vol. of their Memoirs gradually led me to the consideration of ultimate particles or atoms and their combinations. Under the date of Sept 3d, 1803, I find in my notebook 'Observations on the ultimate particles of bodies and their combinations,' in which the atomic symbols I still use [were] introduced."

It was fitting therefore that the 16th IMSC should have marked the bicentenary of this historic event in chemical history. We were pleased to host a small Exhibition of Dalton memorabilia, assembled as part of the major Dalton bicentenary celebrations in Manchester in October 2003. Dr Ron Oliver, formerly of the University of Salford, arranged this Exhibition, and we were grateful to him for agreeing to put it on display in Edinburgh.

Opening of the Thomson Medal Awards Ceremony
16th IMSC Edinburgh 2003

by

Emilio Gelpi
Chairman of the Thomson Medal Awards Committee 2003

11 nominees were presented by various National Representatives for the Thomson Medal Award which I had the privilege to chair. There are 33 member states represented now in the Governing Committee of the IMSS and a total of 24 cast their votes in the first round. As a result of this first round of votes, in compliance with the established rules for these awards, the top nine candidates were selected for the second and final voting round.

The second round produced also a total of 24 votes, with the peculiarity that two countries that had cast their votes in the first round did not vote in the second but there were two more countries that voted for the first time. Thus, the total was again 24 votes counted or 73% of the eligible member states.

After vote counting I informed our IMSS President that the winners in alphabetical order had been Richard Caprioli, Franz Hillenkamp, and Victor Talrose. These three distinguished scientists collected the highest number of points that ranged for all of the nominees from a high of 71 points to a low of 17. Now it is time to express my personal gratitude to all that contributed to these Awards by either submitting nominations or casting their votes after having to put up with my various and annoying reminders. And last but not least my warmest congratulations to the winners who will now receive their Awards from our President. All of them are worthy recipients of such an honour!

Presentation of the Thomson Medals

by

Nico M.M. Nibbering
President of the International Mass Spectrometry Society

Richard,

You have been one of the people most responsible for the revolution of mass spectrometry in biochemistry over the past twenty years. As a biochemist by training you have developed new technologies and applied them to the study of the structure, metabolism, and function of peptides and proteins. Your work has been particularly important in providing a basis for the on-line capabilities of MS so much in use today, such as:

1. Continuous-flow (dynamic) FAB that allowed the power of liquid chromatography, capillary electrophoresis, and flow-injection analysis to be employed for the analysis of complex biological mixtures.

2. Microdialysis/mass spectrometry as an on-line capillary-based flow rate technology that enabled in vivo measurements at extraordinary high sensitivity and drug pharmacokinetic measurements in live animals.

3. Micro-electrospray ionization with which ultra-high sensitivity was achieved in the very low attomole range to the mid to high zeptomole range.

4. Imaging mass spectrometry which you developed for the analysis of the distribution of peptides and proteins in tissue sections as well as other biological samples over the past few years and which presently is applied in colon, breast, and brain cancer research.

It is, therefore, with very great pleasure that I can present to you the Thomson Medal Award in recognition of your scientific achievements in and service to International Mass Spectrometry for which many, many thanks!

Franz,

Your scientific career is almost entirely connected with the combination of lasers and mass spectrometry. Trained in electronic engineering you did radiation and environmental research for your Ph.D. After that, and that was in the mid-sixties of the last century, you had already published a paper about the application of lasers in medicine and biology. Then following a lectureship in Physics you became Professor of Biophysics at the JW Goethe University in Frankfurt and around that time you developed together with the late Professor Raimund Kaufmann the laser microprobe mass analyzer which connected a microscopically focused short-pulse laser beam and a time-of-flight mass spectrometer. The aim was to determine spatially metal cations in biological tissues and this took you in the field of mass spectrometry.

Your work in the mid- to late-seventies of the last century to analyze non-volatile and thermally labile (bio)molecules attracted the attention of the group of the late Professor Hans Beckey and Professor Franz Rollgen of the University of Bonn with whom my group and I had at that time a strong cooperation, in particular, in the development of negative ion field desorption mass spectrometry. In the mid-eighties of the last century you took up the position of Professor for Medical Physics and Biophysics at the University in Muenster and it was there where you together with Michael Karas developed MALDI mass spectrometry now being applied worldwide in laboratories for research in the biological and medical fields and which has given a completely new dimension to mass spectrometry.

It is, therefore, with very great pleasure that I can present to you the Thomson Medal Award in recognition of your scientific achievements in and service to International Mass Spectrometry for which many, many thanks!

Victor,

You have been active now for more than fifty years in mass spectrometry.

You started your scientific career in 1946 in the Institute of Chemical Physics of the Soviet Academy of Sciences and organized forty years later your own Institute of Energy Problems of Chemical Physics.

In 1952 you had already developed an ion trap and studied ion/molecule reactions of methane where you discovered the protonated methane species and named it the methonium ion which afterwards has formed the basis for the method of chemical ionization mass spectrometry. You also found out at that time that exothermic ion/molecule reactions occur without activation and if endothermic they do not occur and you invented the bracketing method of proton affinity determination which afterwards has been applied worldwide in basic studies of gas-phase ion/molecule reactions and nowadays is used for proton affinity measurements of matrices applied in MALDI mass spectrometry. You have built many mass spectrometers and one of them should be mentioned, that is a beam instrument made in the sixties of the last century with which it was possible to lead neutral free radicals through a hexapole magnetic lens and to study their elementary chemical reactions. You measured more than forty rate constants of such reactions that were of paramount importance for atmosphere chemistry.

You were also one of the first who developed the coupling of GC and LC with MS and in the GC/MS you continuously monitored the GC output as several MS traces for the magnet type MS, later called mass fragmentography. For the last seven years you have been active in biological mass spectrometry where you apply your experience and knowledge in chemical kinetics and physics. Together with the Uppsala group you have applied high explosives as matrices in plasma desorption mass spectrometry to increase the yield of multiply charged ions and with success. At present you are working in the Mass Spectrometry Facility headed by Professor Al Burlingame at the University of California in San Francisco, to understand better the physicochemical mechanism of laser desorption of biomolecules.

Victor, you are a great scientist and, therefore, it is an extremely great pleasure for me to present to you the Thomson Medal Award as recognition of your important scientific achievements and your significant contributions to International Mass Spectrometry for which many, many thanks!

Opening of the Curt Brunnée Award Ceremony 16th IMSC Edinburgh 2003

by

Yuzo Nakagawa
Chairman of the Curt Brunnée Award Committee 2003

Ladies and Gentlemen,
I have the great pleasure to welcome you and open the Curt Brunnée Award Ceremony.

This Award is named after the former Director of Research and Development of the Finnigan Corporation, which now is named Thermo Finnigan Co., and sponsors the Award. This Award shall be given for the fourth time to an individual under the age of 45 years old who has made outstanding contributions to the development of instrumentation for mass spectrometry.

Previous winners of the Curt Brunnée Award are Professors G. Brenton, M. Guilhaus, and Scott A. McLuckey.

For the search of a mass spectrometrist as described, an international selection committee was appointed by the International Mass Spectrometry Society (IMSS) in close cooperation with the National Organizing Committee of this IMSC.

The selection committee consisted of the designee of the President of the IMSS, the chair of the IMSC Edinburgh, three representatives from each of the geographical regions of the IMSS nominated by three regional representatives on the Executive committee of the IMSS, and the Curt Brunnée Award Committee chair. The selection committee was chaired by Dr Yuzo Nakagawa, Japan.

The nominations of two candidates were received before November 30, 2002, and each nomination document was very carefully read and considered by 12 committee members. We had a very hard time to choose one, because both nominees are excellent scientists. After voting, we decided on one winner—Dr Michisato Toyoda.

Dr Michisato TOYODA, 31 years old, Assistant Professor at the Department of Physics, Graduate School of Science, Osaka University. He graduated at Osaka University in 1994, received his Doctor of Philosophy under the direction of the late Prof. T. Matsuo and Prof. I. Katakuse in 2000. Since 1998 he had been working in the group of Prof. I. Katakuse and received a Research Award from the Mass Spectrometry Society of Japan in 2002. 10 days ago he joined the group of the Prof. Peter J. Derrick, Univ. of Warwick as a Research Associate.

Dr Toyoda has greatly contributed to the progress in the instrumentation of mass spectrometry, especially in the development of a multi turn type mass spectrometer.

His outstanding contribution is introduced by himself. Now I would like to say to him in Japanese "GOJUSHOH OMEDETOH GOZAIMASU," which means: Congratulations!

Report by the Retiring President of the International Mass Spectrometry Society

Introduction

After a deliberate process of preparation over 6 years, the International Mass Spectrometry Society (IMSS) was established by ratification of its Constitution by the representatives of 32 countries, acting as the Governing Committee of the IMSS, at the 15th International Mass Spectrometry Conference (IMSC), held in Barcelona in 2000.

However, in this report the History, Organization, Aims, and Membership of the IMSS will not be repeated, as they have been described extensively in the Journal of Mass Spectrometry **36** (2001) 119–123 and in the Proceedings of the 15th IMSC, which have been published as Advances in Mass Spectrometry by Wiley in 2001. They can further be found at the IMSS web site: http://www.imss.nl.

New Institutional Members

Serbia and Montenegro has become an Institutional Member of the IMSS.
Ireland has been invited to become an Institutional Member of the IMSS.

Conference Committee Rules and Regulations

The Treasurer of the IMSS, Jan Vink, and the undersigned have set up "Conference Committee Rules and Regulations" which subsequently have been discussed at length with John Monaghan, the Vice-President of the IMSS for the 16th IMSC and after some modifications have been established as a contract between the IMSS and the 16th IMSC. They concern the composition and appointment of the committee, which is organizing an IMSC, its tasks, powers, and finances and the 50/50 sharing of any credit balance between the IMSC and IMSS. The Conference and Committee Rules and Regulations have been endorsed by the Governing Committee of the IMSS in its meeting on Sunday 31 August, 2003 in Edinburgh, Scotland and thus can be used for future IMSCs

Thanks

Thanks go to:

- The **Organizing Committee of the 16th IMSC** including members of the **British Mass Spectrometry Society** [Prof. J.J. Monaghan (Chair), Dr A.E. Ashcroft (Vice-Chair), A.K. Upton (Secretary), Dr J. Oxford (Treasurer), Committee Members: Prof. A.G. Brenton, Dr C. Eckers, Dr H. Evans, Prof. S.J. Gaskell, Dr N. Haskins, Dr G.J. Langley, Dr M.A. McDowall, Dr F. Mellon, Prof. F.S. Pullen, Prof. J.H. Scrivens];
- The **Officers of the Society** for the period 2000–2003, E. Gelpi, Vice-President (Society) and Chair Thomson Medal Award Committee; J.J. Monaghan, Vice-President (Conference) and Chair 16th IMSC; J. Vink, Treasurer; J. Traeger, Secretary; A.E. Ashcroft, Representative, Region A; Y. Nakagawa, Representative, Region B and Chair Curt Brunnée Award Committee; M.N. Eberlin, Representative, Region C and Chair JMS Awards Committee;
- The Past-President of the Society, R.G. Cooks;
- **Award Sponsors:** ThermoFinnigan for Sponsoring the Curt Brunnée Award; John Wiley & Sons for Sponsoring the Journal of Mass Spectrometry Award Symposium;
- The IMSS web masters Peter Jacobs of NV Organon, Oss, The Netherlands until the beginning of July 2002 and from then on Dr Fred Mellon of the Institute of Food Research, Norwich, UK.

New Executive Committee of the IMSS

The Governing Committee of the IMSS in its meeting on Sunday 31 August, 2003 in Edinburgh, Scotland has elected the following Officers of the Society for the period 2003–2006:

J.J. Monaghan, President; J. Traeger, Vice-President (Society); Z. Herman, Vice-President (Conference); J. Vink, Treasurer; A.E. Ashcroft, Secretary; J. Grotemeyer, Representative, Region A; I. Katakuse, Representative, Region B; M.N. Eberlin, Representative, Region C.

17th IMSC in Prague, Czech Republic in 2006

Professor Zdeněk Herman: Chairman of the Conference (Vice-President of the IMSS for the 17th IMSC).

Dr Vladimír Havlíček: Treasurer and Vice-Chairman of the Conference Petr Halada: Secretary.

Committee members: Josef Čáslavský, Michal Holčapek, Jan Hrušák, Josef Chmelík, Karel Lemr, Lenka Kolářová, Marek Kuzma, Petr Man, Petr Novák, Petr Pompach, Miroslav Šulc, Zbyněk Zdráhal.

Venue of the 18th IMSC

Five proposals were received to host the 18th IMSC in 2009, which were Sydney (Australia), Lyon (France), Bremen (Germany), Torino (Italy), and Stockholm (Sweden). The Governing Committee of the IMSS in its meeting on Sunday 31 August, 2003 in Edinburgh, Scotland selected Bremen, Germany as venue for the 18th IMSC under the chair of J. Grotemeyer.

IMSS Budget

IMSS is a low budget organization. It is committed to sponsor the Thomson Medal Awards program at the 16th IMSC and future IMSCs, to pay also to maintain the IMSS web site and to pay legal and banking costs. It has no other expenses and has not paid anything for office staff, travel, or any other expenses. Further details are contained in the financial report of the Treasurer.

Mode of Operation

The IMSS is operating in an entirely open fashion. Information on its aims, operations, and past actions is available for the asking.

Long Term Planning

In the past three years the finances of the IMSS have grown steadily by almost equal contributions from institutional members and personal members. Expenses until now have been very limited (see comments above on the IMSS budget) because for a healthy and strong IMSS first a solid financial basis should be built up. This is necessary for the continuity of the IMSS and its IMSCs and makes it possible to start with organizing IMSCs financially independently. Soon the time will come to spend part of the finances of the IMSS to fulfill its aims, that is to promote mass spectrometry worldwide. For this it is recommended to install a committee consisting of a few Executive Committee and Governing Committee Members of the IMSS to make proposals for spending money in a fair and balanced way.

The issue of the 2 year versus 3 year cycle of future IMSCs was discussed fruitfully in the Governing Committee of the IMSS in its meeting on Sunday 31 August, 2003 in Edinburgh, Scotland. Hopefully the decision for the 2 year cycle of future IMSCs, for which many representatives were in favour, can be made ultimately at, if not earlier than, the 17th IMSC in Prague, Czech Republic in 2006.

The undersigned wishes the new Executive Committee all success in its efforts to make the IMSS an increasingly flourishing Society to keep up with the fast developments in our exciting field of mass spectrometry.

Nico M.M. Nibbering
October, 2003

Report on the 16th International Mass Spectrometry Conference

31 August–5 September 2003, Edinburgh, UK

The Conference, held at the Edinburgh International Conference Centre, was chaired by Professor John Monaghan of the University of Edinburgh. The Organising Committee was composed of 14 members of the British Mass Spectrometry Society, assisted by members of the IMSS Executive Committee and the IMSS Governing Committee. Dr. Alison Ashcroft of the University of Leeds chaired the international Scientific Organising Committee.

The total number of delegates reached 1434, representing a total of 43 countries. The highest number of delegates was, as might be expected, from the UK (447), followed by the USA (190), Germany (101), Sweden (87), Japan (64), France (49), Netherlands (49), Sweden (49), Switzerland (49), Italy (46), and Belgium (40). Financial support from several organisations enabled a number of free registrations to be made available to students and academics.

Scientific Programme

More than 1000 Abstracts were received and the final Scientific Programme contained 6 Plenary Lectures, 18 Keynote Lectures, 83 contributed Oral presentations, and around 800 Poster presentations (split into 4 daily sessions of 200 because of space limitations within the Conference Centre). Also integral to the science on view at the Conference were the numerous Workshops and Vendor Seminars held each afternoon.

Authors of Plenary and Keynote Lectures were invited to submit their full manuscripts for publication in the Proceedings of the Conference and we are grateful that most of them did so. Abstracts from the authors of all the Oral and Poster presentations are provided in a CD that is supplied with this volume of the Proceedings.

Social Programme

For most of the delegates the conference started on Sunday 31 August with a mixer reception held in the magnificent Main Hall of the Royal Museum of Scotland. An Exhibitors' Wine Reception was held within the Conference Centre on Tuesday 2 September. The Conference Dinner was held on Thursday 4 September at Dynamic Earth, a newly opened Exhibition Centre intended to teach people about their own planet Earth. Several trips and visits within Edinburgh and further afield were arranged for Accompanying Persons and for those delegates who had reached scientific saturation.

Excellence Awards

No report of the Conference would be complete without mention of the award of the 2002 Nobel Prizes in Chemistry to John Fenn and Koichi Tanaka and their Plenary Lectures on Friday morning provided a fitting climax to the week.

Three scientists were awarded the Thomson Medals for "achievement in and service to international mass spectrometry." The medals were presented to Richard B. Caprioli, Vanderbilt University, Franz Hillenkamp, University of Muenster, and Victor Talroze, Russian Academy and the University of California, San Francisco. Professor Richard Caprioli was recognised for his work on the application of novel MS techniques to the study of the structure, metabolism, and function of peptides and proteins in biological systems. Professor Franz Hillenkamp was recognised for work on laser mass spectrometry, particularly on MALDI. Professor Victor Talroze received his medal for his many achievements in ion chemistry and physics. The Thomson Medals were first given in 1985 at the IMSC in Swansea, UK and have been given previously to the following scientists: J.H. Beynon, K. Biemann, M.T. Bowers, C. Brunnée, R.G. Cooks, C. Djerassi, J.B. Fenn, D.E. Games, D.F. Hunt, K.R. Jennings, A.G. Marshall, H. Matsuda, F.W. McLafferty, N.M.M. Nibbering, A.O.C. Nier, H. Schwarz, and J.F.J. Todd.

The Curt Brunnée Award, for achievements in instrumentation by a scientist under 45 years of age, was made to Dr Michisato Toyoda of Osaka University for his work on the design and use of multi-turn time-of-flight mass spectrometers. Previous winners were Gareth Brenton (1994), Michael Guilhaus (1997), and Scott McLuckey (2000).

Five Journal of Mass Spectrometry Awards, recognising the best contributions to the scientific programme of the Conference by young scientists, were presented to Filip Cuykens, Zoe A. Harvey, Lenka Kolorova, Youri O. Tsybin, and Tohru Yamagaki.

Commercial Participation

There was a strong commercial presence at the Conference with 55 exhibitors occupying space within all parts of the Conference Centre.

Closing Remarks

The 17th International Mass Spectrometry Conference, to be chaired by Professor Zdeněk Herman, will be held in Prague, the Czech Republic in 2006.

I would like to add my personal thanks to all who helped make the 16th IMSC a success in so many different ways.

John J. Monaghan

Chairman 16th IMSC
School of Chemistry,
University of Edinburgh
October, 2003

Advances in Mass Spectrometry, Volume 16
A.E. Ashcroft, G. Brenton and J.J. Monaghan (Editors)
© 2004 Published by Elsevier B.V.

CHAPTER 1

Drug Discovery and Development in the Post-Genome Era. Can We Rationally Design Safer Drugs?

Thomas A. Baillie

Department of Drug Metabolism, Merck Research Laboratories, WP75A-303,

West Point, PA 19486, USA

1.1 Introduction

The proteomics and genomics revolution of recent years has provided the pharmaceutical industry with unprecedented opportunities for the identification of new drug targets and for the assessment of drug effect. It has been estimated that the human genome contains approximately 30,000 genes [1] and, assuming historic success rates, if only 10% or so of the human genome is amenable to drug intervention, one would predict that some 3000 new drug targets ultimately may emerge from the human genome project [2]. This prospect, viewed in light of the past two decades' progress in molecular biology, ultra high throughput screening techniques, combinatorial chemistry and rapid analog synthesis, suggests that there may never have been a brighter time for innovation in pharmaceutical research. Yet the discovery and development of novel therapeutic agents has become a more challenging, riskier and expensive proposition than ever before, due to factors such as heightened competition for global market share, the high cost of modern technology, the more stringent regulatory requirements for drug registration, and little change in the relatively high rate of failure ("attrition") in drug development. Collectively, these factors have contributed to the decline in the approval of "new molecular entities" by the U.S. Food & Drug Administration (FDA) from 44 in 1996 to 21 in 2002 [3], despite a doubling in research spending by the pharmaceutical industry over the same period. In terms of the financial risk incurred by

1

drug developers, it is noteworthy that a recent survey from the Tufts Center for the Study of Drug Development [4] estimated that the R&D cost of bringing a new drug to market has increased dramatically over the past 15 years and now stands at around $800 million, a significant fraction of which may be attributed to previous development failures [5]. As a result of these rising costs, drug spend in most healthcare markets worldwide is increasing at a rate of 15 to 20% per year, a figure which is widely viewed as non-substainable, particularly in a depressed economic climate. Not surprisingly, the need to contain such escalating costs, through realization of improved efficiencies in drug discovery and decreased attrition rates in drug development, has become a major driver in today's pharmaceutical R&D environment.

In considering possible reasons for what some claim to be a decline of innovation in an industry with an enviable record of scientific achievement, it has been argued that the "easy" drug targets already are addressed by therapeutic agents which are both safe and effective, such that the industry is now faced with developing treatments for much more complex disease states with multiple causes [6], such as Alzheimer's disease, asthma, diabetes, cancer and obesity. At the same time, the full impact of the genomics revolution remains to be realized in the drug discovery process; pharmaceutical scientists are still primarily gathering information on the patterns of gene expression changes elicited by drug candidates, and are only beginning to understand how these changes in the expression levels, not simply of single genes but of gene clusters, translate to drug effect, drug toxicity and to potential intersubject variability in these outcomes. Similarly, the emerging field of systems biology, which attempts to integrate genomics, proteomics and metabonomics data in developing an understanding of biological pathways and networks, remains largely untapped in terms of its potential to influence drug discovery [7,8]. Realistically, therefore, several years may elapse before these exciting new areas of science have matured to the point where drug discovery can capitalize fully on the explosion of knowledge derived from today's human genomics and proteomics research. Indeed, the historically cyclical nature of new drug discovery, based not only upon scientific and technological breakthroughs, but also upon the development of improved therapies which truly are differentiated from their predecessors, has been highlighted in a recent article by Schmid and Smith [2].

As indicated above, rates of attrition during the drug development process have remained high over the years, notably at the preclinical stage where compounds most often fail due to unacceptable toxicity in animals, although inappropriate pharmacokinetics, metabolic properties and physicochemical characteristics contribute to the drop-out rates. While attrition during clinical development is much lower than during preclinical evaluation, the financial consequences of losing a drug candidate escalate dramatically as development proceeds from initial safety and tolerability (Phase I) studies through large-scale clinical trials (Phase II/III).

Consequently, it is imperative that new chemical entities being considered for development be evaluated critically at the preclinical stage, and that liabilities that might preclude successful development of the compound are uncovered at as early a stage as possible so that scarce development resources are not wasted on flawed candidates. Failure of drug candidates, which do enter clinical development, usually result either from a lack of desired efficacy in man or from an unacceptable side-effect profile. Current research on the identification of appropriate "biomarkers" of drug effect and toxicity offers considerable promise for minimizing the risk of such failures in development, and has the potential to accelerate the overall drug development and registration process [9,10]. Nevertheless, the prospect of a drug candidate causing rare, but serious, adverse events in human subjects, either during clinical development or even following regulatory approval, remains a concern, and underscores the need for a fuller appreciation of the multiplicity of factors which may contribute to foreign compound-induced toxicities. A series of highly publicized product withdrawals over the past few years, brought about by concerns over the safety of the drug itself or as a consequence of interactions with co-administered therapeutic agents, serves to further highlight the need for a high degree of safety, in addition to efficacy, in a new drug product.

1.2 Reactive metabolites in drug-mediated toxicity

From the foregoing discussion, it will be evident that safety assessment has become an important focus area in the pharmaceutical industry where significant gains could accrue (in terms of reduced attrition) from an improved understanding of basic mechanisms of toxicity, together with the development of predictive toxicology models. While many approaches to this objective currently are being pursued, one promising avenue deals with minimizing the formation of chemically reactive metabolites which, in certain cases, are believed to mediate the toxic effects of their respective parent compounds through covalent modification of key structural or functional proteins [11]. With recent advances in the sensitivity and versatility of analytical techniques (notably those based upon mass spectrometry), and the wider use of radiolabeled compounds early in the drug discovery process, it has become easier to detect, identify and quantify the covalent adducts to peptides and (in favorable cases) proteins to which these short-lived reactive intermediates give rise. However, not all reactive metabolites are toxic [11,12], and the identities of target proteins for toxicity largely remain elusive [12,13]. In light of these and other uncertainties associated with exposure to biological reactive metabolites, it seems prudent in the drug discovery process to select candidates that have a low propensity to undergo metabolic activation to reactive electrophiles. This is particularly important in light of the fact that some drug-modified proteins are believed to be immunogenic in nature and to cause idiosyncratic reactions in humans that are not effectively predicted from traditional toxicology studies in animals [14–16]. The objective of minimizing metabolic activation, in turn, requires some appreciation of the mechanisms by which certain functional groups in a new chemical entity may be metabolized to reactive species, so that appropriate chemical modifications can be made to the lead structures.

At Merck Research Laboratories, a strategy has been adopted whereby drug candidates being considered for development are evaluated for their ability to form chemically reactive, electrophilic metabolites, both in vitro and in vivo, in an effort to minimize the risk of advancing compounds that carry a risk of causing idiosyncratic toxicities in man [17]. While the experimental protocol involves the preparation of a radiolabeled analog of the compound-of-interest for covalent binding studies, "trapping" experiments with nucleophiles such as glutathione (GSH) and cyanide also comprise an important element of the in vitro evaluation process since the lifetimes of the reactive species in aqueous media usually are too short to permit direct analysis. Indeed, such trapping approaches routinely are applied in early-phase drug discovery programs to evaluate the propensity of a new structural template to undergo metabolic activation. Structural characterization of the trapped adducts, using a combination of tandem liquid chromatography-mass spectrometry (LC-MS/MS) and high-field nuclear magnetic resonance (NMR) techniques, provides an indirect insight into the nature of the reactive intermediates, and thereby affords the medicinal chemist a rationale for minimizing the potential for metabolic activation through modification of the original structure. In the course of such work, it is important to have the ability to detect, identify and quantify, in an unbiased fashion, metabolites of drug candidates that may arise from metabolic activation pathways. Frequently, such metabolites (or their adducts with GSH or cyanide) may be predicted based on literature precedents, but in many cases unexpected products are formed that reveal the operation of novel metabolic transformations. Hence, the analytical requirements for such studies call for a comprehensive survey of all drug-related materials present in the aqueous biological sample, whether derived from an in vitro incubation or from an animal dosed with the drug candidate. It should be borne in mind that products of metabolic activation pathways can be present at very low levels and, therefore, high sensitivity and specificity of detection are prerequisites for success. Of the arsenal of analytical techniques currently available to the drug metabolism scientist, LC-MS/MS, in one or other of its many guises, is unrivaled in its versatility and overall suitability for such demanding applications, and likely will remain so for the foreseeable future [18]. The following examples, which are taken from recent work performed at Merck Research Laboratories, serve to illustrate the above approach to the issue of reactive intermediates in drug discovery and development, and also underscore the power of mass spectrometry for the detection of novel types of drug metabolites.

Table 1.1: Metabolic activation of Compounds I–V as measured by covalent binding to liver microsomal preparations from rat and human in the absence and presence of GSH. Experimental details are provided in [18]

Compd.	Structure*	Covalent binding to liver microsomal protein (pmol equiv/mg protein/1-h incubation[†])			
		Human		Rat	
		No GSH	5-mM GSH	No GSH	5-mM GSH
I	R–O–⬡	3870 ± 303	647 ± 64	1490 ± 133	325 ± 16
II	R–O–⬡ (F, F)	1690 ± 315	472 ± 76	841 ± 94	120 ± 47
III	R–O–pyridyl (N)	911 ± 109	294 ± 18	535 ± 21	139 ± 50
IV	R–O–pyridyl (N, Cl)	303 ± 81	100 ± 19	190 ± 43	54 ± 17
V	R–O–pyridyl (N, CF$_3$)	88 ± 4	27 ± 11	111 ± 21	24 ± 5

*For all compounds, the tritium label was placed at the same position within the functional group labeled 'R.'

[†]Parallel experiments suggested that irreversible binding of radioactivity to microsomal protein at time zero and in the absence of an NADPH regenerating system was < 5 pmol equiv/mg protein/1-h incubation in all cases.

1.3 Metabolic activation of aryloxy-substituted leads in a drug discovery program. LC-MS^n on a 3-dimensional ion trap

In the course of evaluating the metabolic fate of selected lead compounds in a new drug discovery program, it was found that a series of aryloxy-substituted derivatives of a common structural template underwent metabolism in rat liver microsomal preparations to chemically reactive species which could be trapped

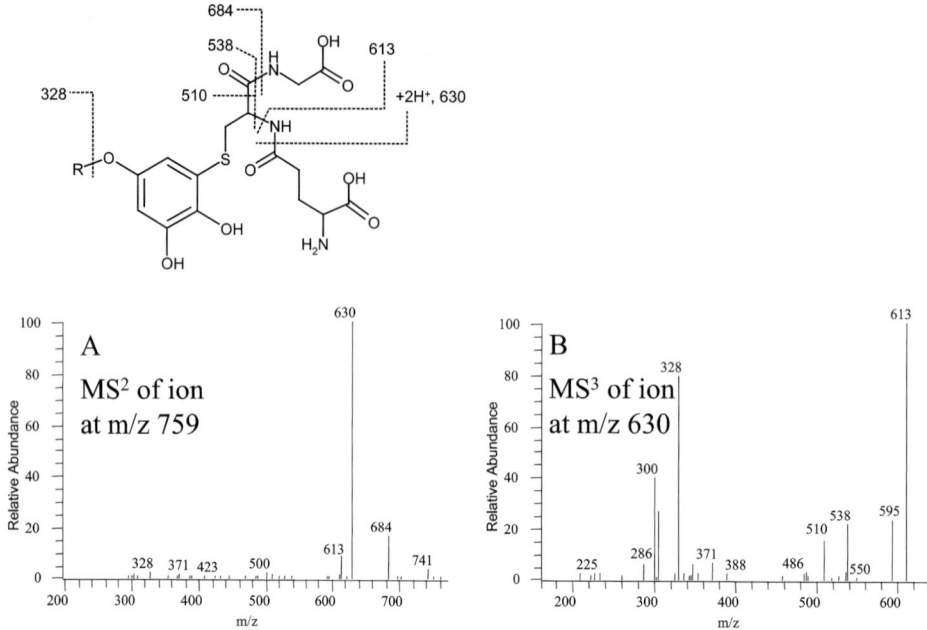

Figure 1.1 (A) MS^2 (products of m/z 759) and (B) MS^3 (products of m/z 630) spectra of a GSH adduct of Compound I (Table 1.1) detected in rat liver microsomal incubations conducted in the presence of GSH. Relevant fragment assignments rationalizing the phenyl ring as the site of metabolic activation are also depicted. The exact regiochemistry of GSH addition is not known. Adapted from [18].

by GSH [18]. The adducts were detected readily by LC-MS analysis of post-incubation samples using a Thermo-Finnigan LCQ-DECA XP^{TM} ion trap system, and were subject to MS^3 or MS^4 in order to obtain more detailed structural information. For the prototype compound in the series (Compound I, Table 1.1), two metabolites were detected with MH$^+$ ions at m/z 759, reflecting an addition to the parent molecule (MW $= 421$) of 337 Da. Collision-induced dissociation (CID) of the m/z 759 species afforded, in each case, an MS^2 spectrum with product ions at m/z 684 (-75 Da) and 630 (-129 Da), both of which are characteristic of GSH conjugates [19], while the MS^3 spectrum obtained from activation of the m/z 630 ion yielded information on the drug residue, notably the prominent m/z 328 ion indicative of an unchanged 'R' moiety (Fig. 1.1).

These results demonstrated that Compound I had undergone metabolic activation via oxidation of the aromatic ether substituent, resulting in addition to the ring of two oxygen atoms (32 Da) and a molecule of GSH (307 Da), accompanied by the elimination of two hydrogens (2 Da), to afford two isomeric GSH conjugates. This information was of key importance, in that it indicated that the aryl ether

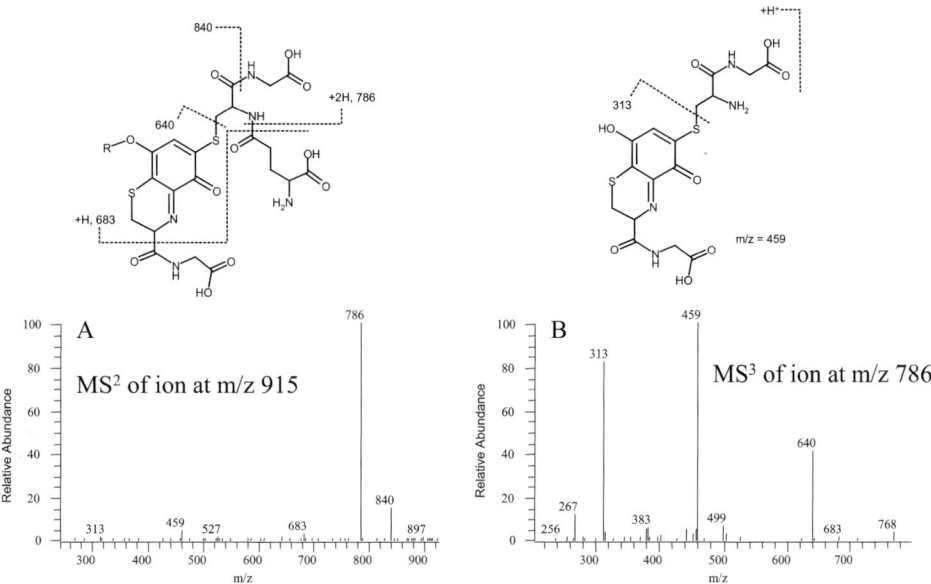

Figure 1.2 (A) MS^2 (products of m/z 915) and (B) MS^3 (products of m/z 786) spectra of a GSH adduct of Compound I (Table 1) detected in human liver microsomal incubations conducted in the presence of GSH. Relevant fragment assignments rationalizing the phenyl ring as the site of metabolic activation are also depicted. The exact regiochemistry of GSH addition is not known. Adapted from [18].

moiety was the primary site of metabolic activation, and suggested that sequential cytochrome P-450-mediated oxidations of the parent compound led to a reactive *ortho*-quinone intermediate which underwent capture by GSH through Michael addition reactions [18].

Interestingly, the above GSH adducts were not detected in parallel incubations with human liver microsomal preparations, which instead afforded conjugates whose MH$^+$ species appeared at m/z 915 (addition of 493 Da to the parent structure). The MS^2 and MS^3 spectra obtained from CID of m/z 915 (MH$^+$) and 786 ([MH$^+$−129]), which are reproduced in Fig. 1.2, again point to the aryloxy moiety as the site of metabolic transformation.

Moreover, the data suggest that the same *ortho*-quinone-GSH adducts detected in the rat liver preparations must have been formed in the human liver experiments, and subjected to a further two-electron oxidation and conjugation with a second molecule of GSH. It is proposed that the resulting *bis*-adducts, following a third oxidation and hydrolytic removal of a γ-glutamyl residue from the second GSH moiety, spontaneously cyclize with elimination of the elements of H$_2$O to yield the observed products. This type of behavior is consistent with what has been reported on the biological reactivity of a series of nephrotoxic polyphenolic GSH conjugates [20].

Based upon the above findings, it was surmised that the aryloxy moiety in Compound I represented a potential liability for this compound, and possibly for structurally related derivatives with similar aryloxy substituents. Incubation of a tritium-labeled derivative of Compound I, labeled in the 'R' group, with both rat and human liver microsomal preparations confirmed that metabolic activation led to appreciable levels of covalently bound drug–protein adducts, an undesirable feature of any new series of lead compounds. Extension of these studies to derivatives in which the phenoxy substituent in Compound I was replaced by a variety of aryloxy moieties designed to be resistant to metabolic oxidation (Compounds II–V, Table 1.1) led to progressively lower levels of GSH adducts (as detected by LC-MS) and decreased levels of covalent binding to microsomal protein (as detected by radioactivity measurements). Ultimately, Compound V was deemed acceptable from a drug metabolism perspective and was taken forward into clinical development. This example serves to highlight the value of the MS^n capability of the 3D ion trap, which played a key role in rapidly pinpointing the site of metabolic activation in this structural series.

1.4 Metabolism of α-substituted piperidine derivatives to ring-contracted pyrrolidines. Accurate mass measurement on a hybrid Q-TOF instrument

Piperidine derivatives are encountered widely in medicinal chemistry since this heterocycle often enhances drug potency due to the introduction of basicity, lipophilicity or both. However, the piperidine moiety frequently serves as a metabolic "soft spot" in that oxidation may occur on one or other of the carbon atoms α to the ring nitrogen, leading to the formation of lactam or ring-opened products. In addition, iminium ions can be formed from N-alkylpiperidines, and these electrophilic species have been postulated to play a role in the toxicity of piperidine-containing xenobiotics [21]. In a drug discovery program at Merck Research Laboratories, early candidates containing an N-methylpiperidine ring system were found to be subject to the latter route of bioactivation, as evidenced by efficient trapping experiments using an equimolar mixture of unlabeled and stable-isotope-labeled ($^{13}C^{15}N$) cyanide as the nucleophile. By this approach, characteristic "doublet" ions, separated by 2 Da, were evident in the mass spectra of the resulting cyano adducts which thus facilitated their detection by LC-MS/MS techniques. In an effort to address this metabolic deficiency, a group of 2,2,6,6-tetramethyl-N-methylpiperidine derivatives was prepared with the intent of blocking α-carbon oxidation and thus preventing iminium ion formation. While in vitro metabolic

Figure 1.3 Structure of the 2,2,6,6-tetra-methyl-N-methylpiperidine derivative (Compound VI). Adapted from [22].

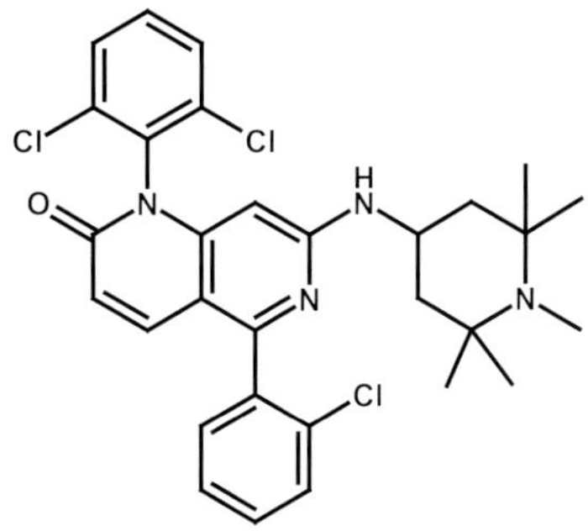

VI

studies with these compounds confirmed that iminium ions no longer were generated, they revealed, surprisingly, the formation of a series of novel products which differed in mass from their respective parent compounds by −56 Da [22].

Elucidation of the structures of these novel products was accomplished largely through the use of accurate mass measurements, conducted on a Micromass Q-TOF II^{TM} hybrid tandem instrument operated in the positive ion ESI LC-MS/MS mode with a minimum resolution of 6000 (fwhm). In the case of parent Compound VI (Fig. 1.3), the MH$^+$ species gave a measured m/z ratio of 569.1669, corresponding to the elemental composition $C_{30}H_{32}N_4OCl_3$ (+4.8 ppm), while the corresponding value for the novel metabolite was 513.1010 ($C_{26}H_{24}N_4OCl_3$, −1.1 ppm), denoting a loss of the elements C_4H_8 (56 Da) from the parent structure.

Collisional activation of the metabolite MH$^+$ ion at m/z 513 afforded a series of structurally informative fragments. Notable among these ions was m/z 416.0125 ($C_{20}H_{13}N_3OCl_3$), which also was present in the MS/MS spectrum of the parent molecule and was attributed to cleavage of the bond between C-4 of the piperidine ring and the exocyclic nitrogen, with charge retention on the tetracyclic core (Fig. 1.4).

The presence of this key fragment demonstrated that the metabolic transformation of interest must have occurred on the substituted piperidine ring. Ions at m/z 496.0752 (MH$^+$–NH$_3$), 457.0411 (MH$^+$–C$_4$H$_8$) and 440.0130 (MH$^+$–C$_4$H$_8$–NH$_3$) afforded an insight into the nature of the modified heterocycle, and strongly suggested that ring contraction of the original 2,2,6,6-tetramethyl-N-methylpiperidine had occurred to yield a 2,2-dimethylpyrrolidine derivative (Fig. 1.4). Finally, a sample of this metabolite was isolated from incubation media, purified and subjected to analysis by ^{1}H NMR at 600 MHz, which corroborated the structure deduced from the high resolution mass spectral analysis through two-dimensional correlation spectroscopy and decoupling experiments [22].

This example illustrates a number of points. From an analytical perspective, it is essential to have in place methodologies which reveal not only expected products

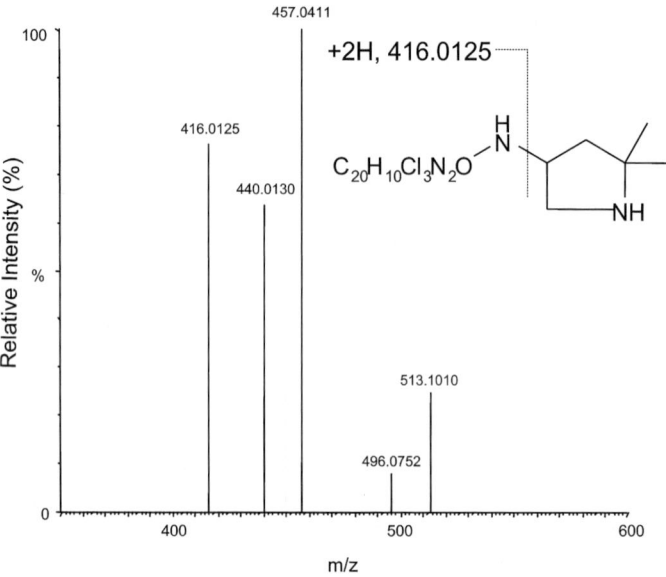

Figure 1.4 Accurate mass MS^2 spectrum of the piperidine ring-contracted metabolite of Compound VI obtained by CID of the MH$^+$ ion at m/z 513.1010. The origin of the key fragment ions is discussed in the text. Adapted from [22].

of metabolism, but unanticipated metabolites that may derive from novel pathways of biotransformation. In the case of Compound VI, it appears that N-demethylation of the parent structure was followed by N-oxidation of the resulting piperidine to afford the corresponding hydroxylamine. Further one-electron oxidation to a nitroxyl radical likely precipitated homolytic C–C bond cleavage and contraction of the original 6-membered ring to a five-membered heterocycle. From a biological perspective, the results of these studies raise interesting questions about the nature of the intermediates involved in this pathway, and raise the possibility that iminium ions may not be the sole reactive species generated during the metabolism of substituted piperidine derivatives. Finally, elucidation of the structure of the ring-contracted metabolite by MS/MS analysis was greatly facilitated by the availability of accurate mass data from the Q-TOF^{TM} instrument, which routinely provides masses of precursor and product ions to an accuracy of 5 ppm or less, more than adequate for the unambiguous assignment of elemental composition to parent and product ions from low molecular weight drug molecules and their metabolites [23].

1.5 Metabolic activation of a pyrazinone-containing thrombin inhibitor. LC-MS/MS analysis on a linear ion trap

A further example of the rich metabolic chemistry associated with 6-membered heterocycles, and of the power of LC-MS/MS combined with high-field NMR techniques, was encountered in studies of a series of inhibitors of the enzyme thrombin, a key element in the blood coagulation cascade. The structural series of interest contained a pyrazinone ring system as the central core (Fig. 1.5), and metabolism studies were performed early in the course of preclinical evaluation both to identify metabolic "soft spots" in the lead compounds, and to screen for possible reactive metabolite formation. The first thrombin inhibitor candidates from this series, while potent and selective against their pharmacological target, were found to suffer from rapid clearance in animal models, raising the prospect of inadequate pharmacokinetics in humans. This high clearance proved to be a consequence of rapid metabolism at the benzylic center adjacent to the monosubstituted aromatic ring, which led to hydroxylated products that were essentially inactive as inhibitors of thrombin. Fluorination at this center to block oxidation

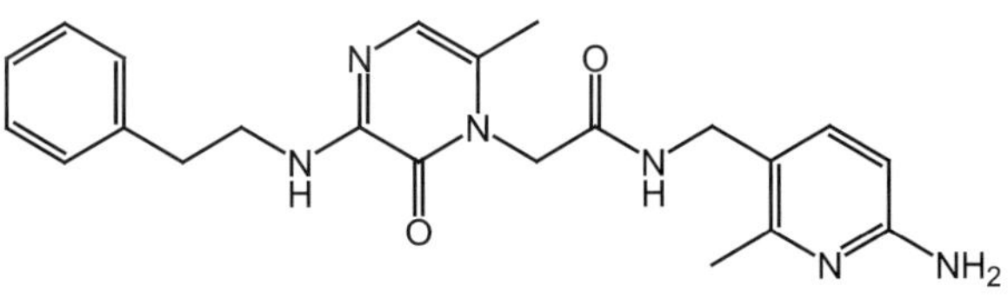

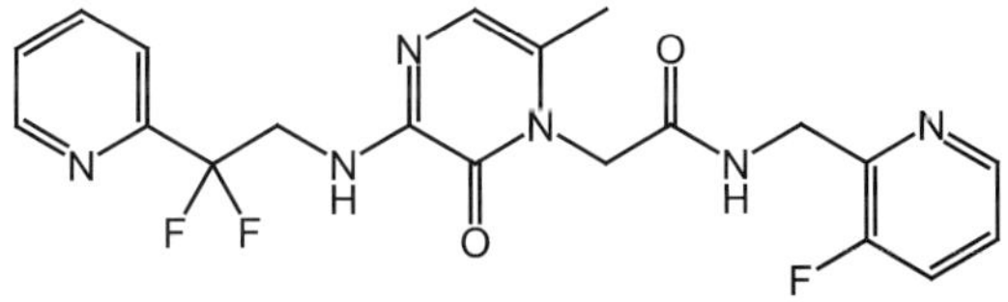

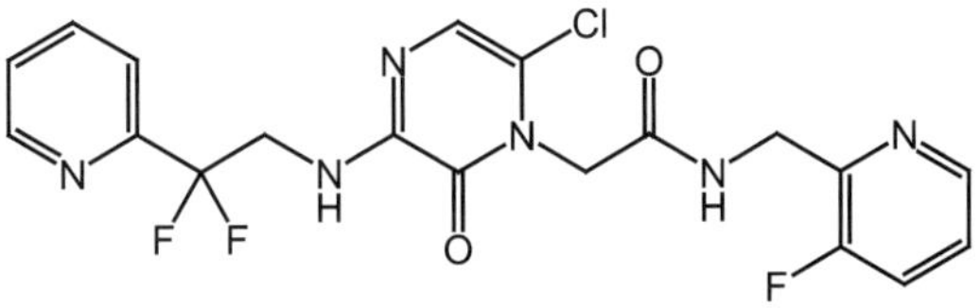

Figure 1.5 Structures of thrombin inhibitors (Compounds VII–IX) referred to in the text.

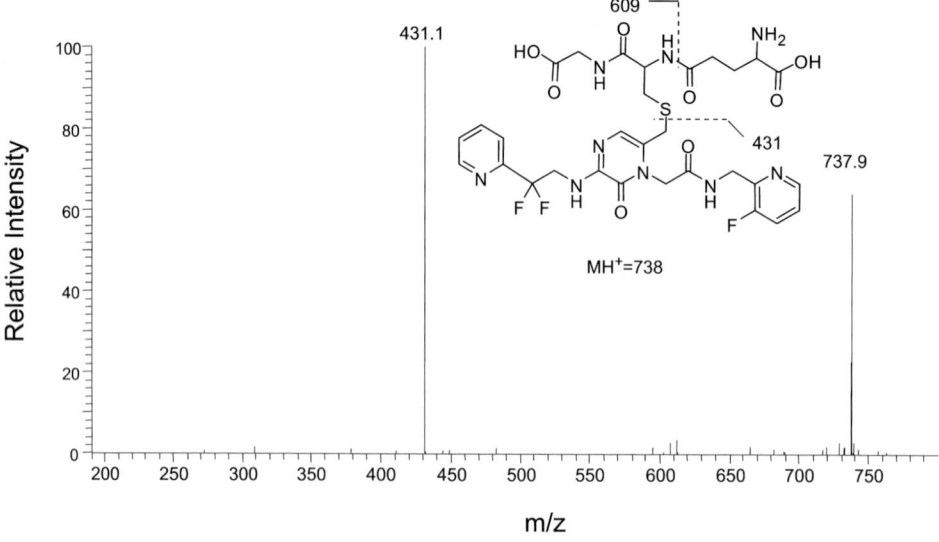

Figure 1.6 MS² spectrum of the 6-methyl-GSH conjugate of Compound VIII, obtained by CID of the MH⁺ ion at *m/z* 738. The spectrum was recorded on a Thermo-Finnigan LCQ™ 3D ion trap instrument. Adapted from [24].

resulted in a second generation of compounds (e.g., Compound VIII, Fig. 1.5) that exhibited significantly decreased clearance in animals, with corresponding increases in elimination half-life. However, these gains in pharmacokinetics were offset by "switching" of metabolism from the benzylic position to the pyrazinone ring system, which generated chemically reactive, electrophilic species that bound covalently to cellular proteins. Thus, in the case of Compound VIII, in vitro trapping experiments with GSH led to the identification of two isomeric adducts with MH⁺ ions at *m/z* 738, whose MS² spectra (obtained on a Thermo-Finnigan LCQ™ ion trap instrument) are reproduced in Figs. 1.6 and 1.7. One adduct, subsequently identified as the product of attachment of GSH at the 6-methyl group, yielded an abundant product ion at *m/z* 431 (MH⁺–GSH), but gave few structurally informative fragments upon MS³ analysis in the trap. Similarly, the second adduct, the product of an unusual pyrazinone ring cleavage reaction with rearrangement to form a substituted imidazole, afforded an abundant product ion at *m/z* 609 (MH⁺–γGlu), but again failed to yield structurally informative ions upon further collisional activation. As a result, the adducts were isolated, purified and their structures determined by a combination of ¹H and ¹³C NMR techniques [24]. These analyses revealed that Compound VIII was subject to two distinct metabolic activation pathways, both of which centered on the 6-methylpyrazinone moiety. It was concluded, therefore, that replacement of the 6-methyl group with an electron-

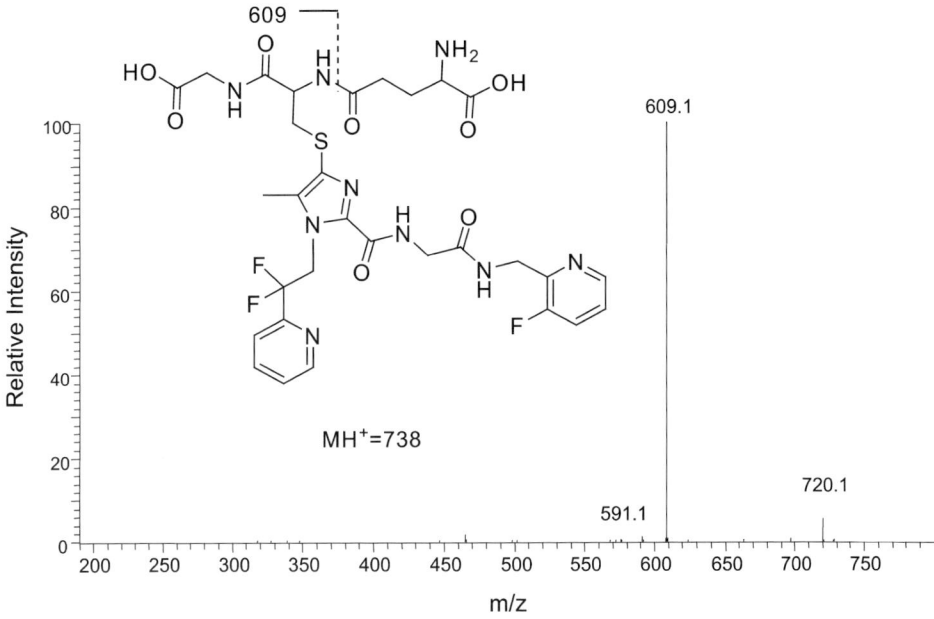

Figure 1.7 MS^2 Spectrum of the rearranged GSH conjugate of Compound VIII, obtained by CID of the MH$^+$ ion at m/z 738. The spectrum was recorded on a Thermo-Finnigan LCQ^{TM} 3D ion trap instrument. Adapted from [24].

withdrawing substituent should suppress metabolic activation of this core structure. Based on this consideration, the 6-chloro analog (Compound IX) was prepared and shown to retain high potency and selectivity for its therapeutic target. As predicted, this compound proved to be relatively resistant to metabolic activation, and only trace amounts of GSH adducts were detected following incubation of Compound IX with rat and human liver microsomal preparations. In light of its favorable pharmacological, pharmacokinetic and metabolic profile, Compound IX, therefore, was advanced into development.

As noted above, MS/MS analysis of the two GSH adducts derived from Compound VIII, performed on a conventional (3D) ion trap mass spectrometer, yielded few fragment ions and, therefore, the CID spectra were of limited value for structure elucidation purposes. Presumably, the fragment ions observed (Figs. 1.6 and 1.7) resulted from low energy pathways of the type commonly observed in ion traps, although in the case of these particular GSH adducts, even MS^3 spectra proved to be relatively uninformative. Recently, hybrid quadrupole linear ion trap (Q-q-Q$_{LIT}$) have become commercially available, and it became of interest to evaluate one such system (MDS Sciex Q TRAP^{TM}) for the characterization of GSH conjugates, using the adduct depicted in Fig. 1.6 as the model compound. The instrument in question, which was based on the Sciex API2000 platform and

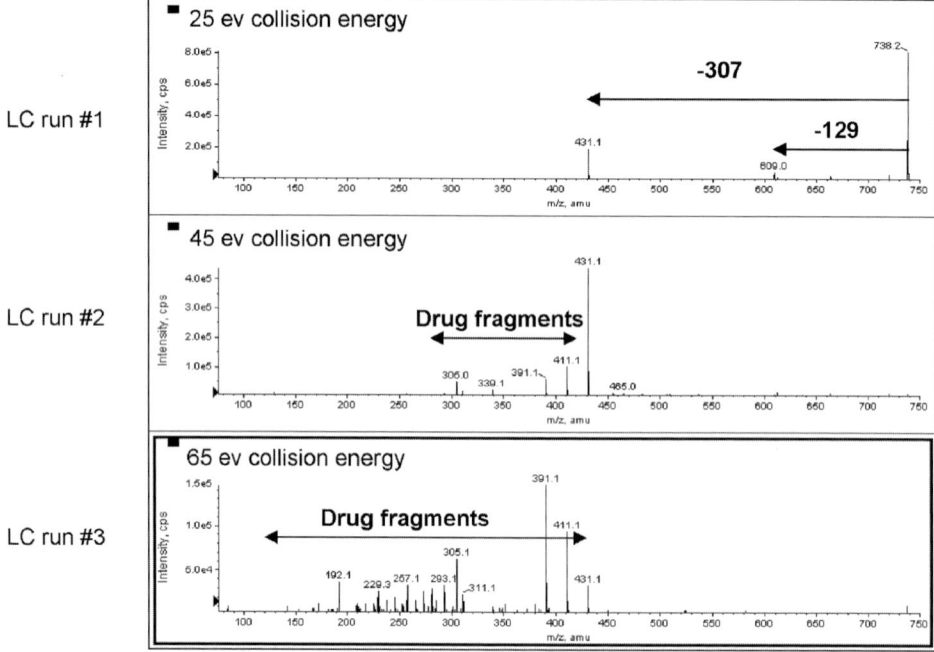

Figure 1.8 MS^2 spectra of the 6-methyl-GSH conjugate of Compound VIII depicted in Fig. 1.6, obtained by CID of the MH$^+$ ion at m/z 738 at different collision energies. The spectra were recorded on an MDS Sciex Q TRAP^{TM} instrument.

equipped with an ESI source operated in the positive ion mode, was configured such that the RF/DC mass filter (Q1) was tuned to allow passage of the proto-nated molecule (m/z 738) into the flow-through collision cell (q2). Trapping and mass analysis of fragment and residual precursor ions occurred in Q3, which was operated as a low pressure linear ion trap (LIT) with axial ion ejection [25,26]. Since the fragmentation mechanism of this hybrid instrument is identical to that in a standard triple quadrupole mass spectrometer, CID spectra from the LIT tend to be closely similar to those obtained from triple quadrupoles which often provide product ion spectra that are richer in information content than those obtained via resonance excitation in a 3D ion trap. Moreover, the high trapping efficiency of the LIT translates to particularly high sensitivities for product ion scanning [26], and the lack of an inherent low mass cut-off results in CID spectra that cover a wide mass range. Indeed, the LIT retains all of the conventional capabilities of a stan-dard triple quadrupole mass spectrometer (including precursor ion scans, constant neutral loss scans and selected reaction monitoring) with no loss in performance, and, therefore, represents an attractive platform for drug metabolism studies [27].

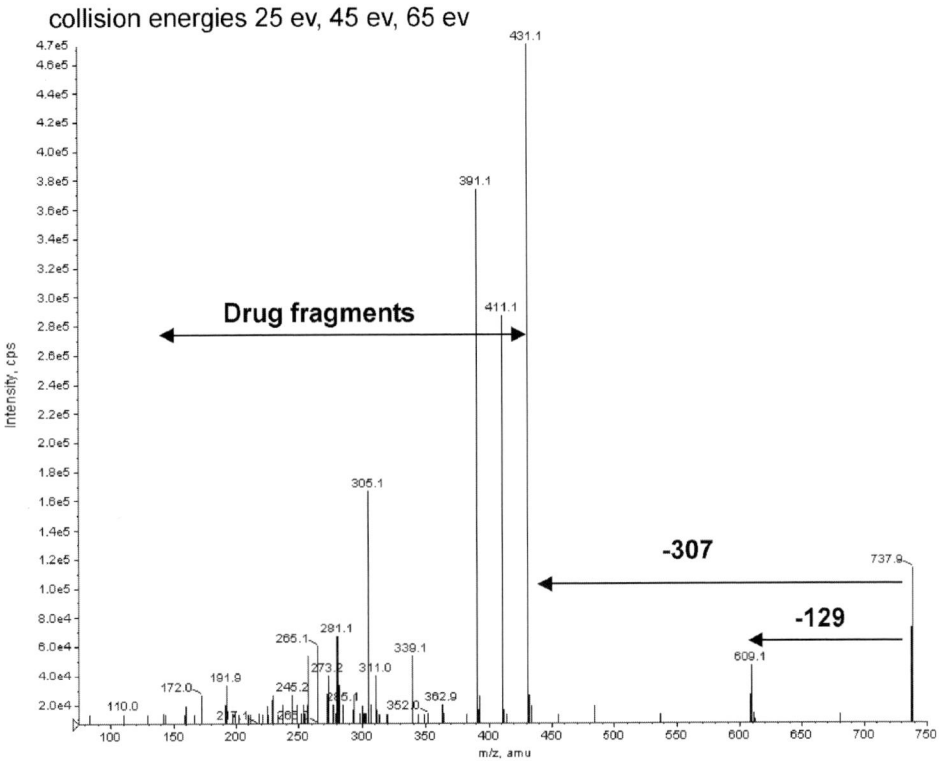

Figure 1.9 Composite MS^2 spectrum of the 6-methyl-GSH conjugate of Compound VIII depicted in Fig. 1.6, obtained by CID of the MH$^+$ ion at m/z 738 at 25, 45 and 65 eV collision energies. The spectrum was recorded on an MDS Sciex Q TRAP^{TM} instrument using 'AutoFrag' software.

When the GSH adduct shown in Fig. 1.6 was subjected to LC-MS/MS analysis on the Q TRAP^{TM}, CID spectra were obtained at three collision energies (25, 45 and 65 eV$_{lab}$) and these are shown in Fig. 1.8. Clearly, appreciable fragmentation occurs at the two higher collision energies, potentially affording a wealth of structural information. A useful feature of the software provided with the Q TRAP^{TM} (termed AutoFrag) allows CID spectra of a given analyte to be recorded at different collision energies 'on the fly' and presented as a composite data set. The result in the case of the model GSH conjugate is depicted in Fig. 1.9, which displays fragments characteristic of both GSH and drug residues, thereby enhancing the overall information content of the spectral data. An alternative approach to this same objective involved CID of the doubly charged ion ([M + 2H]$^{2+}$) at m/z 369.5, which yielded a spectrum of singly charged product ions characteristic of both elements of the adduct structure (data not shown).

1.6 Conclusions

The role of mass spectrometry in the pharmaceutical industry has evolved significantly over the past two decades from its traditional function in the characterization of drug candidates and their impurities to the analysis of combinatorial libraries and the identification of potential new drug targets through proteomics approaches. Also, there is a growing role for mass spectrometry, coupled with powerful bioinformatics tools, in the identification of biomarkers of drug action and toxicity and in the emerging field of systems biology. The value of LC-MS/MS technology for both qualitative and quantitative applications in drug metabolism, pharmacokinetics and biochemical toxicology cannot be overstated, and future developments in hybrid mass spectrometer technology seem likely to extend the already impressive capabilities of LC-MS/MS techniques in terms of speed of analysis, and sensitivity and selectivity of detection. When applied in support of fast-paced drug discovery programs, it should be possible to implement routine LC-MS/MS screening approaches to detect and identify products of metabolic activation, such that information can be provided in a timely fashion to the medicinal chemists charged with lead compound optimization. In that regard, one mechanism by which drug candidates may elicit serious toxicities can be effectively addressed, thereby enhancing the prospect of rationally designing safer therapeutic agents.

From a global perspective, there is no doubt that the pharmaceutical industry faces many challenges in the years ahead due to economic and market forces, the prevailing regulatory climate and socio-political concerns. However, the legacy of the human genome project, together with the remarkable advances in both science and technology that have accompanied our entry into the 21st century, bode well for the future of an industry that has contributed so profoundly to human health.

Acknowledgments

I acknowledge the contributions of my colleagues in the Department of Drug Metabolism at Merck Research Laboratories whose work is described in this article. I am especially grateful to Drs. Rick King, Carmen Fernandez-Metzler and Christine Dieckhaus who provided unpublished data from experiments with the linear ion trap mass spectrometer.

References

1. T. Reiss, Drug discovery of the future: the implications of the human genome project. *Trends Biotechnol.* **19** (2001) 496.

2. E.F. Schmid and D.A. Smith, Discovery, innovation and the cyclical nature of the pharmaceutical business. *Drug Discov. Today* **7** (2002) 563.

3.   M.B. McLellan, Commonwealth Club Speech, U.S. Food and Drug Administration, 2003 (available online at: http://www.fda.gov/oc/speeches/2003/commonwealth0609.html).

4.   K.I. Kaitin, "A methodology for counting costs for pharmaceutical R&D". Presented at the *Crossroad of Biotechnology 2001 Symposium, Montreal, Canada, Tufts Center for the Study of Drug Development*, October 9, 2001.

5.   I. Wilding, Injecting innovation into the drug development process. *Script*, October, 2002.

6.   Editorial, Bigger isn't always better. *Nature* **418** (2002) 353.

7.   J. Boguslavsky, Delivering on the promise of systems biology. *Genomics & Proteomics* **3** (2003) 18.

8.   C.M. Henry, Systems biology. *Chem. Eng. News* **81** (2003) 45.

9.   A.M. Thayer, Biomarkers emerge. *Chem. Eng. News* **81** (2003) 33.

10.   T. Koppal, Benchmarking with biomarkers. *Drug Discovery & Development* **6** (2003) 34.

11.   A. Parkinson, "Biotransformation of xenobiotics". In: *Casarett and Doull's Toxicology: The Basic Science of Poisons*, C.D. Classen (Ed.), McGraw–Hill, New York, 2001, p. 133.

12.   S.D. Nelson and P.G. Pearson, Covalent and noncovalent interactions in acute lethal cell injury caused by chemicals. *Annu. Rev. Pharmacol. Toxicol.* **30** (1990) 169.

13.   N.R. Pumford and N.C. Halmes, Protein targets of xenobiotic reactive intermediates. *Annu. Rev. Pharmacol. Toxicol.* **37** (1997) 91.

14.   J.P. Uetrecht, New concepts in immunology relevant to idiosyncratic drug reactions: the "danger hypothesis" and innate immune system. *Chem. Res. Toxicol.* **12** (1999) 387.

15.   D.J. Naisbitt, S.F. Gordon, M. Pirmohamed, and B.K. Park, Immunological principles of adverse drug reactions: the initiation and propagation of immune responses elicited by drug treatment. *Drug Saf.* **23** (2000) 483.

16.   B.K. Park, D.J. Naisbitt, S.F. Gordon, N.R. Kitteringham, and M. Pirmohamed, Metabolic activation in drug allergies. *Toxicology* **158** (2001) 11.

17.   D.C. Evans, A.P. Watt, D.A. Nicoll-Griffith, and T.A. Baillie, Irreversible binding of drug-related material to protein. An industry perspective on minimizing the potential for drug bioactivation in drug discovery and development, 2003, submitted for publication.

18.   K. Samuel, W. Yin, R.A. Stearns, Y.S. Tsang, A.G. Chaudhary, J.P. Jewell, T. Lanza, L.S. Lin, W.K. Hagmann, D.C. Evans, and S. Kumar, Addressing the metabolic activation potential of new leads in drug discovery: a case study using ion trap mass spectrometry and tritium labeling techniques. *J. Mass Spectrom.* **38** (2003) 211.

19. T.A. Baillie and M.R. Davis, Mass spectrometry in the analysis of glutathione conjugates. *Biol. Mass Spectrom.* **22** (1993) 319.

20. T.J. Monks and S.S. Lau, Biological reactivity of polyphenolic-glutathione conjugates. *Chem. Res. Toxicol.* **10** (1997) 1296.

21. N. Castagnoli Jr., J.M. Rimoldi, J. Bloomquist, and K.P. Castagnoli, Potential metabolic bioactivation pathways involving cyclic tertiary amines and azaarenes. *Chem. Res. Toxicol.* **10** (1997) 924.

22. W. Yin, G.A. Doss, R.A. Stearns, A.G. Chaudhary, C.E. Hop, R.B. Franklin, and S. Kumar, A novel P450-catalyzed transformation of the 2,2,6,6-tetramethyl piperidine moiety to a 2,2-dimethyl pyrrolidine in human liver microsomes: characterization by high resolution quadrupole-time-of-flight mass spectrometry and ^{1}H-NMR. *Chem. Res. Toxicol.* **31** (2003) 215.

23. G. Hopfgartner, I.V. Chernushevich, T. Covey, J.B. Plomley, and R. Bonner, Exact mass measurement of product ions for the structural elucidation of drug metabolites with a tandem quadrupole orthogonal-acceleration time-of-flight mass spectrometer. *J. Am. Soc. Mass Spectrom.* **10** (1999) 1305.

24. R. Singh, M.V. Silva Elipe, P.G. Pearson, B.H. Arison, B.K. Wong, R. White, X. Yu, C.S. Burgey, J.H. Lin, and T.A. Baillie, Metabolic activation of a pyrazinone-containing thrombin inhibitor. Evidence for novel biotransformation involving pyrazinone ring oxidation, rearrangement, and covalent binding to proteins. *Chem. Res. Toxicol.* **16** (2003) 198.

25. J.W. Hager, A new linear ion trap mass spectrometer. *Rapid Commun. Mass Spectrom.* **16** (2002) 512.

26. J.W. Hager and J.C.Y. Le Blanc, Product ion scanning using a Q-q-Q$_{linear\ ion\ trap}$ (Q TRAP^{TM}) mass spectrometer. *Rapid Commun. Mass Spectrom.* **17** (2003) 1056.

27. Y.-Q. Xia, J.D. Miller, R. Bakhtiar, R.B. Franklin, and D.Q. Liu, Use of a quadrupole linear ion trap mass spectrometer in metabolite identification and bioanalysis. *Rapid Commun. Mass Spectrom.* **17** (2003) 1137.

Advances in Mass Spectrometry, Volume 16
A.E. Ashcroft, G. Brenton and J.J. Monaghan (Editors)
© 2004 Published by Elsevier B.V.

CHAPTER 2

Proteomics and Mass Spectrometry in Medicine

Denis F. Hochstrasser [1,3,4,*], Garry Corthals [1], Catherine Zimermann [1],
Alexander Scherl [1], Lydie Bougueleret [4], Jacques Colinge [4],
Alexandre Masselot [4], Pierre-Alain Binz [2], Markus Muller [2], Amos Bairoch [2,3,4],
Ron D. Appel [2,3,4], Jean-Charles Sanchez [1], and Keith Rose [4]

[1] Biomedical Proteomics Research Group, Central Clinical Chemistry Laboratory,
Department of Clinical Pathology, Geneva University Hospital,
CH 1211 Geneva 14, Switzerland

[2] Swiss Institute of Bioinformatics, Geneva, Switzerland

[3] Medical and Sciences Faculties, Geneva University, Switzerland

[4] GeneProt Inc., CH 1217 Meyrin, Switzerland

Abstract

Diseases are often, if not always, due to both genetic predisposition and the environment. The latter plays a major role and modifies drastically the effects of this genetic predisposition. It acts on the epigenetic network. Proteins are essential components of this network and, therefore, are excellent disease markers. They can be used to establish disease diagnosis, prognosis and as a guide for disease treatment. Several of them are even therapeutic agents. The range of protein concentrations in body fluids such as blood exceeds twelve logs and their chemical diversity is tremendous. Their number is still unknown even in samples such as blood. Both the concentration and number fluctuate and thus represent a dynamic biological system. The large concentration range and great number of different proteins require that we use a number of molecular analysis tools, including large-scale approaches, to unravel human proteome complexity, to discover new biomarkers

*E-mail address: denis.hochstrasser@medecine.unige.ch (D.F. Hochstrasser).

and most importantly to find new therapeutic agents. Today mass spectrometry (MS) plays a central and essential role in proteomics. Matrix assisted laser desorption ionization (MALDI)-MS/MS and electrospray ionization (ESI)-MS/MS are complementary techniques generating a large amount of data after gel and liquid chromatography separation and fractionation of patient samples. The development of the molecular scanner and of nonredundant MS strategies complements new software to identify and better score proteins (OLAV) or to visualize gels and even tissues by mass spectrometry. The study of plasma, spinal fluid or nucleoli proteins unravels the tremendous biological complexity of human subproteomes. The ongoing development of a unified protein database "Uniprot" is more than ever critical to capture, collect and share this substantial and rapidly growing body of new data. It should help to partially bridge the existing tremendous gap between data and knowledge.

2.1 Introduction

The text below is divided into five parts. The first part highlights the unforeseen complexity of biological samples. It demonstrates the theoretical analytical challenge to study human plasma or serum by multiple separation methods and mass spectrometry. The second part summarizes the industrial approach to tackle this challenge by using multidimensional chromatography and ESI-MS/MS and the results obtained identifying and partially characterizing numerous plasma proteins. The third part presents an innovative method to analyze numerous proteins in parallel. It is called the molecular scanner, where gels or tissue slices are electroblotted through a porous trypsin membrane onto a peptide-collecting hydrophobic sheet which captures the peptides produced; this sheet is then directly scanned by MALDI-TOF MS equipment. An example of analysis of a bacterial sample is given, which demonstrates the power of this relatively simple technique. The fourth part presents some examples of proteomic studies in fundamental and clinical medicine: the analysis of human cell nucleoli and of cerebro-spinal fluid and serum of patients suffering cerebral strokes. The fifth part underscores the need for very powerful bioinformatics in proteomics and, particularly, in the field of mass spectrometry. Finally, a provocative conclusion states further that the progresses of mass spectrometry are essential for the future of molecular medicine.

2.2 Proteome complexity

The concentrations of known plasma proteins vary from millimolar for albumin, down through high micromolar for immunoglobulins as a group, to femtomolar for tumor necrosis factor—a range of 12 logs. As dying cells often release their contents into the blood circulation, it might be expected that most soluble proteins with reasonably long half-lives might be found in plasma or serum. If, for example, a cell, which contains three copies of a rare transcription factor dies, it might release the three copies into the seven liters of blood or four liters of plasma. Therefore, this transcription factor will be found in yoctomolar concentration, just below the Avogadro number. It is possible that most human proteins could be found in blood at concentrations varying from millimolar to yoctomolar, thus extending the gap between albumin and tumor necrosis factor from the known 12 to 21 logs. Many genes produce not one but multiple mRNAs and final proteins by several mechanisms which have been well characterized, including gene splicing, mRNA editing and modifications at the co- and posttranslational level. It is anticipated that on average in our species one gene produces five or six final protein products. If the human genome contains 35,000 genes, one might expect more than 200,000 different proteins in a human body. Many of them should be found in blood but at an extremely low concentration. There may be a log–linear or log–log relationship between the number of proteins in blood and their relative concentration: a few at the millimolar level, several dozen at the micromolar level, several hundred at the nanomolar level, a few thousand at the picomolar level, tens of thousands at the femtomolar level, and more than a hundred thousands at the attomolar level or below. Our previous work on the plasma proteome at the University Hospital in Geneva and at Geneprot confirms this hypothesis so far.

A crude two-dimensional gel electrophoresis approach provides a simultaneous detection of proteins in the order of magnitude of 4 logs of concentration. About a hundred different proteins (gene products) can be detected and identified loading unfractionated plasma on a two-dimensional gel [1]. Extensive fractionation of large pooled plasma samples by multiple orthogonal chromatography offers a detection and identification sensitivity at the mid to low picomolar concentration range extending the number of identified proteins to about one thousand. The protein concentration range and chemical diversity represent a formidable analytical and especially MS challenge. The best mass spectrometers have sensitivities at the attomolar level. But often, the loading volumes do not exceed nanoliters resulting in sensitivity no better than nanomolar, which is above the concentration of interesting and physiologically relevant molecules. Adequate massive prefractionation and concentration is, therefore, mandatory.

2.3 Large-scale industrial approach

As stated above, the need of adequate and massive prefractionation of body fluids led Geneprot to develop an industrial approach [2]. Pooled patient samples are analyzed on a very large scale (up to 2.5 liters in the case described briefly below). Abundant proteins such as albumin and immunoglobulins are removed by

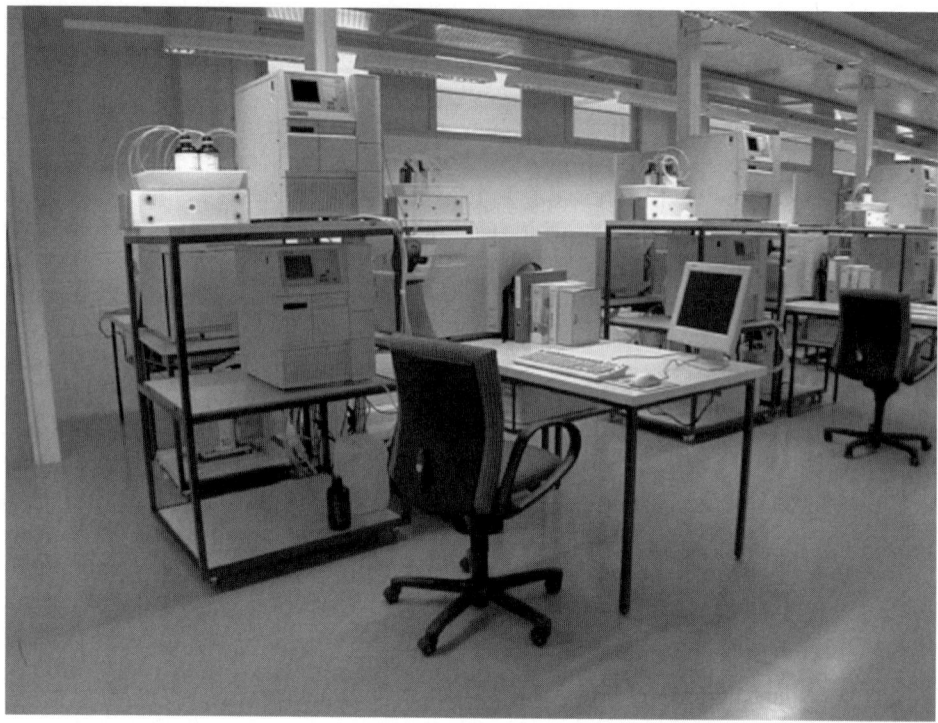

Figure 2.1 GeneProt's analytical units consist of two HPLC machines coupled to an ion-trap mass spectrometer.

highly specific affinity columns. The resulting depleted samples are separated by ion exchange and two reverse phase chromatography steps and finally analyzed by MALDI-MS and ESI-MS-MS. Multiple analytical units each consisting of two automatic high performance liquid chromatography (HPLC) machines "feeding" alternatively every half hour an Ion-Trap MS are used to analyze in parallel multiple final fractions of patients' samples (Fig. 2.1). After the data have been processed by the identification software and entered in the proteomics database, differential analysis can be performed. The final results of the multidimensional separation of samples from the patients' pool and the control pool are represented on vector graphs that allow the comparison of different indicators such as the number of runs in which a protein (or a peptide) was identified, its best score, or the peptide coverage (Fig. 2.2). This offers a convenient visual tool to verify the quality of the results with high accuracy. In a study performed by GeneProt, more than 3 million spectra were acquired. More than 320,000 peptides were manually validated. More than 18,000 peptides were identified including many redundant sequences. 755 distinct proteins were identified with high confidence. An additional 180 proteins would require further validation and about 100 MS/MS spectra matched DNA

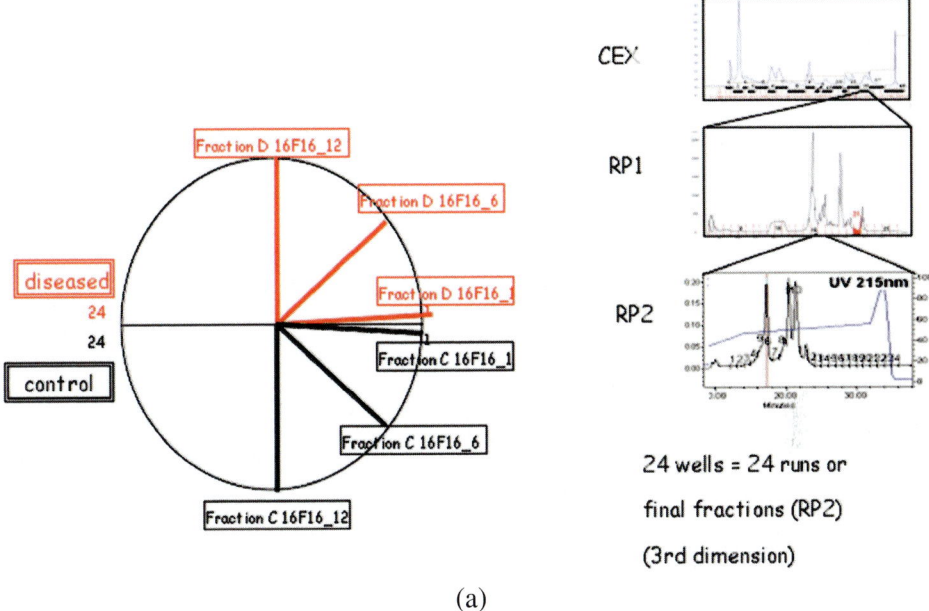

(a)

Figure 2.2 Each final fraction is represented on a table (b) (see overleaf) by a vector: 24 vectors in the half circle Diseased and 24 vectors in the half circle Control (a). The *y* coordinate of the table represents the ion exchange fraction number and the *x* coordinate, the first reverse phase fraction numbers. Two half circles represent on the *x*−*y* coordinate the diseased and control final MS score and reverse phase two fraction numbers. The vector angles from 3 to 9 o'clock represent the reverse phase two fraction numbers of the control samples and counter clockwise from 3 to 9 o'clock the diseased reverse fraction numbers. The length of the vector is proportional to the MS/MS identification score. Similar graphs have been developed to display the percentage of gene coverage by identified peptides.

sequences, where no genes were found to our knowledge, representing either artifacts, known genes absent from our databases or novel genes. To summarize, the first proteome analyzed at GeneProt revealed roughly a thousand distinct proteins with many novel polypeptides (publication submitted).

2.4 The molecular scanner

Innovative ideas could facilitate large-scale approaches. Highly parallel methods would improve the analytical throughput of fractionated samples. The concept of the molecular scanner is to maximize the benefit of the parallel protein transfer during a Western blot procedure and of the imaging power and potential of a mass

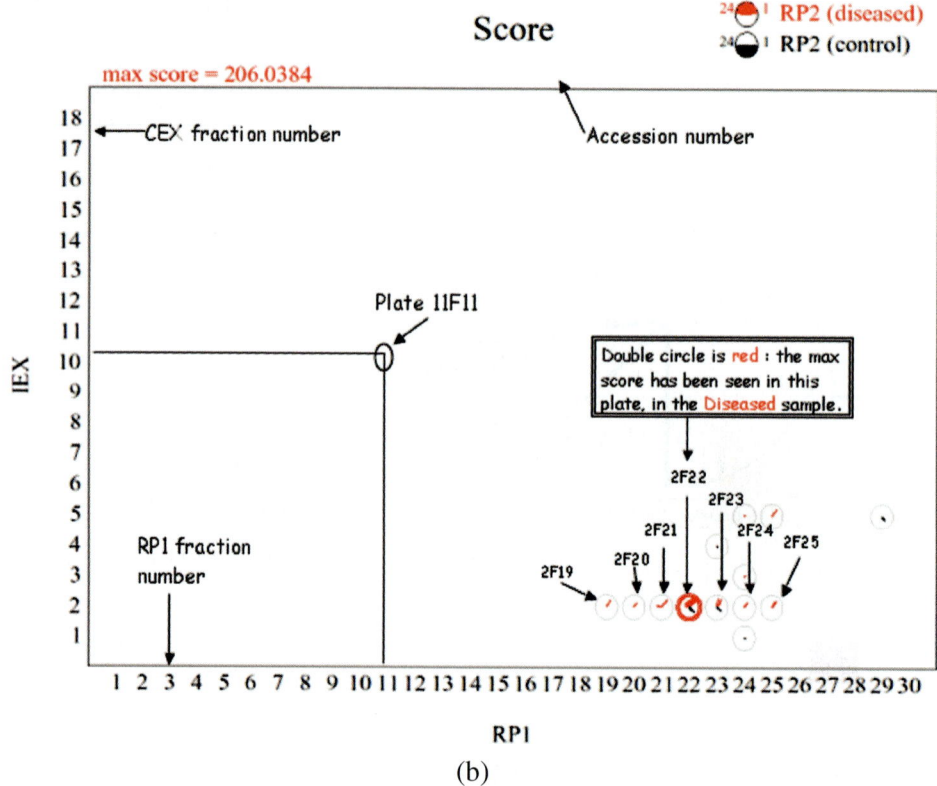

Figure 2.2 (Continued.)

spectrometer [3,4]. An enzymatic membrane was developed where, for example, trypsin is covalently attached to a porous surface. The enzymatic membrane is then intercalated between a gel or a fine tissue section and a collecting membrane. The attached enzyme processes most proteins from the gel or the tissue during the electro-transfer through the membrane onto the collecting surface. The final support that contains the chemically modified (digested) peptides is covered with an appropriate matrix and scanned by the UV or IR laser of the MALDI-MS or better the MALDI-MS/MS instrument. Software tools reconstruct the image of the original gel or tissue with multidimensional views and provide for the automatic identification and partial characterization of the proteins separated in the gel or present in the tissue section (Fig. 2.3). Data correlation algorithms take advantage of the spatial resolution and distribution of the MS or MS/MS signal, significantly enhancing the signal-to-noise ratio. Scanning bacterial samples that are being separated by one- and two-dimensional gel electrophoresis revealed the sensitivity and the power of this technique [5]. Clinical applications are currently being devel-

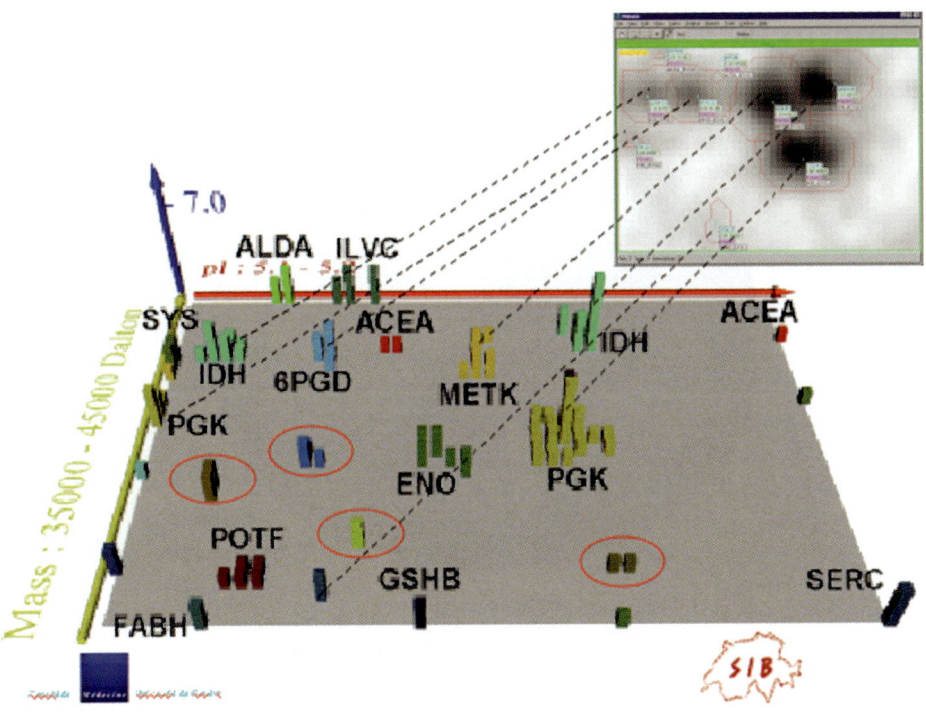

Figure 2.3 The top image represents a portion of a two-dimensional gel electrophoresis analysis of *E. coli* sample scanned with the molecular scanner. The spots' intensities are related to the ion counts on the MS instrument and could be seen by "MS" staining. The lower image is a three-dimensional view of the top image, where the separated proteins are automatically identified and shown with their MS peak intensities. The circles highlight clustered peptides with no identification by peptide mass finger-printing.

oped to diagnose proteinuria and solid tumors from urine and biopsy specimens, respectively.

2.5 Conventional proteomic studies in fundamental and clinical medicine at the University of Geneva

Nucleoli are dense bodies in the nuclei of cells. Their known function is the biogenesis of ribosomes, although it is apparent that they must have numerous other functions. They play a role in cell cycle regulation, cell aging and mRNA export [6–8]. They are enlarged or increased in number in several categories of cancers. They are involved in certain viral infections such as herpes [9]. Two recent independent studies gave new insights into the human nucleoli proteome [10,11].

Functional classification of nucleolar proteins

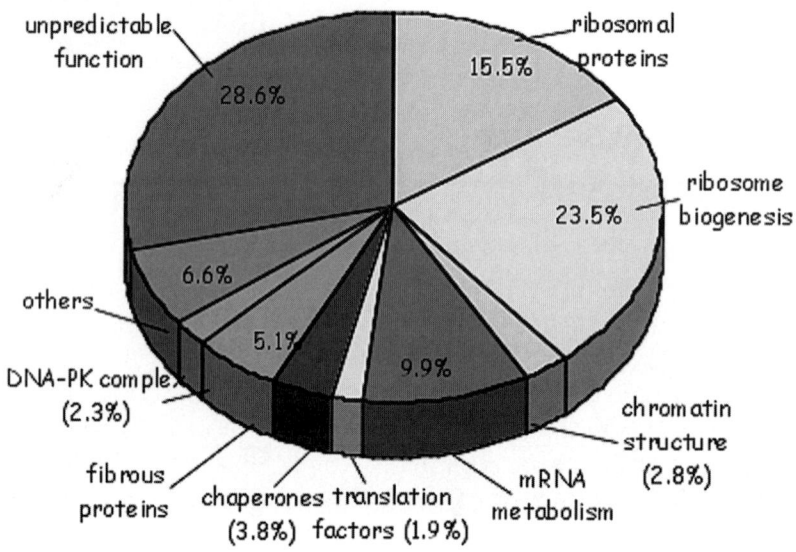

unpredictable function 28.6%

ribosomal proteins 15.5%

ribosome biogenesis 23.5%

others

6.6%

5.1%

9.9%

DNA-PK complex (2.3%)

fibrous proteins

chaperones (3.8%)

translation factors (1.9%)

mRNA metabolism

chromatin structure (2.8%)

A. Scherl. et al. 2002 Mol Biol Cell 13 (11), 4100 - 4109

Figure 2.4 Functional classification of nucleolar proteins.

To establish the proteome of nucleoli, nuclei from human HELA cells were pre-fractionated. Nucleoli were separated from nucleoplasm and analyzed by one- and two-dimensional gel electrophoresis. Bands and spots were excised, the contained proteins digested by trypsin and eluted. The final peptide mixtures were analyzed by MALDI-MS and ESI-MS/MS. In these two studies, more than 300 proteins were found in nucleoli. More than 50 new proteins were discovered, 8 being hypothetical so far. Thirty percent of the nucleoli proteins have unknown function. From the list of known proteins having known activities, several new functions could be attributed to nucleoli (Fig. 2.4).

There are no sensitive and specific stroke markers available in clinical laboratories. The hallmark of stroke is the death of brain tissue due to ischemic events following the formation or the migration of a blood clot or in other cases the rupture of blood vessels and bleeding inside the brain. These two etiologies require precise diagnosis as the first one mandates for its treatment rapid clot lyses. In the second situation, a treatment to destroy blood clots would potentially kill the patient. The most serious brain damage that can occur is death. The body fluid that would collect in the first instance from the released proteins from dead brain tissue is obviously cerebro-spinal fluid. Consequently, we analyzed and compared the cerebro-spinal fluid of control and deceased patients and found several potential diagnostic biomarkers [12]. We then developed immunoassays and measured their blood level. H-FABP protein was found to be up to now the best marker for strokes with a sensitivity of 0.68 and a specificity of 0.99 in a small collection of patient samples.

2.6 Bioinformatics

The large-scale industrial proteomics approach, the molecular scanner concept and improvements in the "classical" proteomics workflow generate a huge amount of MS data, requiring improved software. At GeneProt, scientists designed algorithms to infer peptide charge states a posteriori from the MS/MS mass list [13]. These algorithms can be used with limited resolution instruments like quadrupole-ion traps to reduce computation time and/or increase peptide identification confidence. They also developed an entirely new algorithm (OLAV) to identify proteins by MS/MS [14,15] that performs significantly better than widely used systems like Mascot [16] or Sequest [17]. By applying signal detection theory principles, they defined a large family of peptide scoring functions to match theoretical and experimental MS/MS spectra. These scoring functions make use of probabilistic models aimed at capturing key features of peptide fragmentation spectra. In particular, the prevalence of each ion type, the relation between peak intensity and ion type and the detection of successive fragments are modeled. Fragment amino acid composition is also taken into account to a limited extent. Classical sequence analysis tools like Markov chains and hidden Markov models [18] are used. In addition to reporting peptide and protein matches, OLAV is able to estimate peptide matched P-values, i.e., confidence levels, and it implements a two-pass search close to what has been recently described by Craig and Beavis [19].

OLAV has been extensively used at GeneProt with various ion trap, MALDI TOF/TOF, and Q-TOF instruments on complex biological samples. It allowed automation of peptide/protein identification in an industrial proteomics setting. An alpha version will be accessible soon through ExPASy web server (www.expasy.org). OLAV web interface is an information-rich user interface displaying the number of peptides found per protein; the details of each peptide match with color codes representing peak intensities. It is possible to reject certain peptide identifications via the interface with protein scores recomputed automatically. Other functionalities of OLAV include the display of multiple interpretations of spectra, the ability to search several databases simultaneously plus the possibility to include additional sequences entered manually with integrated display of the results, and the possibility to set variable modifications on specific amino-acids to

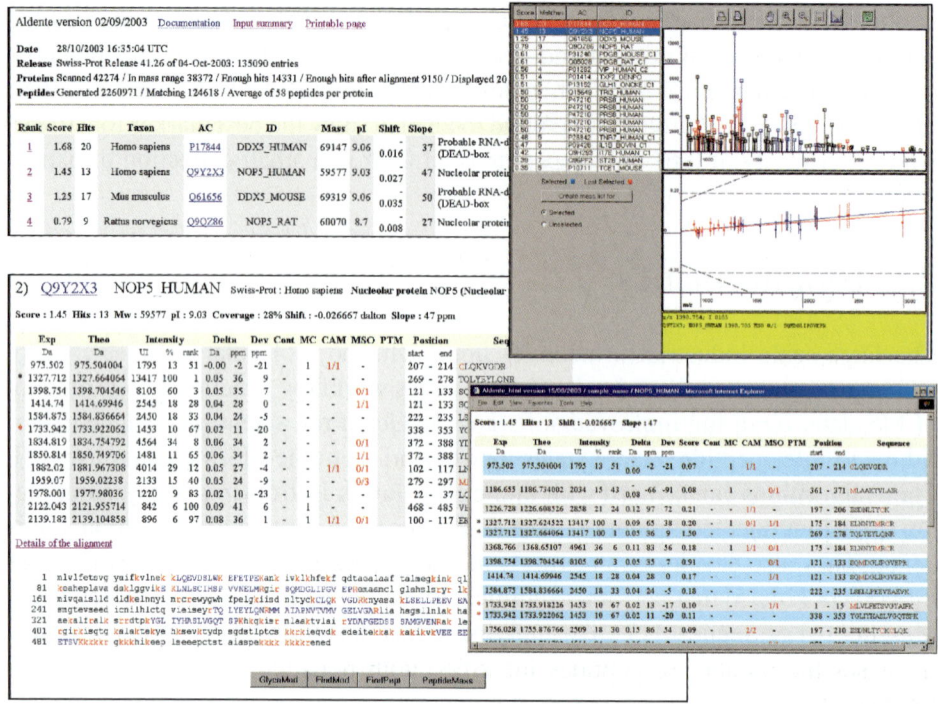

Figure 2.5 Web page showing Aldente software user interface.

better exploit annotations available in databases. The search engine can be run on parallel computers if large databases are searched.

GeneProt's scientists also developed a peptide mass-fingerprinting version of OLAV concepts [20] which exhibits performance superior to ProFound [21] and MSA [22]. Similarly, scientists at the Swiss Institute of Bioinformatics have developed an innovative peptide mass-fingerprinting tool named Aldente [23]. This tool is rapid and highly flexible. It takes advantage of the Hough transform to recalibrate spectra automatically and, therefore, allows easy validation of protein mixtures in the identification process. It considers protein modifications as annotated in the SwissProt database as well as user defined modifications. It is in a final testing phase and will replace soon PeptIdent on the ExPASy server (www.expasy.org). It is cross-linked to the characterization tools of the ExPASy server (Fig. 2.5).

Imaging software tools enhance greatly the power of the molecular scanner process [24]. Three-dimensional views of the combined mass spectra acquired over the collecting membrane display mass distribution and clearly separate data from noise. From the same spectra using data correlation and filtering algorithms, it is possible to attribute peptides to different proteins and, therefore, to enhance peptide fingerprinting capability. The expected workflow with the molecular scanner is to scan the membrane three times. A first rapid scan provides a rough image representing the density of ions detected per pixel. It is like an MS "stain." The second, slower, path is used to obtain precise measurements of peptide masses only where signals were detected in the first pass. The number of laser shots is then much greater to get better precision. Proteins are identified in real time by peptide fingerprint recognition software. If peptides are clustered together, but no protein could be identified, a third scan is used to obtain MS/MS data and consequently more specific data.

Nonredundant MS workflow is a computer-driven peptide selection for further MS/MS analysis. The information resulting from the peptide mass fingerprint and early MS/MS acquisition is used to drive further MS/MS acquisition. It is particularly suited to MALDI ionization, because data acquisition can be paused during data analysis. The computer selects which peptides should be fragmented in the collision cell for MS/MS analysis in a result-dependent manner. After identification of 2 or 3 peptides from one protein, all potential precursors are written on an exclusion list and further MS/MS acquisition is limited on either modified peptides from the same protein or peptides from another protein. Consequently, more proteins are identified with less sequence coverage. This acquisition strategy also increases the number of peptides identified with posttranslational modifications (PTM) [25].

All approaches above require efficient access to nonredundant well-annotated protein and gene databases. The ongoing fusion of SwissProt from the Swiss Institute of Bioinformatics (SIB) and the European Bioinformatics Institute (EBI), the complement of SwissProt, TrEMBL from the EBI and the Protein Information Resource (PIR) from the National Biomedical Research Foundation (NBRF) into a united protein database of protein sequence and functions called UniProt is welcome [26]. It will provide standardized free-access to protein data and information for scientists, and also software. Every UniProt entry will be a central hub for the data available on the protein; it should prove to be a very valuable resource. Its combination with other resources will allow the development of new tools to automate and improve the process of annotation. It is believed that this collaborative effort will reduce the gap that exists today between the amount of data and related knowledge.

2.7 Conclusions

Proteomics, mass spectrometry, and bioinformatics will play a primary role in laboratory medicine. They are essential complements to genomic studies to unravel biological complexity, to discover proteins and protein modifications related to health and diseases. Separation sciences, mass spectrometry, and bioinformatics are the backbone and spinal cord of most proteomic processes. Recent developments such as the molecular scanner or powerful visualization, and identification and annotation software will undoubtedly make proteomics progress to become essential in laboratory medicine and clinical pathology. As a somewhat provocative statement, the recent and expected future progress of mass spectrometry is essential for the future of molecular medicine.

References

1. G.J. Hughes, S. Frutiger, N. Paquet, F. Ravier, Ch. Pasquali, J.-Ch. Sanchez, R. James, J.-D. Tissot, B. Bjellqvist, and D. Hochstrasser, Plasma protein map: an update by microsequencing. *Electrophoresis* **13** (9–10) (1992) 707–714.

2. K. Rose, "Industrialization of proteomics: scaling up proteomics processes". In: *Nature Encyclopedia of the Human Genome*, Nature Publishing Group, London, 2003, pp. 435–439.

3. W. Bienvenut, J.-Ch. Sanchez, A. Karmime, V. Rouge, K. Rose, P.-A. Binz, and D.F. Hochstrasser, Toward a clinical molecular scanner for proteome research: parallel protein chemical processing before and during westernblot. *Anal. Chem.* **71** (1999) 4800–4807.

4. P.-A. Binz, M. Müller, D. Walther, W. Bienvenut, R. Gras, C. Hoogland, G. Bouchet, E. Gasteiger, R. Fabbretti, S. Gay, P. Palagi, M.R. Wilkins, V. Rouge, L. Tonella, S. Paesano, G. Rossellat, A. Karmime, A. Bairoch, J.-Ch. Sanchez, R.D. Appel, and D.F. Hochstrasser, A molecular scanner to highly automate proteomic research and to display proteome images. *Anal. Chem.* **71** (1999) 4981–4988.

5. M. Muller, R. Gras, W.V. Bienvenut, D.F. Hochstrasser, and R.D. Appel, Visualization and analysis of molecular scanner peptide mass spectra. *J. Amer. Soc. Mass Spectrom.* **13** (2002) 221–231.

6. M. Carmo-Fonseca, L. Mendes-Soares, and I. Campos, To be or not to be in the nucleolus. *Nat. Cell Biol.* **2** (2000) E107–E112.

7. M.O. Olson, M. Dundr, and A. Szebeni, The nucleolus: an old factory with unexpected capabilities. *Trends Cell Biol.* **10** (2000) 189–196.

8. T. Pederson, The plurifunctional nucleolus. *Nucleic Acids Res.* **26** (1998) 3871–3876.

9. J.A. Hiscox, The nucleolus—a gateway to viral infection? *Arch. Virol.* **147** (2002) 1077–1089.

10. J.S. Andersen, C.E. Lyon, A.H. Fox, A.K. Leung, Y.W. Lam, H. Steen, M. Mann, and A.I. Lamond, Directed proteomic analysis of the human nucleolus. *Curr. Biol.* **8** (12) (2002) 1–11.

11. A. Scherl, Y. Couté, C. Déon, A. Callé, K. Kindbeiter, J.C. Sanchez, A. Greco, D. Hochstrasser, and J.J. Diaz, Functional proteomic analysis of the human nucleolus. *Mol. Biol. Cell.* **13** (2002) 4100–4109.

12. C.G. Zimmermann-Ivol, P.R. Burkhard, J. Le Floch-Rohr, L. Allard, D.F. Hochstrasser, and J.C. Sanchez, Fatty acid binding protein as a serum marker for the early diagnosis of stroke: a pilot study. *Mol. Cell Proteomics*, October 26 (2003).

13. J. Colinge, J. Magnin, T. Dessingy, M. Giron, and A. Masselot, Improved peptide charge state assignment. *Proteomics* **3** (2003) 1434–1440.

14. J. Colinge, A. Masselot, M. Giron, T. Dessingy, and J. Magnin, OLAV: towards high-throughput mass spectrometry data identification. *Proteomics* **3** (2003) 1454–1463.

15. J. Colinge, J. Magnin, and A. Masselot, "A systematic statistical analysis of ion trap tandem mass spectra in view of peptide scoring". In: R. Page, G. Benson (Eds.), *Proceeding of the Workshop on Algorithms in Bioinformatics (WABI)*, Budapest, September 2003, LNBI, Springer.

16. D.N. Perkins, D.J. Pappin, D.M. Creasy, and J.S. Cottrell, Probability-based protein identification by searching sequence databases using mass spectrometry data. *Electrophoresis* **20** (18) (1999) 3551–6357.

17. M.J. MacCoss, C.C. Wu, and J.R. Yates III. Probability-based validation of protein identifications using a modified SEQUEST algorithm. *Anal. Chem.* **1** (74) (2002) 5593–5599.

18. R. Durbin, S. Eddy, A. Krogh, and G. Hutchinson, *Biological Sequence Analysis*, Cambridge Univ. Press, Cambridge, 1998.

19. R. Craig and R.C. Beavis, A method for reducing the time required to match protein sequences with tandem mass spectra. *Rapid Commun. Mass Spectrom.* **17** (2003) 2310–3216.

20. J. Magnin, A. Masselot, C. Menzel, and J. Colinge, OLAV-PMF: a novel scoring scheme for high-throughput peptide mass fingerprinting. *J. Proteome Res.*, in press.

21. W. Zhang and B.T. Chait, ProFound: an expert system for protein identification using mass spectrometric peptide mapping information. *Anal. Chem.* **72** (11) (2000) 2482–2489.

22. G. Bejerano, Y. Seldin, H. Margalit, and N. Tishby, Markovian domain fingerprinting: statistical segmentation of protein sequences. *Bioinformatics* **17** (10) (2001) 927–934.

23. M. Tuloup, C.P.-A. Hoogland, P.-A. Binz, and R.D. Appel, "New peptide mass fingerprinting tool on ExPASy: ALDentE". In: *Proc. Swiss Proteomics Society Congress, Applied Proteomics, Lausanne, 3–5 December 2002*, Fontis Media (ISBN 2-88476-003-2), 2002, pp. 93–96.

24. M. Muller, R. Gras, P.A. Binz, D.F. Hochstrasser, and R.D. Appel, Improving protein identification for a molecular scanner experiment with human plasma by using correlations between spectra. *Proteomics* **10** (2002) 1413–1425.

25. A. Scherl, P. Francois, V. Converset, M. Bento, J.A. Burgess, J.-C. Sanchez, D.F. Hochstrasser, J. Schrenzel, and G.L. Corthals, Non-redundant mass spectrometry: a strategy to integrate MS acquisition and analysis. *Proteomics*, in press.

26. B. Boeckmann, A. Bairoch, R. Apweiler, M.-C. Blatter, A. Estreicher, E. Gasteiger, M.J. Martin, K. Michoud, C. O'Donovan, I. Phan, S. Pilbout, and M. Schneider, The SWISS-PROT protein knowledgebase and its supplement TrEMBL in 2003. *Nucleic Acids Res.* **31** (2003) 365–370.

Advances in Mass Spectrometry, Volume 16
A.E. Ashcroft, G. Brenton and J.J. Monaghan (Editors)
© 2004 Published by Elsevier B.V.

CHAPTER 3

The Origin and the Future of Macromolecule Ionization by Laser Irradiation

Koichi Tanaka

Shimadzu Corporation, 1 Nishinokyo-Kuwabaracho, Nakagyo-ku, Kyoto 604-8511, Japan

Starting in 1984, a Shimadzu Corporation Central Research Laboratory development team set about the challenge of developing an instrument for analyzing macromolecules such as proteins. At the time, the conventional consensus among chemists was that, "laser ionization of macromolecules exceeding a molecular weight of 10,000 is impossible." However, not being a chemist, I was unaware of this widely held belief.

Although the Nobel Prize was awarded for the "development of soft desorption ionization methods for mass spectrometric analyses of biological macromolecules," this was only one part of the work involved in developing the mass spectrometer. If other aspects of the technology had not been prepared, this success could not have been realized.

A co-researcher of mine, Yoshikazu Yoshida, developed a reflectron for a time-of-flight mas spectomter. At the same time, he also developed a time-to-digital converter (TDC) circuit to measure the time-of-flight of ions with 1 ns accuracy, and proposed the method for efficient laser heating of a sample in a mixture using ultra fine metal powder (UFMP) as a matrix. Another co-researcher, Dr Tamio Yoshida, invented a method for delaying the extraction of ions after the completion of ion generation, in order to prevent loss of mass resolution. Co-researcher, Yutaka Ido, invented a method for further accelerating ions after mass separation, as well as a method for converting ions to electrons, and also notably developed a detection mechanism that fully achieves sensitivity when macromolecules are ionized.

Finally, co-researcher, Satoshi Akita, developed an analog-to-digital converter (ADC) circuit making use of the "pipeline method," which made high speed data acquisition possible. He also developed a high-speed signal accumulation circuit.

33

Back in those development days, I spent my time repeatedly changing the UFMP-retained organic solvent and its concentration on a trial-and-error basis to gather even the smallest possible improvement for high mass ion formation and detection. Normally, acetone was used as a solvent, but one day in 1985, I made a fortunate mistake coupled with some very coincidental actions:

(1) I used glycerin instead of acetone.
(2) I decided to use it as a matrix solution instead of throwing it away because I thought that UFMP was too expensive to waste.
(3) Although glycerin gradually evaporates in a vacuum, I thought it a good idea to boost vaporization by continuous laser irradiation.
(4) As I was eager to see some kind of result, I monitored the TOF spectrum, which led to my first sight of a macromolecular ion signal.

Following this "monumental blunder," I worked to optimize parameters such as concentration until a measurement of a molecular weight of 35,000 was achieved in 1985, and we succeeded in measuring ions with a mass number exceeding 100,000 in 1987.

Even now, the principle of the ionization method has not been proven. Nevertheless, I, as an engineer employed at a private enterprise, turned to the task of developing this discovery into a product, even though the principle was not fully understood, because we believed that if the technology could be of practical use, this was the correct course of action.

In May of 1987, our five team members made the first presentation of our achievements outside of the company at the conference of the Mass Spectrometry Society of Japan, held in Kyoto. The presentation elicited some response but, on the whole, very few people could foresee that in the future this technology would take a major role in science.

In September of the same year, the team participated in the 2nd Japan–China Joint Symposium on Mass Spectrometry held in Takarazuka, Japan, using a poster-based English presentation. It was here that a participant from the USA, Dr Robert J. Cotter, took a great interest in the high-mass ion data provided by me, and soon joined with Dr Catherine Fenselau to disseminate the results and spectra to researchers in Europe and America.

Meanwhile, Associate Professor Takekiyo Matsuo (deceased) of Osaka University ardently recommended me to recompose the report as a formal English-language paper, which we did in 1988. And this is how the contribution of this team became recognized as one of the sources of the MALDI technique.

Based on this technology, first Laser Desorption TOF-MS was released as a product; however, this was commercially unsuccessful with only one sale made, to the City of Hope in the USA. Though the instrument offered a unique capability, its practical application was still limited at that time.

Following this, the technology developed by our team was improved upon by numerous individuals in Europe and the USA, and has become a technology benefiting untold people around the world. Notably, the efforts of Professors, Dr Michael Karas and Dr Franz Hillenkamp, deserve high praise, together with the research results achieved by thousands of other researchers.

The advancement of this macromolecule ionization technology was not due to the efforts of a single genius, but rather, was achieved through the teamwork of five people that provided the backbone of this triumph; a success born from the intense efforts of numerous people striving to develop better technology. The need for highly qualified chemists to push the frontiers of research is ever present. However, the procedure that I developed went beyond the conventions of chemistry. Had I stayed within those conventions, this unique discovery would not have been possible, which proves that it is not always necessary to possess highly specialized knowledge or academic credentials to make a contribution to the advancement of chemistry.

Advances in Mass Spectrometry, Volume 16
A.E. Ashcroft, G. Brenton and J.J. Monaghan (Editors)

CHAPTER 4

Current Challenges in Proteomics: Mining Low Abundance Proteins and Expanding Protein Profiling Capacities

David W. Speicher*, KiBeom Lee, Hsin-Yao Tang,
Lynn Echan, Nadeem Ali-Khan, Xun Zuo, and Peter Hembach

The Wistar Institute, Philadelphia, PA, USA

Abstract

Proteomics has rapidly grown into a major commercial and research enterprise with great prospects for dramatically advancing our knowledge of basic biological and disease processes. The most common type of proteomics analysis involves quantitative protein profile comparisons of two or more experimental states of cells, tissues, or organisms by either "top-down" (separation and analysis of intact proteins) or "bottom-up" (separation and analysis of peptides) methods. Ideally, global profiling experiments should be capable of quantitatively comparing the majority of proteins present in complex proteomes such as mammalian cells and tissues. Unfortunately, no single current technology platform can reproducibly separate and quantitatively compare more than about 1000 to 2000 proteins, and throughput for all current quantitative methods is relatively low. Hence, two related challenges are to obtain more comprehensive coverage of proteins in complex proteomes, and to improve detection and analysis of low abundance proteins in cells, tissues, and biological fluids. Some of the most promising strategies for substantially extending protein profiling capacities utilize emerging methods for subdividing complex proteomes. For quantitative comparisons, an effective initial prefractionation step should be rapid, high recovery, and capable of slicing the

---

*Corresponding author. The Wistar Institute, 3601 Spruce Street, Philadelphia, PA 19104, USA. E-mail address: speicher@wistar.upenn.edu.

proteome into a moderate number of well-resolved fractions with minimal cross-contamination of proteins between adjacent fractions. One such promising prefractionation method recently developed and optimized in our laboratory is microscale solution isoelectrofocusing (MicroSol-IEF), which separates complex proteomes based upon pl by utilizing partitions between separation chambers that contain covalently-bound immobilines. Complex human samples such as cancer cells and biological fluids have been fractionated into well-resolved pools based upon the protein pl's. A commercial version of this method, the ZOOM IEF Fractionator (Invitrogen Corp.) is now routinely used to subdivide proteomes either with commercially-available membrane partitions that fractionate proteomes into up to five pH ranges, or with custom-made membranes that fractionate proteomes into up to seven pH ranges. ZOOM IEF is compatible with most down stream analysis methods including 2-D gels and non-gel methods. It is an important early separation mode in several alternative 3-D and 4-D protein profiling strategies capable of analyzing far more proteins than previous methods. Unlike conventional 2-D gels, this prefractionation method separates proteins effectively regardless of protein size. Hence, when narrow range 2-D gels are used as the primary protein profiling method for separating proteins less than 100 kDa, very large proteins can be separated and analyzed on large pore 1-D gels for proteins larger than 100 kDa. Alternatively, ZOOM IEF pools can be digested directly with trypsin and analyzed by LC/LC-MS/MS. However, the most promising method is a "batch 2-D array/pixelation" strategy, where ZOOM IEF is combined with short 1-D SDS gels. Each lane of the 1-D gel is diced into uniform slices to form a batch 2-D separation or array, where each point in the array (pixel) represents a discrete pH range and size range of the proteome. Each sample is subjected to high sensitivity LC-MS/MS analysis to identify as many proteins as possible. One of the key advantages of this method compared with bottom-up methods is that increasing the number of fractions in each separation dimension results in a substantial increase in the total number of proteins detected and, therefore, improved detection of lower abundance proteins.

4.1 Introduction

Proteomics, a new approach to biology and biomedical research, is less than a decade old but has been growing at an incredible rate. The birth of this new research paradigm directly resulted from two key advances, successful genome projects that provided complete genome sequences, and development of high sensitivity mass spectrometry techniques and instruments for analyzing proteins and peptides. Although there is not complete consensus concerning what proteomics encompasses, it is most commonly considered to include any large-scale protein-based systematic analysis of the entire proteome or a defined subproteome from a cell, tissue, or entire organism.

The proteome, or complete complement of proteins present at a given time in a cell, tissue, or entire organism, is constantly changing in response to environmental signals. Because most proteomics studies are discovery-based rather than hypothesis driven, they are not constrained by prior knowledge. Proteomics, therefore, provides opportunities for unprecedented progress in biomedical and biological research. However, to fully realize this potential, more powerful analysis methods are critically needed for the very complex proteomes encountered with human and other higher eukaryote cells, tissues, and biological fluids. Although the human genome contains less than 50,000 genes, the number of biologically unique protein components that can be expressed by this genome probably exceeds 2 million. This extensive molecular diversity is primarily due to alternative gene splicing and extensive posttranslational modifications of most proteins [1]. A single human cell line is likely to express more than 20,000 functionally distinct protein components with a range in abundance of at least 10^5 to 10^6. Biological fluids, such as plasma or serum, may have even more protein components than cells or tissues and definitely have far greater ranges in protein abundances, which are typically estimated to be in the range of 10^{10} to 10^{12} or greater [2].

The three major types of proteome analysis include: (1) protein profile comparisons, where quantitative changes in proteins between two or more experimental conditions are compared; (2) compositional analysis, which seeks to identify all proteins present in a sample without considering abundance level; and (3) analysis of protein–protein interactions. Protein profiling and compositional analysis, the most common types of proteome analyses, usually utilize essentially the same separation and analysis tools. The latter type of study simply lacks the requirement for quantitation, which simplifies the analysis.

One descriptive way of categorizing alternative protein profiling methods is top-down versus bottom-up (Table 4.1). Top-down methods separate and quantitate intact proteins using one or more separation modes, while bottom-up methods fragment the protein mixture early in the analysis process with separation and quantitation occurring at the peptide level. Bottom-up methods usually utilize trypsin to fragment the proteins and stable isotope incorporation into one of the comparative samples to obtain relative abundance changes. Absolute quantitation of protein abundance levels cannot be readily obtained unless synthetic peptide standards with stable isotopes are utilized, an approach that has been termed AQUA for absolute quantitation [3]. This is at least initially an expensive low throughput process that is most suited to quantitation of a few select proteins of interest or specific known posttranslational modifications. However, similar to antibody arrays, large reagent libraries can be built up over time.

Regardless of the protein profiling method used, the scope of an experiment can be either global or targeted. Global profiling experiments attempt to characterize as many proteins as possible in a proteome, while targeted studies focus on

Table 4.1: Alternative protein profiling methods

Type	Examples	Most common quantitation basis	Capable of absolute quantitation	Discovery based
Top-down	2-D gels	Noncovalent stains	Yes	Yes
	2-D DIGE*	Covalent fluorescent	No	Yes
	Protein LC/LC or IEF/LC	UV absorbance	Yes	Yes
	SELDI-MS[†]	MS signal intensity	No	Yes
	Antibody arrays	Various	Yes	No
Bottom-up	MudPIT[‡] (LC/LC-MS/MS)	Stable isotopes	No	Yes
	ICAT[§]	Stable isotopes	No	Yes
	AQUA[¶]	Stable isotopes	Yes	No

*2-D differential gel electrophoresis [7].
[†] Surface enhanced laser desorption/ionization mass spectrometry [8].
[‡] Multidimensional protein identification technology [9,10].
[§] Isotope coded affinity tag [11].
[¶] Absolute quantitation [3].

a specific subproteome such as an organelle, cellular machine, subcellular fraction, functional activity such as proteolytic activity, posttranslational modification, etc. [4–6]. If the subproteome of interest can be reproducibly isolated, the greater simplicity of these subproteomes more closely matches the analytical capacity of current profiling methods. In contrast, the complexity of most unfractionated higher eukaryote proteomes far exceeds the capacity of any current protein profiling method.

Each of the protein profiling methods in Table 4.1 have strengths and weaknesses that are often complementary. As a result, certain protein changes may be easily detected by Method A and undetectable by Method B, but other protein changes will be invisible to Method A and easily detected by Method B. Most important, although each method will tend to examine a different subset of a complex proteome, none of these methods have been shown to reproducibly detect and quantitate more than a few thousand proteins. Hence, most current protein profiling methods can investigate most or at least a substantial portion of prokaryote proteomes or yeast extracts. But when applied to most eukaryote samples that typically contain more than 20,000 protein components, these methods will at best look at the "tip of the iceberg"; that is, about 10% or less of the total number of proteins present.

Recent advances in mass spectrometry instrumentation, software, and methods strongly suggest that capacities and throughput of proteome analytical tools are likely to continue to improve for some time. However, the most promising short term strategy to achieve more comprehensive global analysis of proteomes is to utilize methods that will subdivide the proteome into a modest number of well resolved fractions. A modest number of fractions is desirable because none of the current profiling methods is very high throughput, with the possible exception of antibody arrays, which are currently constrained by the limited availability of suitable high affinity, specific reagents. Therefore, it becomes prohibitively time consuming to apply existing profiling methods such as MudPIT, ICAT, 2-D gels or 2-D DIGE to large numbers of fractions. Also, if global quantitative comparisons are sought, the separation must be of very high resolution. Extensive cross distribution of a large number of proteins between multiple fractions will: (1) minimize the potential benefit of simplifying the sample, and (2) complicate or totally abolish the ability to make quantitative comparisons unless comparative samples are prelabeled and mixed before fractionation.

As a result of the constraints discussed above, global proteome analysis studies are increasingly turning to some type of sample prefractionation strategy to obtain more comprehensive analysis of complex proteomes and to enhance detection of lower abundance proteins. Most protein profiling methods already utilize at least two dimensions of separation and, therefore, prefractionation adds a third dimension that ideally should be orthogonal to the two (or more) subsequent separation steps. Prefractionations that are orthogonal to both 2-D gel and non-gel based separation strategies include: sequential sample extraction with increasingly stronger solubilization solutions, subcellular fractionation, selective removal of the most abundant protein components, and fractionation of eukaryotic cell extracts using different chromatographic techniques [12–16]. However, all of these are relatively low resolution with substantial cross-contamination of many proteins between two or more fractions. Hence, these methods do not provide optimal simplification of fractions, and more importantly, quantitative comparisons become difficult or impossible, especially if the distribution of incompletely resolved proteins varies between samples as is usually the case for low resolution separations.

The highest resolution methods available for protein separations are electrophoretic methods, particularly isoelectrofocusing under denaturing conditions and separation by size in the presence of SDS. The primary strength of 2-D PAGE is that it efficiently combines these two orthogonal separation methods. It is, therefore, not surprising that various preparative IEF methods are being used for proteome prefractionation. Although preparative IEF separations are not orthogonal to the first dimension analytical IPG separation of 2-D PAGE, there are nonetheless at least two advantages of prefractionating complex proteins prior to using a series of narrow range gels. First, sample is conserved compared with loading replicate aliquots of an unfractionated sample on narrow range 2-D gels. More importantly, the simplified samples permit much higher relative loads of prefractionated samples with better resolution and more consistent spot recovery [17–19]. While preparative IEF fractionation is advantageous when combined with 2-D PAGE, it is even more powerful when combined with non-2-D gel methods.

The preparative IEF methods that have been used to fractionate complex proteomes utilize one of two separation principles; i.e., either "free solution IEF," where soluble carrier ampholytes are used to create and maintain pH gradients across a single focusing chamber or "multi-compartment solution IEF," where pH selective partitions with protein-scale pores divide a series of tandem chambers containing the sample.

The most common free solution IEF method, the Rotofor (BioRad Laboratories) separates proteins in a rotating chamber [20,21], which is either 60 or 18 ml. Bulk liquid can move throughout this single chamber, although the rotation and a series of screens made of woven polymers are used to minimize mixing of focused proteins due to convection currents. The protein mixture is dispersed throughout the chamber and soluble ampholytes produce the pH gradient when an electrical field is applied. The Rotofor approach has been widely applied for initial protein purification under either native or denaturing conditions. A second free solution IEF method is free flow electrophoresis (FFE), which has been used to separate intact cells, organelles, and complex protein mixtures [22–26]. A commercial version of this device is the ProTeam™ FFE system (Tecan Group Ltd.). This complex device uses a very thin (0.2 mm) rectangular separation chamber. When this device is used in the free solution IEF mode, the sample flows perpendicular to the electrodes and carrier ampholytes create the pH gradient between the electrodes. Laminar flow is used to both maintain the separation and collect the sample at the opposite end of the separation chamber.

The second type of preparative IEF separation, which uses multiple sample compartments that restrict bulk liquid flow was originally described by the Righetti group [27]. The original complex device used multiple chambers separated by partitions or membranes composed of immobiline/acrylamide gels cast in glass fiber filters for mechanical support. Each membrane contained immobilines that buffered at specific pHs. When an electrical field was applied, proteins migrated into chambers containing membranes with pHs that bracketed the pIs of the separated proteins. A commercial device called Isoprime for "**Iso**electrofocusing" and "**Pr**eparative **I**soelectric **M**embrane **E**lectrophoresis" (Amersham Biosciences) has large chambers, requires cooling, and uses peristaltic pumps to recirculate samples through the separation chambers. It was initially used primarily for final stage purification of individual proteins under native conditions [28,29] and is difficult to adapt to denaturing conditions for proteome studies. Recently, a modified IsoPrime-type apparatus called a "multicompartment electrolyzer (MCE)" (Proteome Systems) was developed that is somewhat simpler and smaller than the Isoprime and, therefore, easier to use with denaturing conditions [30].

Separation devices that use immobiline buffered membranes to effect protein separation are capable of very high resolution separations because membrane partitions can be selectively made with precise pH's and proteins with pl's differing by as little as 0.01 pH units can be separated [28]. However, both devices described above are relatively complex, expensive, and yield purified fractions with large volumes; that is, about 30 ml for the Isoprime and 5 ml for the MCE. Hence the fractionated samples are usually very dilute unless a very large amount of initial sample was used, and effective concentration of samples in urea, detergents and ampholytes in high yield is usually difficult.

4.2 The MicroSol IEF (ZOOM IEF) Fractionator

A simpler, smaller volume device using immobiline membranes was developed in our laboratory [31] at approximately the same time that the MCE Fractionator was developed. This **microscale solution isoelectrofocusing** (MicroSol-IEF) device utilizes 500 µl sealed separation chambers in a compact unit without external mixing or cooling. The acrylamide/immobiline partition membranes have also been improved by using large pore acrylamide to improve recovery of large proteins and 0.67 mm hydrophilic porous polyethylene instead of glass fiber filters for mechanical strength [18,19]. A commercial version of this device with 700 µl separation chambers, the ZOOM IEF Fractionator (Invitrogen Corp.) is flexible and easy to use (Fig. 4.1).

The ZOOM IEF Fractionator can be used to separate microgram to milligram amounts of complex proteomes including tissue or cell extracts and biological fluids such as serum or plasma. It contains seven separation chambers but the actual number of fractions and the pH ranges of each fraction can be easily tailored to different experimental needs. Either a spacer with an open hole or an immobiline/acrylamide partition disc can be placed between separation chambers and between the terminal separation chambers and electrode chambers. Hence, there is flexibility in the number and pH ranges of fractions obtained depending upon the number of partition membranes and the buffering pH of these membranes. Similarly, the volumes of individual fractions can equal individual chamber volumes (700 µl or less) or two or more chambers can be connected by spacers to increase volumes of select fractions. Pre-made partition membranes are available with six different buffering pHs, which can be used to divide proteomes into five fractions (Fig. 4.1C). Alternatively, partitions of any pH can be custom prepared in the laboratory [17–19,31]. A membrane configuration frequently used in this laboratory to divide mammalian proteomes into seven pH ranges is shown in Fig. 4.1D. In this

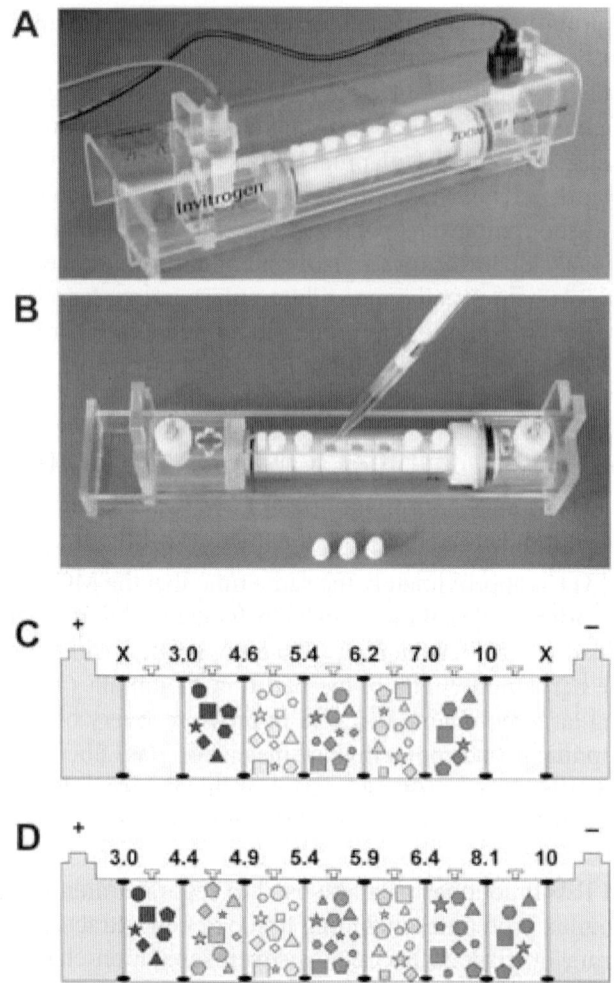

Figure 4.1 The ZOOM IEF Fractionator. (A) An assembled unit. (B) Sample being loaded into the central three chambers. (C) A schematic of a unit utilizing the six commercially available partition membranes to define five pH ranges in tandem 700 µl chambers. The pHs of partition membranes are shown above the unit. X indicates use of an open spacer instead of an immobiline/acrylamide membrane. (D) A schematic of a unit utilizing eight custom made membranes to define seven pH ranges.

scheme the pH 4.4 to pH 6.4 range is divided into four 0.5 pH fractions because this region of many proteomes is populated with the largest number and mass of proteins.

Although this device can be used with different separation buffers, the most appropriate buffers for most proteome study prefractionations are the same as those used for IPG gels. Specifically, a preferred buffer is: 8M urea, 2M thiourea, 4% CHAPS, 1% DTT, and 0.2–0.5% soluble ampholytes. There is substantial flexibility in how protein extracts are loaded into the device. For example, in the seven fraction separation scheme depicted in Fig. 4.1D, the initial extract can be loaded into any single chamber, several chambers, or all separation chambers. The advantages of different loading strategies appear to be sample dependent. For example, it may be advantageous to load an entire sample such as serum into the single chamber that will contain albumin, the most abundant protein in most biological fluids. However, if a sample has a relatively high initial ionic strength, diluting it in a larger volume to decrease the initial ionic strength will be advantageous. Also, if the available sample is too dilute to fit into one or a small number of chambers, it is usually advisable to load it into more chambers rather than risk sample losses, proteolysis, etc. during a concentration step. However, some samples show poor focusing in the terminal chambers when the sample is loaded into all chambers. This is probably because ampholytes migrate more rapidly than proteins and the terminal chambers reach pH extremes before proteins with pIs outside the terminal chamber pH ranges can migrate out of the chamber. Some of these proteins may not be stable at pH extremes, causing aggregation and precipitation. Therefore, as a first approximation, it is generally advisable to avoid loading samples in the most acidic and most basic separation chambers.

The maximum amount of a sample that can be effectively separated into five or seven fractions using the schemes shown in Fig. 4.1 is similarly sample dependent. At a minimum, at least 1–3 mg of most types of protein extracts should be effectively separated by the ZOOM IEF Fractionator, and much higher loads are feasible for at least some types of samples. For example, at least 25 mg or more of human or murine serum can be loaded into a single central chamber and efficiently separated.

In summary, the ZOOM IEF Fractionator has a number of advantages over the MCE and Isoprime instruments for prefractionation of complex proteomes. It is simple, compact, inexpensive, and highly versatile. Most importantly, it uses small sample volumes and produces very well resolved fractions with minimal cross contamination between fractions and minimal sample losses. As a result, fractions do not typically need to be concentrated before using them for down stream analyses. Due to these advantages, it is a critical first step in several alternative 3-D and 4-D comprehensive protein profiling methods that have been developed in this laboratory for analyzing mammalian cell extracts and biological fluids as described below.

4.3 Analysis of mammalian cellular and tissue proteomes using alternative 3-D protein profiling strategies

The ZOOM IEF Fractionator is compatible with any down stream protein profiling method that can use denatured proteins, including schemes that use 2-D gels or MS-based protein profiling methods (Fig. 4.2).

4.3.1 ZOOM IEF and slightly overlapping narrow pH range 2-D gels

As discussed above, there are several advantages of prefractionating complex proteomes prior to use of narrow range 2-D gels, especially the ability to use much higher protein loads to increase detection of lower abundance proteins [17–19,31]. In addition, very large proteins can be effectively separated and recovered due to

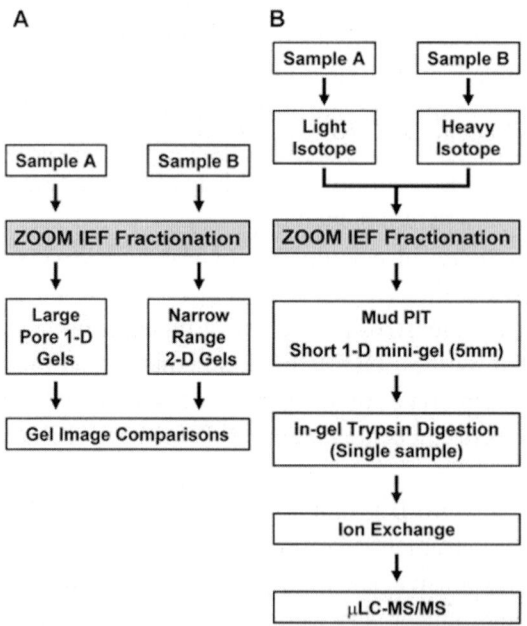

Figure 4.2 Scheme for two alternative 3-D protein profiling methods. (A) Two or more samples are separately prefractionated using ZOOM IEF followed by parallel analysis of each fraction on large pore 1-D gels for proteins > 100 kDa and slightly overlapping narrow pH range 2-D gels for proteins < 100 kDa. (B) Integration of ZOOM IEF prefractionation with a MudPIT-type (LC/LC-MS/MS) bottom-up protein profiling approach.

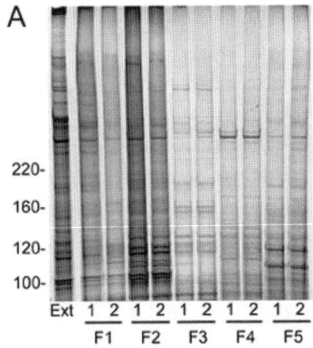 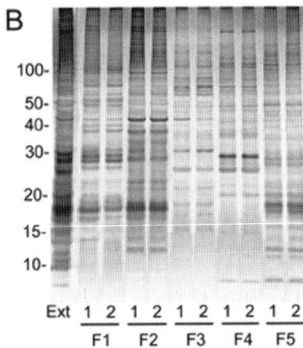

Figure 4.3 Evaluation of replicate prefractionations of a breast cancer cell extract. The separation scheme shown in Fig. 4.1C was used. (A) Large proteins on a NuPAGE 3–8% Tris-acetate gel. F1, F2, etc. are the ZOOM IEF fraction numbers. (B) Evaluation of lower molecular weight proteins on a NuPAGE 10% Bis–Tris gel. Gels were silver stained.

the large pore size of the acrylamide (3–4%) in the partition membranes and the lower local protein concentrations of focused proteins compared with IPG gels. As a result, very large proteins that are poorly recovered on conventional 2-D gels can be detected and compared using 1-D gels (Fig. 4.2A) in parallel with very narrow pH range 2-D gels to analyze ZOOM IEF fractions.

Proteins in the 100–500+ kDa range are consistently recovered from replicate prefractionation runs (Fig. 4.3A). Of course some high molecular weight proteins appear to be present in multiple adjacent fractions, but many large proteins are extensively and variably glycosylated, which results in charge trains that will sometimes bridge two or more pH ranges. Also, a previous study showed that 1-D gels could detect multiple quantitative changes in proteins > 100 kDa when two closely related breast cancer cell lines were compared, but most of the bands on 1-D gels of unfractionated extracts contained multiple proteins [17]. Clearly many of these distinct proteins with the same molecular weight are likely to migrate to different pH ranges. Hence, the use of simple high throughput large pore 1-D gels to detect and compare proteins > 100 kDa after ZOOM IEF separation is a useful complement to parallel analysis of fractions on narrow range 2-D gels. In addition, 1-D gels of the lower molecular weight range (Fig. 4.3B) provides a convenient rapid method to initially survey the quality of each prefractionation and the relative amounts of protein in different fractions (Fig. 4.3B) prior to running more laborious 2-D gels on each fraction. After fractions are surveyed by 1-D gels, the sharpness of separation boundaries can be rapidly surveyed using 11 cm IPG gels that are 3 pH units wide followed by a second dimension run on Bio-Rad Criterion gels. These gels are as high throughput as mini-gels but with substantially higher resolution. Finally, large very narrow range 2-D gels are run to detect and compare as many proteins as possible. For optimal separation, the IPG gels should be

Table 4.2: Scheme for producing high resolution high throughput slightly overlapping 2-D gels using commercially available IPG strips

Fraction	F1	F2	F3	F4	F5	F6	F7
ZOOM pH range	3.0–4.4	4.4–4.9	4.9–5.4	5.4–5.9	5.9–6.4	6.4–8.1	8.1–10
Commercial 24 cm IPGs	3.0–6.0	4.0–5.0	4.5–5.5	5.0–6.0	5.5–6.7	6.0–9.0	7.0–10
pH (trimmed to 18 cm)	3.0–5.25	4.25–5.0	4.75–5.5	5.25–6.0	5.7–6.6	6.2–8.45	7.75–10
IPG separation (cm/pH)	8	24	24	24	20	8	8

as long as possible and only slightly wider than the pH ranges of each fraction. This ensures maximum separation distances without having any of the fraction run off the narrow range gels. Also, the larger the second dimension gel the more time consuming and expensive it is to run, and, therefore, using a gel that is much larger than the pH range of the fraction is inefficient. However, large IPG strips that directly match either the five or the seven fraction pH ranges shown in Fig. 4.1 with slight overlap are not commercially available. Hence, the pH ranges shown in Fig. 4.1D were selected in part because these pH ranges could be indirectly matched with commercially available IPG strips to produce optimal separation distances and maximum throughput. This is accomplished by using 24 cm long IPG strips that are trimmed after focusing to the next size smaller second dimension gel (18 cm) as illustrated in Table 4.2 because more 18 cm gels can be run and stained per day than 24 cm gels and at a lower cost. The images from these slightly overlapping 2-D gels can then be combined to produce a composite 2-D gel image with an effective IEF separation distance between pH 3 to 10 of about 86 cm.

The effectiveness of this approach is illustrated in Fig. 4.4 which shows more than 1000 reproducible spots from a single ZOOM IEF fraction. Overall about 5000 to 8000 spots can be detected by the combined ZOOM IEF/2-D + large pore 1-D PAGE method described here.

4.3.2 ZOOM IEF and MudPIT (LC/LC-MS/MS)

The major challenge in combining ZOOM IEF prefractionation with MudPIT type analyses is preparation of the fractions for trypsin digestion. Although denaturation of the sample in detergents or urea is typically used prior to solution digestion with trypsin, the ZOOM IEF fractions also have very high concentrations of CHAPS, DTT and soluble ampholytes. These components can interfere with either trypsin digestion or chromatography and mass analysis. Hence, the simplest method for

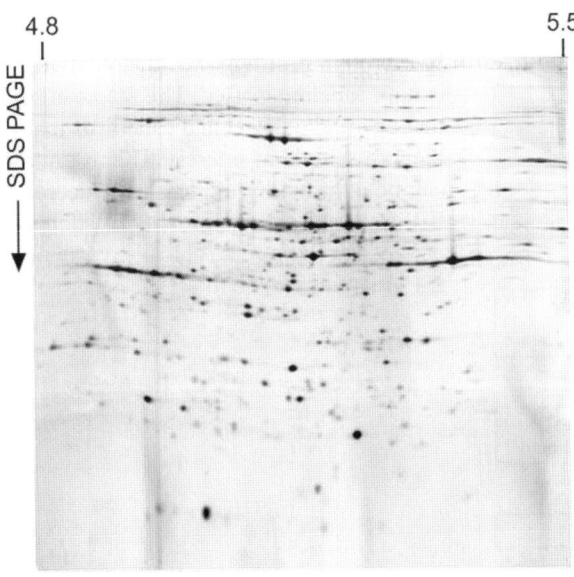

Figure 4.4 Analysis of a ZOOM IEF fraction on an ultra-narrow pH 2-D gel. Fraction 3 of a human breast cancer cell extract was focused on a 24-cm pH 4.5–5.5 IPG strip. After focusing the region below pH 4.8 which did not contain proteins was cut off and the trimmed IPG strip was separated in the second dimension of an 18 × 19 cm SDS gel using a Protean II electrophoresis unit.

cleaning up fractions for trypsin digestion is to run the sample into an SDS gel for a short distance followed by Colloidal blue staining. The entire protein band is then excised and digested with trypsin in-gel followed by a conventional MudPIT-type separation using ion exchange chromatography and μLC-MS/MS analysis (Fig. 4.2B).

4.3.3 The 2-D array/pixelation method

As discussed above, bottom-up LC/LC-MS/MS methods such as MudPIT and ICAT have received much attention over the past several years, and there are high expectations that they will become high throughput and/or substantially sur pass 2-D gels in the number of proteins that they can detect. However, to date these hopes have not been realized. Two problems with bottom-up approaches are: (1) fragmentation of proteins greatly reduces the extensive diversity in physical properties of the proteins to tryptic peptides that have a much more limited range of physical properties, and (2) tryptic peptides of major proteins are usually distrib-uted throughout the peptide population rather than simply overloading one region of a multidimensional separation.

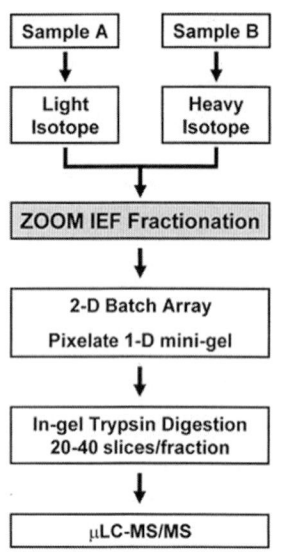

Figure 4.5 The 2-D array/pixelation scheme for comprehensive protein profiling. For quantitative comparisons, stable isotope labeling is used. This step can be deleted for studies that attempt to detect and catalog as many proteins as possible in a sample of interest. The resolution of the 2-D array can be adjusted by increasing or decreasing the number of IEF fractions and the number of slices used to divide each lane of the SDS gel. Typical experiments use five to seven IEF fractions and 10 to 40 slices per gel lane.

A top-down method developed in our laboratory that shows great promise is a batch 2-D protein array followed by systematic analysis of each component of this array using trypsin digestion and μLC-MS/MS (Fig. 4.5). This method can either be used with stable isotope labeling for quantitative comparisons or without a label if the goal is simply to identify as many proteins as possible in a proteome such as plasma. In this scheme, samples are prefractionated by ZOOM IEF followed by running each fraction on a 1-D SDS PAGE mini-gel for a short distance, e.g., typically 2 to 4 cm. The gel is stained with Colloidal Coomassie and each lane is then divided into uniform width slices, e.g., 10, 20, or 40. The result is a 2-D array of fractions or gel slices that is conceptually equivalent to a low resolution 2-D gel (Fig. 4.6). That is, each point or pixel on the array contains a group of proteins in a gel slice with a known pI range and a known molecular weight range. Each slice is then digested with trypsin and analyzed by μLC-MS/MS. The total number of proteins that can be detected by this method is dependent upon: (1) the number of quality MS/MS spectra the LC-MS/MS system can obtain from very complex mixtures and the resulting number of proteins that can be identified, (2) the degree of redundancy between pixels in the 2-D array, and (3) the number of divisions used in each dimension.

Maximizing the number of ZOOM IEF fractions is beneficial because there is little overlap between fractions. Increasing the number of gel slices increases the number of total proteins identified at least up to 40 slices per lane but of course doubling the number of slices doubles total analysis time. A point of diminishing returns is rapidly reached, and halving the slice width or doubling the gel length is usually only a worthwhile trade-off when the number of new proteins detected is

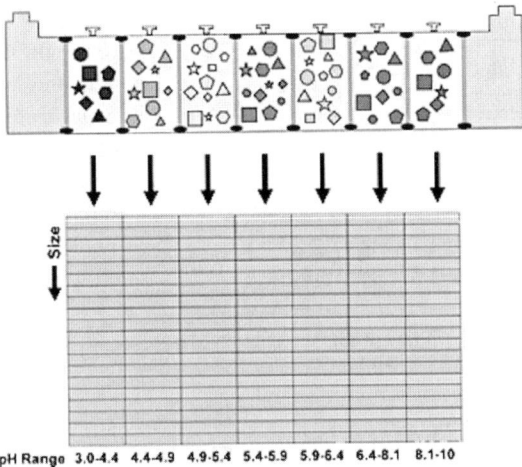

pH Range 3.0-4.4 4.4-4.9 4.9-5.4 5.4-5.9 5.9-6.4 6.4-8.1 8.1-10

Figure 4.6 The batch 2-D protein array. Aliquots of each fraction from IEF fractionation of a proteome are run on an SDS mini-gel. After staining each lane is divided into equal slices. The result is a 2-D protein array, where each point or "pixel" represents a mixture of proteins with defined pI range and M.W. range.

substantial, e.g., at least +30 to 50%. It appears that, in general, the most promising strategy is to slice the proteome into 7 ZOOM IEF fractions and 10 or 20 gel slices per fraction for a total of 70 or 140 fractions, respectively. Sensitive LC-MS/MS systems are capable of detecting up to 50 to 100 or more proteins in a single extended LC run, and initial results suggest that this method is capable of detecting up to 5000 to 8000 proteins from complex proteomes such as human cell extracts and plasma or serum.

4.4 Analysis of human serum and plasma

The 3-D protein profiling methods described above are capable of substantially extending the number of proteins that can be detected in complex proteomes compared with existing methods. However, these approaches still do not have sufficient resolving power for analysis of serum, plasma, and other biological fluids. As discussed above, protein profile analysis of serum by either 2-D gels or non-gel methods is severely limited by a few major proteins that are present at > 1 mg/ml in serum. Albumin alone contributes more than 50% of the total serum protein at 35–45 mg/ml. In contrast, many bioactive proteins and potential serological markers of disease are present in serum at ng/ml to pg/ml or lower levels.

The strategy that has been most frequently used to attempt to detect lower abundance proteins is to deplete major proteins, particularly the two most abundant proteins, albumin, and immunoglobulins. Albumin has usually been depleted

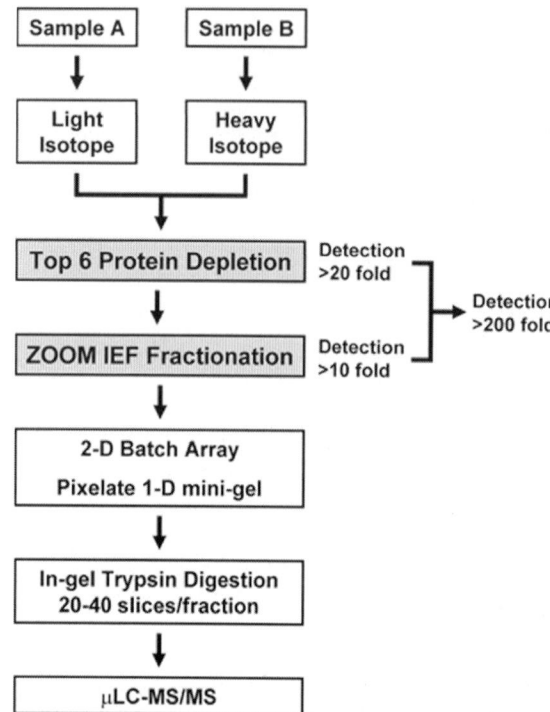

Figure 4.7 A 4-D separation scheme for profiling proteins in serum and plasma. This scheme is similar to the 2-D batch array/pixelation method described in Fig. 4.5 with the addition of affinity depletion of six abundant proteins on a polyclonal antibody column (Agilent).

using various dyes coupled to solid supports and immunoglobulins have been depleted using Protein A or G or a combination of the two proteins. A major concern with such approaches is that all resins, as well as serum albumin and other major proteins, nonspecifically bind some minor proteins. Hence, many minor proteins are usually lost during depletion, and detecting and monitoring such losses is difficult. In addition, these resins usually only remove about 80–95% of the albumin and the residual albumin is still a major component of the sample. However, despite these concerns, even partial removal of these major proteins enhances the ability to detect lower abundance proteins. Specifically, we have found that partial depletion of albumin and IgG using the ProteoPrep Blue Kit (Sigma) enables separation of 5- to 10-fold more serum on 2-D gels. In earlier studies we also showed that IEF prefractionation of whole mouse serum into 4 pH ranges without major protein depletion allowed much higher protein loads and detection of lower abundance proteins [17,18]. About 10- to 20-fold higher detection sensitivity was achieved in those experiments by combining IEF prefractionation with high protein loads on narrow range 2-D gels. Although, these initial attempts did improve protein detection thresholds substantially, further improvements were needed to enable detection of proteins in the low ng/ml range or lower.

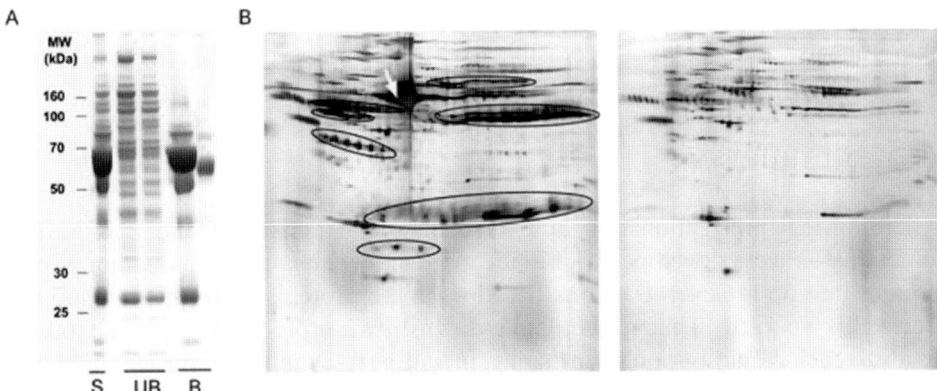

Figure 4.8 Depletion of six abundant plasma proteins using an HPLC polyclonal antibody column. (A) Separation of 15 μl of human plasma on the antibody column. S—unfractionated sample, UB—unbound fractions, B—bound fractions. (B) Comparison of 100 μg undepleted plasma (left panel) with an equivalent amount of depleted serum (right panel) on pH 3-10L 2-D gels. Albumin is indicated with an arrow. The other five proteins, including known proteolytic products and aggregates, are enclosed in ellipses.

Further enhancements in serum/plasma protein detection capacity were achieved by combining immunoaffinity depletion of six abundant blood proteins with the 3-D protein array/pixelation method discussed above. The overall scheme for quantitative comparisons is shown in Fig. 4.7. This method utilizes a 4.6 × 50 mm HPLC column containing polyclonal antibodies to albumin, transferrin, haptoglobin, α-1-antitrypsin, IgG, and IgA which is called the Multiple Affinity Removal System (Agilent Technologies). The results from a typical human plasma sample depletion are shown in Fig. 4.8. The 2-D gel of the depleted plasma sample confirms these proteins are depleted with high efficiency. Since these six proteins represent about 85% of the total protein mass in plasma, much higher protein loads can be applied to 2-D gels and presumably to the ZOOM IEF Fractionator, and of course, these proteins are no longer present in high yield to dominate MS spectra. By combining depletion of these six proteins with fractionation of the resulting depleted serum or plasma sample, an increase in detection sensitivity of at least 200-fold should be readily achievable with 2-D gels.

The loading limit of depleted plasma in the ZOOM IEF Fractionator has not yet been reached and the increase in sensitivity that this strategy produces for the 2-D protein array/pixelation method is being tested. Although certainly depletion of the six most abundant proteins and use of seven ZOOM IEF fractions substantially enhance sensitivity in this current scheme, the wide dynamic range of protein concentrations remains a major challenge. It is already apparent that the next most abundant proteins now dominate MS spectra and 2-D gels, thereby limiting detection capacity. It will be interesting to see how robust immunoaffinity depletion is in routine use and whether it is feasible to extend this approach to additional major serum/plasma proteins.

4.5 Summary

MicroSol-IEF prefractionation and its commercial derivative, the ZOOM IEF Fractionator, were developed to provide rapid, convenient prefractionation of proteomes into up to seven fractions based upon the proteins' pls. This method can be readily coupled to most alternative downstream protein profiling methods to analyze each resulting pool of proteins, including 2-D gels and non-gel methods. Because ZOOM IEF fractionation is a very high resolution method, it is an essential early separation mode in several alternative 3-D and 4-D protein profiling strategies capable of analyzing far more proteins than previous methods. One useful 3-D method is to couple ZOOM IEF prefractionation with use of slightly overlapping very narrow pH range gels to produce a composite IEF separation distance of more than 80 cm. Alternatively, ZOOM IEF pools can be digested with trypsin after cleaning up the fractions in very short 1-D SDS gels followed by conventional LC/LC-MS/MS analysis. However, the most promising 3-D method is a "batch 2-D array/pixelation" strategy, where ZOOM IEF is combined with short 1-D SDS gels. Each lane of the 1-D gel is then diced into uniform slices to form a batch 2-D separation or array where each point in the array (pixel) represents a discrete pH range and size range of the proteome. Each sample is subjected to high sensitivity LC-MS/MS analysis to identify as many proteins as possible. One of the key advantages of this method compared with bottom-up methods is that increasing the number of fractions in each separation dimension results in a substantial increase in the total number of proteins detected and, therefore, improved detection of lower abundance proteins. These methods are capable of detecting more proteins than previous methods and characterization of up to 5000 or more proteins should be feasible with either the 2-D gel or 2-D array approach. Finally, to meet the challenging demands of biological fluid analysis, antibody affinity depletion of six abundant blood proteins is being used in combination with the 2-D array/pixelation method to characterize more blood proteins and to discover novel serological markers of disease.

Acknowledgments

The authors gratefully acknowledge the administrative assistance of Emilie Gross and the graphics assistance of Christine DeLaurentis. This work was supported by NIH Grants CA94360, CA77048, and CA92725 to D.W.S. as well as Institutional support by an NCI cancer center grant (CA10815) and the Commonwealth Universal Research Enhancement Program, Pennsylvania Department of Health.

References

1. A.A. Godley and N.H. Packer, "The importance of protein co- and post-translational modifications in proteome projects". In: M.R. Wilkins, K.L. Williams, R.D. Appel, and D.F. Hochstrasser (Eds.), *Proteome Research: New Frontiers in Functional Genomics*, Springer, Berlin, 1997, pp. 65–91.

2. G.L. Corthals, V.C. Wasinger, D.F. Hochstrasser, and J. Sanchez, The dynamic range of protein expression: a challenge for proteome research. *Electrophoresis* **21** (2000) 1104–1115.

3. S.A. Gerber, J. Rush, O. Stemman, M.W. Kirschner, and S.P. Gygi, Absolute quantification of proteins and phosphoproteins from cell lysates by tandem MS. *Proc. Natl. Acad. Sci. USA* **100** (2003) 6940–6945.

4. L.A. Huber, K. Pfaller, and I. Victor, Organelle proteomics: implications for subcellular fractionation in proteomics. *Circ. Res.* **92** (2003) 962–968.

5. M. Dreger, Proteome analysis at the level of subcellular structures. *Eur. J. Biochem.* **270** (2003) 589–599.

6. D.A. Jeffery and M. Bogyo, Chemical proteomics and its application to drug discovery. *Curr. Opin. Biotechnol.* **14** (2003) 87–95.

7. M. Unlu, M.E. Morgan, and J.S. Minden, Difference gel electrophoresis: a single gel method for detecting changes in protein extracts. *Electrophoresis* **18** (1997) 2071–2077.

8. H. Kuwata, T.T. Yip, C.L. Yip, M. Tomita, and T.W. Hutchens, Direct detection and quantitative determination of bovine lactoferricin and lactoferrin fragments in human gastric contents by affinity mass spectrometry. *Adv. Exp. Med. Biol.* **443** (1998) 23–32.

9. M.P. Washburn, D. Wolters, and J.R. Yates III, Large-scale analysis of the yeast proteome by multidimensional protein identification technology. *Nat. Biotechnol.* **19** (2001) 242–247.

10. M.P. Washburn, R. Ulaszek, C. Deciu, D.M. Schieltz, and J.R. Yates III, Analysis of quantitative proteome data generated via multidimensional protein identification technology. *Anal. Chem.* **74** (2002) 1650–1657.

11. S.P. Gygi, B. Rist, S.A. Gerber, F. Turecek, M.H. Gelb, and R. Aebersold, Quantitative analysis of complex protein mixtures using isotope coded affinity tags. *Nat. Biotechnol.* **17** (1999) 994–999.

12. M.P. Molloy, B.R. Herbert, B.J. Walsh, M.I. Tyler, M. Traini, J.C. Sanchez, D.F. Hochstrasser, K.L. Willams, and A.A. Gooley, Extraction of membrane proteins by differential solubilization for separation using two-dimensional gel electrophoresis. *Electrophoresis* **19** (1998) 837–844.

13. L.A. Huber, C. Pasquali, R. Gagescu, A. Zuk, G.J. Gruenber, and K.S. Matlin, Endosomal fractions from viral K-ras-transformed MDCK cell reveal transformation

specific changes on two-dimensional gel maps. *Electrophoresis* **17** (1996) 1734–1740.

14. B.A. Lollo, S. Harvey, J. Liao, A.C. Stevens, R. Wagenknecht, R. Sayen, J. Whaley, and F.G. Sajjadi, Improved two-dimensional gel electrophoresis representation of serum proteins by using ProtoClear™. *Electrophoresis* **20** (1999) 854.

15. M. Fountoulakis, H. Langen, C. Gray, and B. Takacs, Enrichment and purification of proteins of *Haemophilus influenzae* by chromatofocusing. *J. Chromatogr.* **806** (1998) 279–291.

16. M. Fountoulakis, M.F. Takacs, and B. Takacs, Enrichment of low-copy-number gene products by hydrophobic interaction chromatography. *J. Chromatogr.* **833** (1999) 157–168.

17. X. Zuo, L. Echan, P. Hembach, H.Y. Tang, K.D. Speicher, D. Santoli, and D.W. Speicher, Towards global analysis of mammalian proteomes using sample prefractionation prior to narrow pH range two-dimensional gels and using one-dimensional gels for insoluble and large proteins. *Electrophoresis* **22** (2001) 1603–1615.

18. X. Zuo and D.W. Speicher, Comprehensive analysis of complex proteomes using microscale solution isoelectrofocusing and slightly overlapping narrow range two-dimensional gels. *Proteomics* **2** (2002) 58–68.

19. X. Zuo, P. Hembach, L. Echan, and D.W. Speicher, Enhanced analysis of human breast cancer proteomes using micro-scale solution isoelectrofocusing combined with high resolution 1-D and 2-D gels. *J. Chromatogr. B* **782** (2002) 253–265.

20. M. Bier, N.B. Egen, T.T. Allgyer, G.E. Twitty, and R.A. Mosher, "New developments in isoelectric focusing". In: E. Gross and J. Meienhofer (Eds.), *Peptides: Structure and Biological Functions*, Pierce Chemical Co., Rockford, IL, 1979, pp. 79–89.

21. N.B. Egen, W. Thormann, G.E. Twitty, and M. Bier, "A new preparative isoelectric focusing apparatus". In: H. Hirai (Ed.), *Electrophoresis,* de Gruyter, Berlin, 1984, pp. 547–549.

22. K. Hannig, New aspects in preparative and analytical continuous free-flow cell electrophoresis. *Electrophoresis* **3** (1982) 235–243.

23. G. Weber and P. Bocek, Recent developments in preparative free flow isoelectric focusing. *Electrophoresis* **19** (1998) 1649–1653.

24. G. Weber, D. Grimm, and J. Bauer, Application of binary buffer systems to free flow cell electrophoresis. *Electrophoresis* **21** (2000) 325–328.

25. P. Hoffmann, J. Hong, R.L. Moritz, L.M. Connolly, D.F. Frecklington, M.J. Layton, J.S. Eddes, and R.J. Simpson, Continuous free-flow electrophoresis separation of cytosolic proteins from the human colon carcinoma cell line LIM 1215: a non two-dimensional gel elecrophoresis-based proteome analysis strategy. *Proteomics* **1** (2001) 807–818.

26. J. Bauer, Advances in cell separation: recent developments in counterflow centrifugal elutriation and continuous flow cell separation. *J. Chromatogr. B* **722** (1999) 55–69.

27. P.G. Righetti, E. Wenisch, and M. Faupel, Preparative protein purification in a multi-compartment electrolyser with immobiline membranes. *J. Chromatogr.* **475** (1989) 293–309.

28. P.G. Righetti, E. Wenisch, A. Jungbauer, H. Katinger, and M. Faupel, Preparative purification of human monoclonal antibody isoforms in a multi-compartment electrolyser with immobiline membranes. *J. Chromatogr.* **500** (1990) 681–696.

29. E. Wenisch, P.G. Righetti, and W. Weber, Purification to single isoforms of a secreted epidermal growth factor receptor in a multicompartment electrolyzer with isoelectric membranes. *Electrophoresis* **13** (1992) 668–673.

30. B. Herbert and P.G. Righetti, A turning point in proteome analysis: sample prefractionation via multicompartment electrolyzers with isoelectric membranes. *Electrophoresis* **21** (2000) 3639–3648.

31. X. Zuo and D.W. Speicher, A method for global analysis of complex proteomes using sample prefractionation by solution isofocusing prior to two-dimensional electrophoresis. *Anal. Biochem.* **284** (2000) 266–278.

Advances in Mass Spectrometry, Volume 16
A.E. Ashcroft, G. Brenton and J.J. Monaghan (Editors)

CHAPTER 5

Occurrence and Environmental Behavior of Chiral Compounds: Enantioselective Processes and Source Apportioning

Thomas Poiger*, Markus D. Müller, Ignaz J. Buerge, Marianne E. Balmer, and Hans-Rudolf Buser,

Agroscope, Swiss Federal Research Station for Horticulture, CH-8820 Wädenswil, Switzerland

Abstract

Many organic compounds (natural and synthetic) possess at least one element of asymmetry, leading to the presence of stereoisomers. Whereas synthetic compounds are most often produced as mixtures of stereoisomers, natural compounds typically are single isomer compounds. Stereoisomers have similar (diastereomers) or identical (enantiomers) physical–chemical properties, but often differ significantly with respect to biological properties such as toxicity and environmental behavior. However, the differences in biological properties of individual stereoisomers of synthetic organic compounds are often not well understood. This is, in part, due to lack of awareness of these differences, lack of legal requirements (e.g., in pesticide registration) and limited availability of appropriate analytical methods. Studies on individual stereoisomers of chiral compounds require special (enantioselective) analytical methods. Separation of enantiomers by HPLC, GC, and CE is achieved by addition of so-called "chiral selectors" such as cyclodextrins to the stationary or mobile phase. Separation of chiral compounds into two or more

---

*Corresponding author. Agroscope FAW, Schloss, CH-8820 Wädenswil, Switzerland.
E-mail address: thomas.poiger@faw.admin.ch.

peaks increases the need for very selective detection, particularly in complex matrices. Therefore, enantioselective separation is often coupled to mass spectrometric detection. Because interactions between target compound and chiral selector cannot be predicted, the choice of a suitable selector often is not an easy task and must be done on a case-by-case basis. Single isomer or isomer enriched reference compounds, which are often not commercially available, can either be prepared by stereoselective synthesis or by semipreparative enantioselective (liquid) chromatography. In both cases, the absolute configuration must then be determined by X-ray diffraction or by chiroptical measurements in reference to compounds with known configuration. Several examples from the areas of pesticides and pharmaceuticals research illustrate the general approach for studying the behavior of chiral compounds with isolation and characterization of single isomer compounds, investigation of enantioselective degradation processes using racemic as well as enantiopure compounds, and enantioselective residue analysis in wastewater and surface waters. In this article particular emphasis is put on source apportioning of chiral compounds. Examples include apportioning of use of old (racemic) versus new (enantio-enriched) form of a pesticide, agricultural versus nonagricultural use of pesticides, and discharge of pharmaceuticals via urine (after consumption and metabolization in the body) versus disposal of unused drugs. In these cases, the stereoisomer composition of residues serves as a "built-in" marker for the source of the compound or the processes that affected the compound between its source and its point of occurrence.

5.1 Introduction

Organic compounds with at least one element of asymmetry in their three-dimensional structure are chiral and consist of two or more stereoisomers (enantiomers, diastereomers). Naturally occurring chiral compounds most often are biosynthesized as single isomer compounds. Examples include amino acids and sugars, as well as the corresponding oligo- and polymers (peptides, proteins, and enzymes, cellulose and amylose). Synthetic chiral organic compounds, on the other hand, are very often produced as mixtures of stereoisomers.

Even though the physico-chemical properties of enantiomers are, in general, identical, they may behave very differently in a chiral environment such as in biological systems. In certain areas of research, awareness of the impact of stereochemistry on the biological effects of compounds is higher than in others. In the pharmaceutical sector, for example, most new chiral active ingredients are marketed as single isomer compounds and many racemic compounds presently on the market are being replaced by single isomers ("chiral switch") [1]. In the pesticide sector, a similar trend can be observed, though with a certain delay [2]. The environmental behavior, toxicity, and ecotoxicity of different stereoisomers of chiral compounds can be vastly different. Nevertheless, data on fate and effects of individual stereoisomers are still rare. This is, in part, due to lack of awareness of these differences, lack of legal requirements (e.g., in pesticide registration) and limited availability of appropriate analytical methods.

The last decades have seen a growing perception of possible adverse effects of micropollutants in the environment. To minimize possible risks for human and environmental health associated with micropollutants in the aquatic environment, or to meet generic environmental standards, reduction of contamination at the source is often necessary or desirable. However, depending on the chemical of concern, there are often several sources possible. In case of pesticides, these sources include agricultural use of the compound, but also other sources, such as private households or industry. Agricultural use may give rise to contamination from correct use, but also from accidental input (e.g., overspray of open water, runoff from cleaning of spraying equipment, accidental spills, etc.). If reduction of contamination is necessary, measures should be taken at the predominant source or, if multiple sources are present, where they are most (cost) effective.

Source apportioning is, thus, a key issue. Chiral compounds are particularly interesting in this respect, as they consist of mixtures of isomers which are transported in the environment in the same way (same partitioning behavior, solubility, volatility, etc.), but their composition reflects their initial composition as altered by environmental (biological) processes. They, thus, exhibit something like a "built-in" label. In the following, three case studies of environmental chemistry and source apportioning of chiral compounds will be presented. The chiral switch of the herbicide metolachlor, the first example, occurred during the last few years. During the same time, a field investigation was carried out on the occurrence and enantiomer composition of metolachlor residues in surface waters and it could be shown that residues in the environment quickly responded to the replacement of racemic by enantioenriched metolachlor and that it was possible to verify the chiral switch using enantioselective residue analysis.

The second example of the herbicide mecoprop illustrates many aspects of the environmental behavior of chiral pesticides in soil and how this behavior affects the composition of residues in natural waters. At the same time it illustrates that pesticides also may have significant nonagricultural sources.

A range of micropollutants are regularly detected in ground and surface waters, including industrial and household chemicals, and increasingly also pharmaceuticals and personal care products [3]. The chiral pharmaceutical compound ibuprofen, the third example, was consistently detected in surface water samples collected for the monitoring of pesticides. Its ubiquitous occurrence in natural waters and wastewater, as well as its enantiomer composition and the simultaneous presence of its principal metabolites in raw wastewater indicated human consumption and subsequent excretion, rather than industry, as its source in the environment.

5.2 Enantioselective residue analysis—the key to the assessment of the environmental fate of chiral compounds

Conventional analytical techniques based on gas chromatography (GC), high-performance liquid chromatography (HPLC), or capillary electrophoresis (CE) do not allow distinction between enantiomers. To achieve separations of enantiomers, so-called "chiral selectors" are added to the stationary (in some cases also to the mobile) phase. Cyclodextrins, native or derivatized (alkylated, acylated, silylated) are among the most widely used chiral selectors. The interactions between enantiomers and chiral stationary phase are very specific and difficult to predict as diastereomeric host–guest complexes are formed. As a consequence, the selection of a suitable separation system for chiral compounds is mainly based on trial and error and, therefore, requires a fair amount of intuition and practical experience and access to a wide range of enantioselective HPLC and/or GC columns.

The complexity of enantioselective separations expectedly increases with the number of possible stereoisomers of a compound. In addition, in enantioselective residue analysis, the need for a sensitive and selective detection of the stereoisomers increases, because the detector signal is distributed over several peaks and chances for interferences increase with the number of peaks. Consequently, coupling to mass spectrometry (MS or MS/MS) in many cases is required.

Reference compounds of single stereoisomers are needed to study processes such as enantiomerization/racemization in laboratory experiments. However, such compounds often are not commercially available and have to be prepared, e.g., by semipreparative enantioselective (liquid) chromatography starting with the mixtures. An example of such a semipreparative isolation is shown with the stereoisomers of the selective herbicide metolachlor (details to the compound, see below; structure and absolute configurations see Fig. 5.1). The four isomers can be separated by HPLC using an enantioselective Chiralcel OD-H column (Daicel Chem. Ind., Japan; Fig. 5.2). However, some isomers show partial coelution, so that direct isolation of single isomers with this column from the commercial racemic metolachlor is not straightforward (Fig. 5.2b). Using an nonenantioselective ("achiral") column, metolachlor can be separated into two diastereomeric enantiomer pairs, which can be isolated in small (mg) quantities (Fig. 5.2a). Both fractions, when analyzed using the "chiral" column, can then be separated into the two enantiomers (Fig. 5.2c). Using this sequential isolation procedure, it was possible to obtain all four stereoisomers of metolachlor in good (95–99%) enantiomeric purity [4]. The absolute configuration of each isomer was then assigned based on optical rotation measurements in reference to published data.

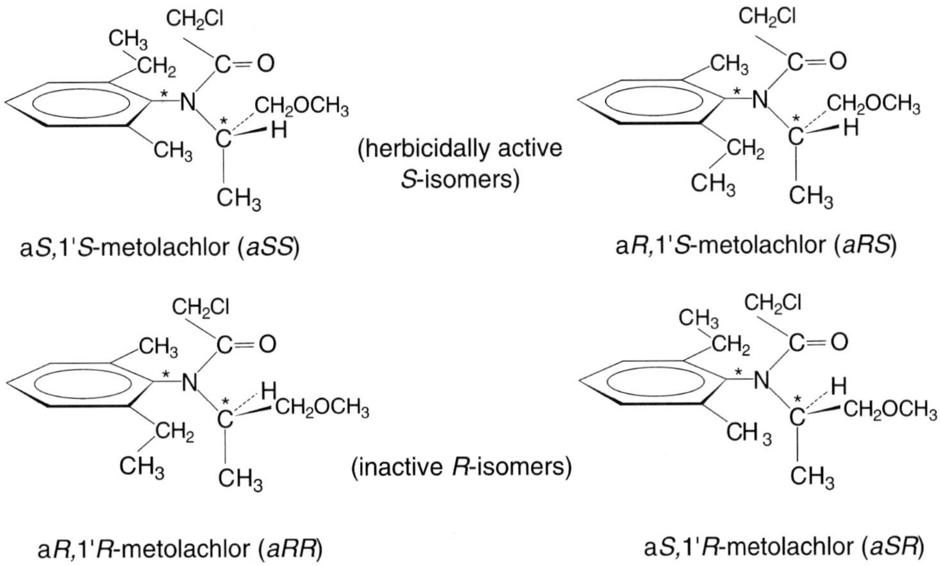

CH$_2$Cl
CH$_3$
CH$_2$
C=O
*N
CH$_2$OCH$_3$
H
CH$_3$
C
CH$_3$

(herbicidally active
S-isomers)

CH$_2$Cl
CH$_3$
C=O
*N
CH$_2$OCH$_3$
H
CH$_2$
C
CH$_3$
CH$_3$

aS,1'S-metolachlor (aSS)   **aR,1'S-metolachlor (aRS)**

CH$_2$Cl
CH$_3$
C=O
*N
H
CH$_2$OCH$_3$
CH$_2$
C
CH$_3$
CH$_3$

(inactive R-isomers)

CH$_2$Cl
CH$_3$
CH$_2$
C=O
*N
H
CH$_2$OCH$_3$
CH$_3$
C
CH$_3$

aR,1'R-metolachlor (aRR)   **aS,1'R-metolachlor (aSR)**

Figure 5.1 Structures (absolute configurations) of the four stereoisomers of metolachlor. The asterisks denote the chiral axis and the asymmetrically substituted C-atom.

Figure 5.2 HPLC chromatograms of *rac*-metolachlor, analyzed using the achiral Hypercarb column (a), and the chiral Chiralcel OD-H column (b), and of an isolate of the first eluted peak on the achiral column, consisting of one enantiomer pair, analyzed using the chiral column (c). Note the diastereomer resolution in panel (a), and the enantiomer resolution in panel (c). Also denoted are the isomer assignments eventually deduced (adapted from Ref. [4]).

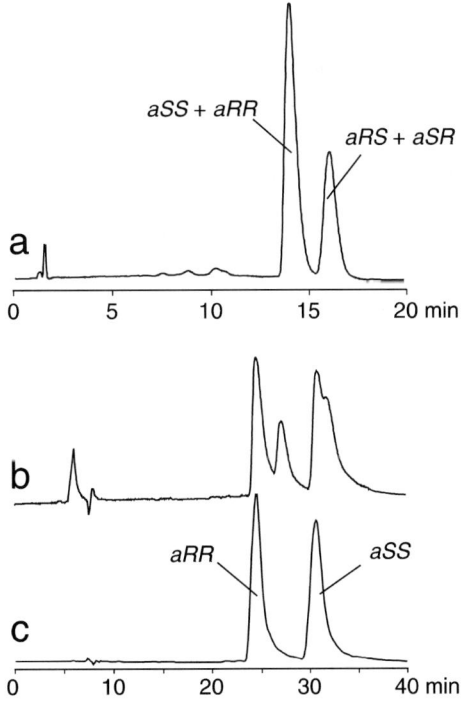

5.3 Metolachlor—monitoring the "chiral switch" in environmental samples

Metolachlor is an important selective herbicide for use in corn to control annual grasses and some broad-leaved weeds [5,6]. The compound is one of the most popular pesticides world-wide with an estimated production in excess of 30,000 t/year [7]. Metolachlor is chiral and—due to the presence of two chiral elements (an asymmetrically substituted carbon and a chiral axis)—consists of four stable isomers (Fig. 5.1). However, most of the herbicidal activity (95%) in racemic metolachlor originates from the two $1'S$-isomers (aSS and aRS, see Ref. [8]). Metolachlor was introduced in 1976 as the racemic compound consisting of equal amounts of active S- (aSS plus aRS) and inactive R- (aSR plus aRR) isomers. Since 1997, metolachlor is being replaced world-wide by S-metolachlor with a content of $\sim 90\%$ S-isomers and $\sim 10\%$ R-isomers and with the same biological effect at 65% of the use rate. The enantioenriched product is manufactured using a novel enantioselective catalytic process [9]. Chromatograms illustrating the different stereoisomer composition of *rac*- and S-metolachlor, analyzed on an enantioselective HRGC column, are shown in Fig. 5.3. In Switzerland, this chiral switch was initiated in 1997 with the registration of S-metolachlor followed by the expiration of the registration for the racemic product. The "chiral switch" from racemic to S-metolachlor, as with other chiral pesticides, not only allows lower application rates, it also reduces the amounts of pesticides released into the environment, prevents deployment of an inactive isomer ("isomer ballast") to the biosphere, and, thus, reduces potential side-effects on non-target organisms. The use of the enantiopure compounds should, thus, lead to lower environmental concentrations and to a changed isomer composition of the residues.

Residues of pesticides in various Swiss lakes are currently being monitored in a survey program initiated by the Federal Office of Agriculture, Berne, Switzerland, to determine concentrations and loads and to correlate these with measures taken to improve farming practices with respect to environmental contamination [10]. Lakes were selected for this program because they are transient reservoirs and reflect the processes in the catchment area, and their composition and properties are less prone to temporal fluctuation than those of rivers and streams (integrating effect). Of particular importance for this study was Lake Baldeggersee. The lake is located in the central region of Switzerland in an area of intense agricultural activities including corn growing, and metolachlor and other pesticides are regularly detected in this lake. Lake Baldeggersee, as most other Swiss midland lakes, is stratified during the warmer season (April to October); in winter (November to March) the lake is mixed. Inputs of the pesticides are expected in spring to early summer into the surface water (epilimnion). Actual pesticide loading to the

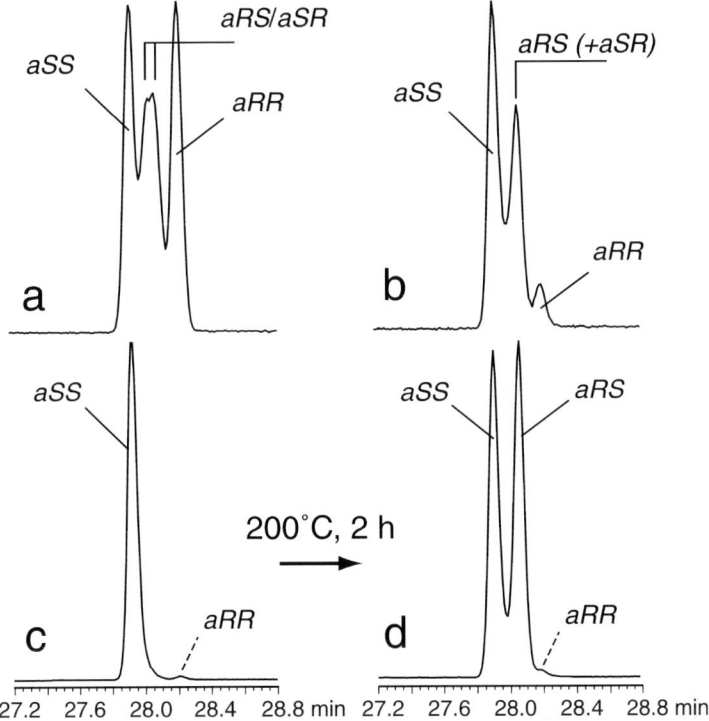

Figure 5.3 HRGC SIM chromatograms (*m/z* 238) of *rac*-metolachlor (a) and
S-metolachlor (b) showing different enantiomer/stereoisomer compositions. Also shown
are chromatograms of the *aSS*-isomer isolated from *rac*-metolachlor by HPLC (c) and a
1:1 mixture of *aSS*- and *aRS*-metolachlor (d), produced from the *aSS*-isomer by thermal
treatment, analyzed using the chiral PS086-BSCD column (adapted from Ref. [4]).

lake (input) depends on a number of parameters, such as the amount of the compounds used in the catchment area, integrated-pest-management (IPM) measures,
and actual weather conditions before, during, and following application (intensity,
duration, and frequency of rainfall events, flooding of fields, etc.).

The time span from 1997 through 2001 covered five application periods of
metolachlor, during which the gradual replacement of *rac*-metolachlor by *S*-meto-
lachlor was expected. Surface water and water from deeper regions (20 m or more)
were analyzed at monthly intervals over a 5-year period. Concentrations of metolachlor during this time were in the range of 3 to 30 ng/L (Fig. 5.4). As for most of
the pesticides observed in the lake, metolachlor showed seasonal changes of concentration and, when analyzed enantioselectively, it showed varying stereoisomer
composition.

Whereas concentrations in the deeper regions of the lake remained more or
less constant, distinct peaks were observed in the epilimnion. These concentration

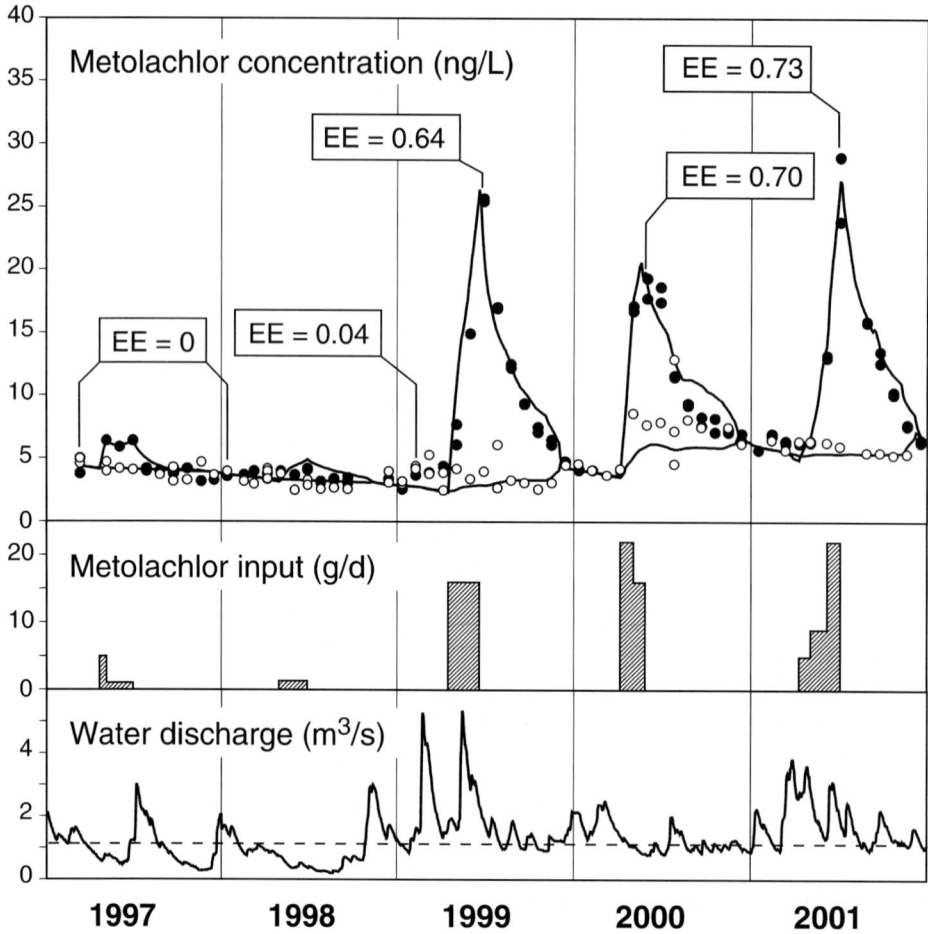

Figure 5.4 Concentration of metolachlor in Lake Baldeggersee (upper panel), simulated metolachlor input to the lake (middle panel), and water discharge from Lake Baldeggersee in 1997 to 2001 (lower panel). Symbols: measured concentrations (●, epilimnion; ○, hypolimnion); simulated data (— concentrations, best fit). The dashed line in the lower panel indicates the long-term average discharge of Baldeggersee. In the upper panel enantiomer compositions (EE) measured in surface water are indicated (adapted from Ref. [22]).

peaks coincided with the application of metolachlor in the catchment area and occurred during spring and early summer at a time when the lake is stratified. During stratification (April–October) flushing of the lake is faster because water exchange to and from the epilimnion (smaller volume) is faster than to and from the whole lake. Because of this rapid elimination (export from the lake), concentrations of metolachlor decreased to almost the same level as before the seasonal input until the lake was vertically mixed again in the fall.

Concentrations (and enantiomer composition) of metolachlor were simulated using a one-dimensional lake model to rationalize the observed behavior and to calculate the input of metolachlor to the lake (for details, see Ref. [11]). In Fig. 5.4, simulated concentrations are plotted together with measured data. The model indicated significant year-to-year variation of metolachlor input to the lake, probably due to strongly varying weather conditions during the time shortly before and after the application of the herbicide (note the different water discharges during these years).

The enantiomer composition, expressed as enantiomeric excess (EE), defined as the excess of the herbicidally active *S*-isomers over the inactive *R*-isomers, is indicated in Fig. 5.4 for selected surface water samples. Residues of metolachlor showed a nearly racemic composition prior to the major input in 1999, followed by a clear change toward a nonracemic composition after this input. The surface water (depth, 2.5 m) during the peak inputs in 2000 and 2001 showed a stereoisomer composition very close to that of *S*-metolachlor and clearly different from that of *rac*-metolachlor. The corresponding chromatograms are plotted in Fig. 5.5.

In terms of absolute concentrations of metolachlor, no direct evidence could be obtained for a replacement of the racemic by the enantioenriched product in the catchment areas of the investigated lakes. In fact, concentrations in 1999 to 2001 were similar and even higher than before, despite the lower application rate of *S*-versus *rac*-metolachlor. This is, in part, due to more surface run-off as a result of intensive rainfall during the application period in these years. In addition,

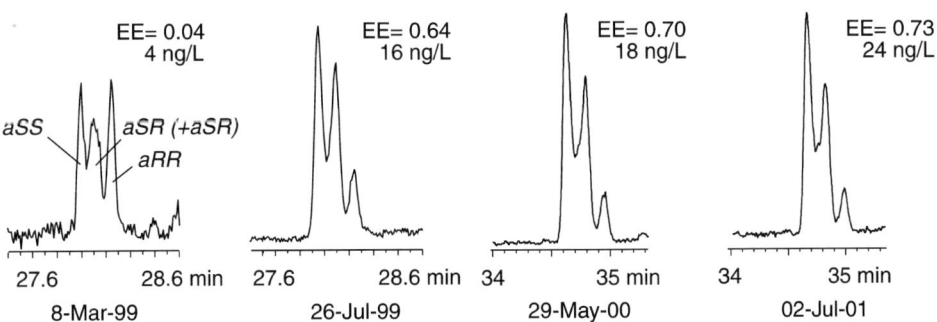

Figure 5.5 EI SIM chromatograms (*m/z* 238) of metolachlor residues in surface water from Baldeggersee, analyzed using an enantioselective PS086-BSCD column, and showing a racemic composition from earlier applications of racemic metolachlor in March 1999, and increasing contributions from *S*-metolachlor following the peak inputs in 1999, 2000, and 2001. Note that different columns were used for analyses resulting in different retention times (adapted from Ref. [22]).

some replacement of other herbicides in favor of metolachlor containing products may have occurred, so that more metolachlor may have been used in the catchment area of the lake. Furthermore, actual herbicide usage depends on the area treated and type of agricultural production (e.g., corn versus other crops) which is also subject to year-to-year variations. Therefore, concentration measurements, at least on a short term basis, could not be used to verify substitution of racemic by *S*-metolachlor.

The enantiomeric composition of the residues, on the other hand, is much less sensitive to these above mentioned factors. The data clearly demonstrate a substantial change of the enantiomeric composition of metolachlor residues in the lakes, reflecting the successive replacement of *rac*-metolachlor by *S*-metolachlor. The data also demonstrate the rapid response of environmental residues to this replacement in terms of enantiomer composition. In the case of metolachlor, the apportioning of racemic versus enantioenriched compound is straightforward, because metolachlor is chirally stable (no racemization or chiral inversion) in soil and surface waters, and is degraded in soil with only slight enantioselectivity. The next example deals with a compound that is degraded in soil with significant enantioselectivity and also undergoes fairly rapid chiral inversion in soil.

5.4 Mecoprop—agricultural versus nonagricultural use of a herbicide

Mecoprop is an important selective herbicide belonging to the family of phenoxyalkanoic acids. Some of these compounds, including mecoprop, are chiral and it was recognized some time ago that for these compounds only the *R*-enantiomers show herbicidal activity [12]. Since 1986 products containing the pure *R*-enantiomers replaced the previously used racemic products in Switzerland and products containing the racemate were deregistered [2]. Nevertheless, there are still many countries, where the racemic products are on the market. The structure and absolute configuration of mecoprop enantiomers are shown in Fig. 5.6.

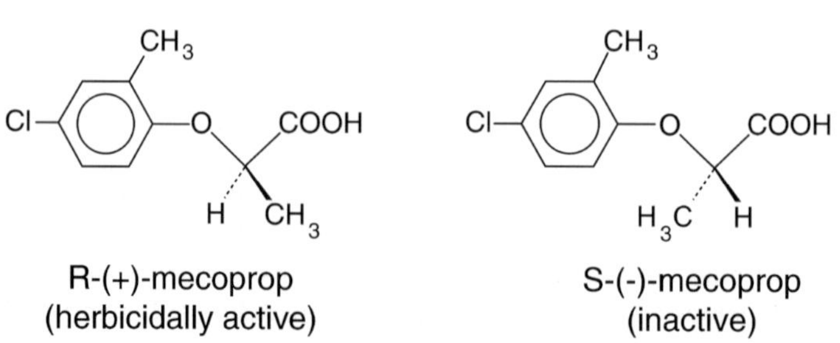

R-(+)-mecoprop
(herbicidally active)

S-(-)-mecoprop
(inactive)

Figure 5.6 Structure and absolute configuration of mecoprop enantiomers.

Mecoprop is degraded in soil with faster dissipation of the *S*-enantiomer, as shown with incubation of the racemic compound (at pH ≤ 7) [13–15]. In addition, incubation of the pure enantiomers in soil revealed some formation and subsequent dissipation of the respective opposite enantiomer [13]. This enantiomerization (or chiral inversion) process, after some time, led to a nonracemic composition of mecoprop in soil (*R* > *S*) independent of the composition at the beginning of the experiment and did not occur in sterilized soil indicating that it is biologically mediated [16]. In another study with alkaline soils (pH ≥ 7.5), mecoprop was shown to be degraded with opposite enantioselectivity [17]. Reevaluation of all these literature data suggested that enantioselectivity of mecoprop degradation in soil may be correlated with soil pH [18]. Residues of mecoprop after application of the enantiopure compound to neutral and acidic soils are thus expected to be enriched in the *R*-enantiomer. If residual mecoprop is subsequently transferred from agricultural fields to surface waters via surface runoff, residues in water are also expected to be enriched with the *R*-enantiomer. As pesticide loss from fields via surface runoff occurs primarily with fresh, mostly unaltered residues, the latter is expected to be true even for alkaline soils.

When water samples from various Swiss lakes were analyzed for the presence of mecoprop, both enantiomers were detected at low levels. However, there were differences in the enantiomer composition between these lakes [19]. In Fig. 5.7 we show GC-MS SIM chromatograms of mecoprop analyzed as methyl ester using an enantioselective GC column, thus demonstrating the presence of both enantiomers in two lakes, Baldeggersee and Greifensee. In Baldeggersee, residues of mecoprop showed a composition with clear excess of the *R*-enantiomer, which is

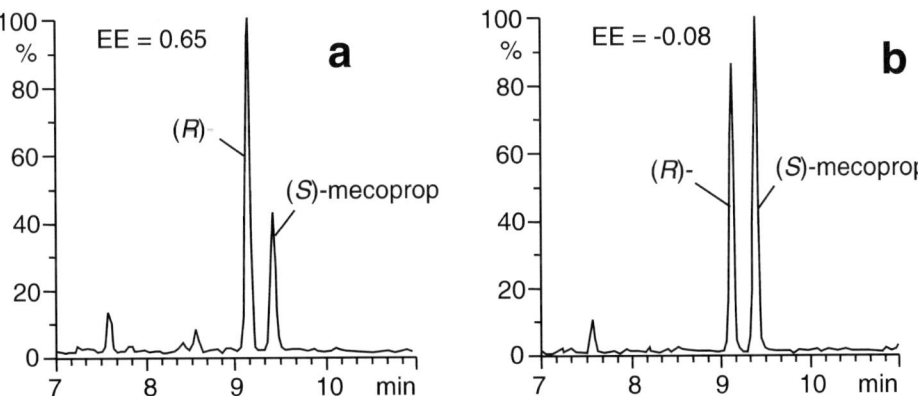

Figure 5.7 EI SIM chromatograms showing elution of (*R*)- and (*S*)-mecoprop (*m/z* 228) as methyl esters from Lakes Baldeggersee (a) and Greifensee (b). Concentrations (sum of *R*- and *S*-enantiomers) in these samples were 20 and 41 ng/L, respectively. Note the different enantiomer composition in the two lakes (adapted from Ref. [19]).

in agreement with the agricultural use of enantiopure (R)-mecoprop and its be-havior in soil as outlined above. In Greifensee, however, mecoprop residues were racemic or even slightly enriched with the S-enantiomer, pointing to an additional ("racemic") source of mecoprop in the tributary of this lake. It was also noted that mecoprop in the tributary of Greifensee was mainly associated with effluents of WWTPs [20,21]. Mecoprop in WWTP effluents was racemic or enriched with the S-enantiomer. The source of mecoprop in wastewater was later identified as an additive to bituminous membranes used for the sealing of flat roofs. The additive, an ester of mecoprop, prevents perforation of the membranes by roots of plants [20]. When the roof is wet, the ester is slowly hydrolyzed and racemic mecoprop is released and discharged during rain events.

Apportioning of these sources of mecoprop (agricultural use versus flat roofs) in lakes is complicated by the fact that the compound is chirally unstable in soil. A quantitative apportioning similar to that done with metolachlor is thus challeng-ing and requires careful evaluation of the impact of possible enantiomerization processes on the enantiomer composition of residues in lakes. Enantiomerization of mecoprop in soil, as discussed above, should have some, but limited influence on the enantiomer composition in surface runoff. Presuming that pH values of soils in the catchment areas of different lakes in the Swiss midland region, on average, are similar, enantiomer composition of mecoprop residues in surface wa-ters originating from agricultural runoff should thus vary little between different lakes.

Laboratory incubations indicated that mecoprop may be chirally unstable in surface waters and that enantiomerization occurs with preferential formation of (S)-mecoprop [19]. Such an enantiomerization in surface waters would have a strong impact on the enantiomer composition of mecoprop, with increasing pref-erence for the S-enantiomer as the residence time in the water body increases. However, mecoprop residues in some lakes with long water residence time such as Baldeggersee (filling time, ≈ 5 years) and Sempachersee (15 years), are enriched with the R-enantiomer, indicating that enantiomerization of mecoprop in natural waters may be negligible compared with other elimination processes. These dif-ferences between field data and data obtained under more artificial conditions in the laboratory may need some further investigation.

Additional indication for the considerable chiral stability in natural waters is obtained from monthly data on mecoprop concentrations and enantiomer compo-sition in Baldeggersee at different depths, determined in 1999 (Fig. 5.8). Meco-prop concentrations in the epilimnion increased during April and May due to seasonal input to this layer after application of mecoprop in the field. Concen-trations in the hypolimnion remained constant, because there is no significant in-put of mecoprop to this compartment and apparently no degradation. Likewise, the chiral composition of mecoprop residues, expressed as enantiomer excess,

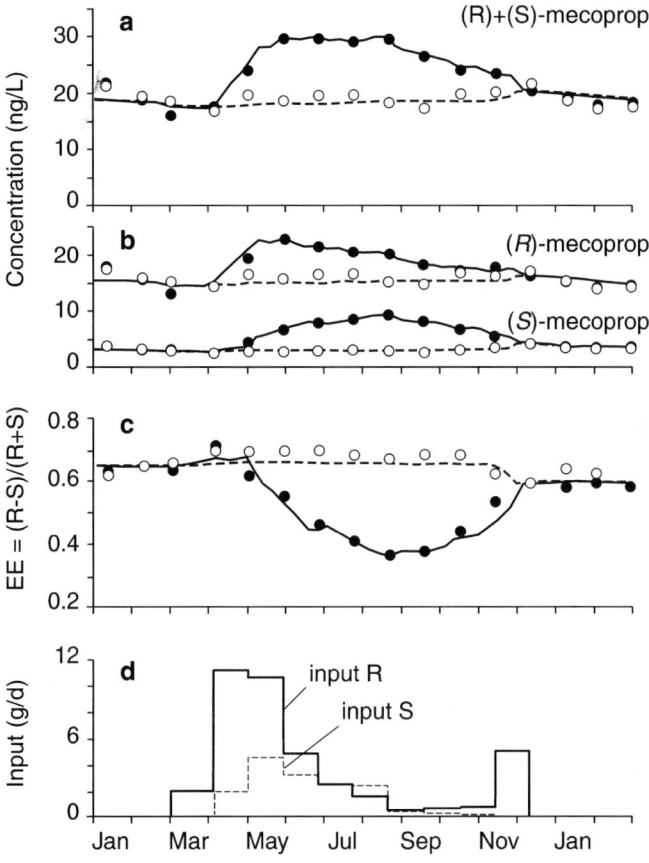

Figure 5.8 Total concentration (a), concentration of *R*- and *S*-enantiomers (b), and enantiomer composition, expressed as enantiomer excess (EE, panel (c)) of mecoprop residues in surface (full circles) and deep water (open circles) from Baldeggersee in 1999 (circles, measured values; lines, modeled data). Modeled monthly input of mecoprop between sampling events is shown on panel (d). Concentration increases due to input only to the surface water (stratified lake). Vertical mixing at the end of the year eliminates concentration differences between surface and deep water.

$EE = ([R] - [S])/([R] + [S])$, in the hypolimnion remained constant (Fig. 5.8c), indicating that mecoprop is also enantiomerically stable in this compartment over a period of several months. In contrast, the enantiomer composition in the epilimnion decreased from May to August, and then increased again from September to December, almost to the same value as in spring. This variation in enantiomer composition could be due to input of mecoprop with varying composition or to microbial conversion or to a combination of both.

To rationalize these findings, a computer model was constructed on the basis of the model used for simulation of metolachlor concentrations and input in the same lake [11,22]. The model included flushing as the only elimination process for mecoprop in the lake and it was assumed that no enantiomerization or degradation of mecoprop occurred during the simulated period (one year). Simulated concentrations of mecoprop enantiomers in the lake were fitted against measured data by variation of the monthly input and the resulting values are plotted in Fig. 5.8. Mecoprop input, calculated with this model, occurred primarily from March to November. There are two input peaks in April/May and November consistent with spring and fall application of mecoprop in agriculture, respectively. During these peaks, input is predominantly from the R-enantiomer. Between these peaks, input seems to be closer to racemic, indicating that mecoprop may be from a different source, likely from runoff of flat roofs during this time. It should be noted that it is possible to rationalize the observed variations of mecoprop concentrations and enantiomer composition solely based on variation of mecoprop enantiomer input to the lake and without introduction of (enantioselective) degradation and/or enantiomerization processes to the model. This result supports the assumption that these processes may be less important or even negligible not only in the hypolimnion, but in the whole lake. It is, however, not a proof for the absence of such processes. Direct evidence for their presence or absence would be difficult to obtain and would require regular determination of mecoprop concentrations and enantiomer composition not only in the lake, but also in tributaries over an extended period of time.

Considering the limitations discussed above, apportioning of the contributions of different sources to the overall input of mecoprop to different lakes, solely based on the enantiomer composition of its residues in these lakes, is somewhat tentative, but will nevertheless be attempted as follows. In Fig. 5.9 we plot enantiomer composition as a function of contribution of roof runoff and agriculture (runoff and drainage from fields, also including input from private households using mecoprop in gardening). Mecoprop from roof runoff (left-hand side in Fig. 5.9) may be racemic or slightly enriched in (S)-mecoprop as observed in many WWTP effluents, with EE values in the range of -0.2 to 0 (see also Refs. [19,20]). Mecoprop from agricultural uses (right-hand side) should be the pure R-enantiomer as applied in the field or a mixture enriched in (R)-mecoprop as observed after some degradation/enantiomerization in soil. The corresponding range of values (shaded area) is fairly large. Therefore, we also plotted a "best estimate" (bold line) considering the observed predominance of (S)-mecoprop in WWTP effluents as well as the fact that surface runoff of mecoprop from fields occurs primarily with fresh, unaltered residues in soil.

Using the diagram in Fig. 5.9, the relative contribution of the two major mecoprop sources can be estimated for different lakes. Sempachersee and Baldeggersee

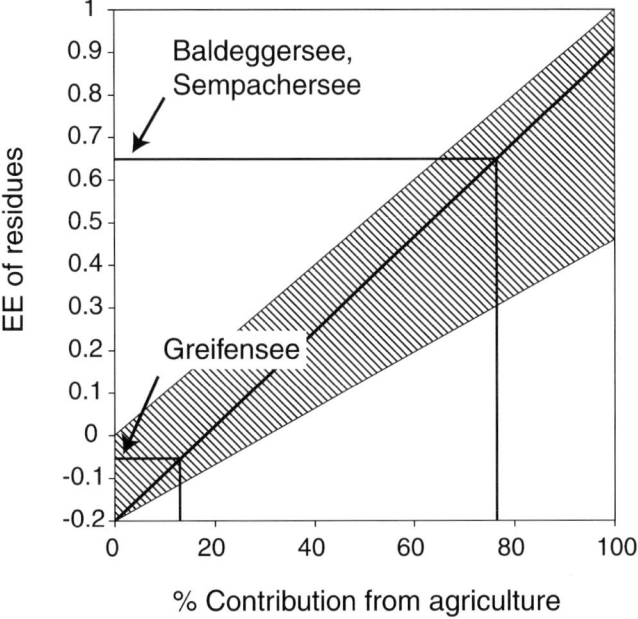

Figure 5.9 Relationship between contribution of different sources (roof runoff versus agricultural) to the overall input of mecoprop to surface waters and enantiomer composition of mecoprop residues. The shaded area represents the range of expected values, the bold line indicates the "best estimate," which is then used to relate the average enantiomer composition of mecoprop residues in selected lakes to the estimated contribution of the different sources.

are located in rural areas receiving little water from residential areas. The enantiomer composition of mecoprop residues, on average, is in the range of EE = 0.6–0.7, corresponding to a relative contribution of agricultural input of $\approx 75\%$. Greifensee, on the other hand, receives much more water from residential areas and EE values of mecoprop residues are in the range of −0.1 to 0, corresponding to only 10–20% agricultural input. The latter finding is consistent with data from continuous monitoring of mecoprop in WWTPs and tributaries of Greifensee [21]. The data, thus, indicate that herbicide input to surface waters from nonagricultural sources may be significant and, in some cases, may exceed that from agricultural applications.

5.5 Ibuprofen—input to surface waters via human medical use

In recent years, reports on the occurrence of pharmaceutical compounds in the environment have received much attention (see, e.g., Ref. [3]). Among the most fre-

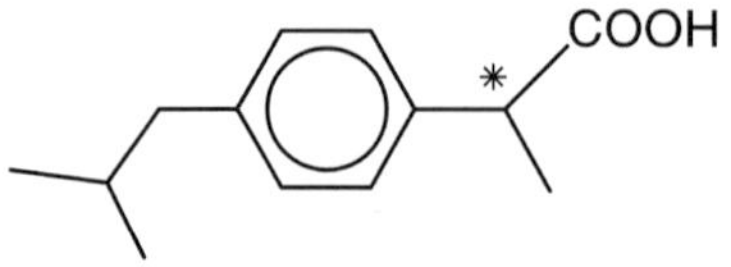

Figure 5.10 Structure of the pharmaceutical drug ibuprofen. The asterisk indicates the chiral center.

quently used pharmaceuticals is ibuprofen (IB; (*rac*)-2-[4-isobutylphenyl]propionic acid, structure see Fig. 5.10), a nonsteroidal antiinflammatory (NSAID), analgesic and antipyretic drug, widely used in the treatment of rheumatic disorders, pain, and fever [23]. It has an estimated annual global production of several kilotons, and it is the third-most popular drug in the world [24]. It is an important non-prescription drug and has a relatively high therapeutic dose (200–1200 mg/day). IB has an asymmetrically substituted carbon atom and is chiral (Fig. 5.10). The desired pharmacological effects reside almost exclusively in the *S*-enantiomer, yet the racemic compound was used as the drug until recently when the enantiopure form of IB has become available as well [1]. It has been shown that in humans and other mammals the inactive (*R*)-(−)-isomer undergoes extensive (unidirectional) chiral inversion to yield the active (*S*)-(+)-isomer [23]. It is excreted to a significant degree (70–80% of the therapeutic dose) as the parent compound (free or conjugated) or in the form of metabolites [23]. Its physico-chemical properties suggest a rather high mobility in the aquatic environment.

IB and its principal metabolites were consistently detected in raw wastewater from various WWTPs [25]. The concentrations of IB in WWTP influents (1 to 3.3 µg/L) correspond to inputs, normalized for the population serviced by these plants, of 0.28–1.1 mg person^{-1} day^{-1}. These inputs are in good agreement with overall estimated consumption of IB, assuming some metabolization in man and degradation in the sewer system. The residual IB in these samples showed a predominance of the *S*-enantiomer (EE = 0.7–0.8; as an example, see Fig. 5.11c), similar to human urine (EE ≈ 0.9; Fig. 5.11b) and in contrast to the drug itself (Fig. 5.11a). The widespread occurrence of IB in WWTP influents, the concurrent presence of the principal human metabolites, and the data on the enantiomer composition of IB thus clearly point to inputs from human therapeutic use of the drug, rather than ("racemic") input from industrial sources or from drugs which were directly disposed of by consumers.

In WWTP effluents, IB was detected at much lower concentrations (≈ 2–81 ng/L) whereas the concentrations of metabolites were below the detection limit, indicating that these compounds are degraded extensively during wastewater treatment (96–99.9% removal of IB). Residual IB in WWTP effluents showed a different enantiomer composition (Fig. 5.11d) than in influents and indicated that (*S*)-IB is somewhat faster degraded than (*R*)-IB. The simultaneously detected pharmaceutical compound clofibric acid (CA), which is more persistent than IB, was essentially not degraded and became enriched in WWTP effluents relative to IB (Fig. 5.11c and 5.11d).

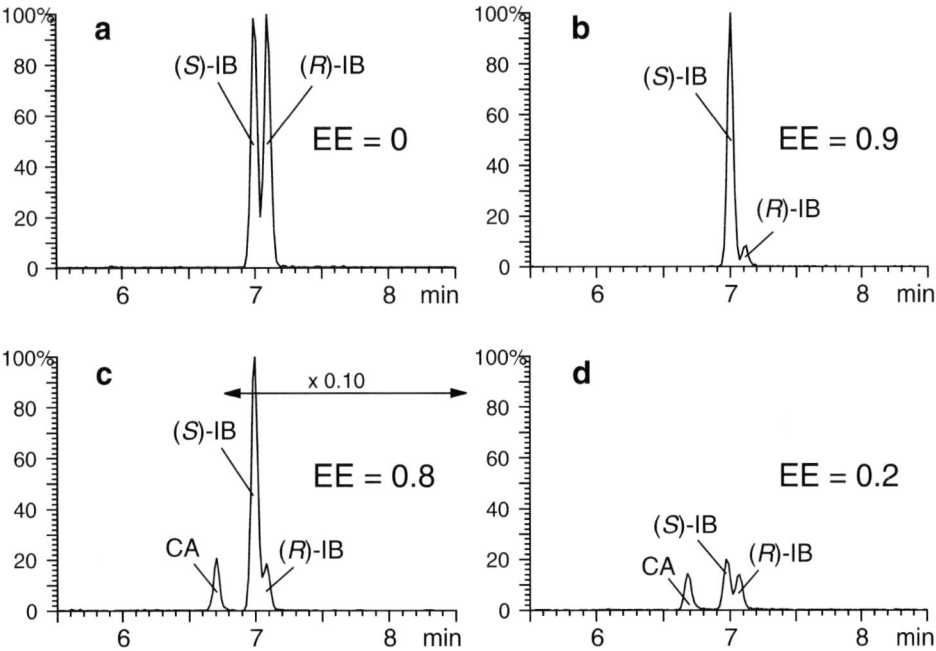

Figure 5.11 EI SRM chromatograms showing elution of ibuprofen (upper chromatograms, transition $220^+ \rightarrow 161^+$) and of ibuprofen and clofibric acid (lower chromatograms, combined trace of transitions $220^+ \rightarrow 161^+$ and $228^+ \rightarrow 169^+$), analyzed as methyl esters on an enantioselective OV1701-DMPen column. Racemic ibuprofen (IB) used as the pharmaceutical drug (a), ibuprofen extracted from urine after use of the drug (b), ibuprofen and clofibric acid (CA) extracted from WWTP influent (c) and effluent (d) (adapted from Ref. [25]).

Even though emissions of IB with treated wastewater are small, IB was consistently detected in water samples from different lakes receiving WWTP effluents at concentrations of up to 8 ng/L. Concentrations found in lakes, in some cases, were higher than expected from the input loads calculated from concentrations in WWTP effluents, pointing to a possible significant contribution of untreated wastewater (such as from overflow during rain events) to overall input. In fact, because of the efficient elimination of IB in WWTPs, even a small fraction of untreated wastewater discharged directly (1–4%), would lead to an input of IB of the same magnitude as from the total amount in WWTP effluents. A recently published study on the occurrence and behavior of caffeine in wastewater and surface water also indicated that compounds which are extensively (> 99%) eliminated in WWTPs are discharged to surface waters primarily via untreated wastewater and could serve as markers for the proportion of untreated versus treated wastewaters [26].

References

1. I. Agranat, H. Caner, and A. Caldwell, *Nat. Rev. Drug Discov.* **1** (2002) 753–768.

2. A. Williams, *Pestic. Sci.* **46** (1996) 3–9.

3. C.G. Daughton and T.A. Ternes, *Environ. Health Perspect.* **107** (1999) 907–938.

4. M.D. Müller, T. Poiger, and H.R. Buser, *J. Agric. Food Chem.* **49** (2001) 42–49.

5. W.A. Ahrens, *Herbicide Handbook*, seventh ed., Weed Society of America, Champaign, IL, 1994.

6. K.K. Hatzios, *Herbicide Handbook*, Supplement to seventh ed., Weed Society of America, Champaign, IL, 1998.

7. U.S. Environmental Protection Agency, Office of Pesticide Programs, Washington, DC (http://www.epa.gov/oppbead1/pestsales) accessed in July 2003.

8. H. Moser, G. Rihs, and H. Sauter, *Z. Naturforsch. B* **37** (1982) 451–462.

9. H.U. Blaser, H.P. Buser, K. Goers, R. Hanreich, H.P. Jalett, E. Jelsch, B. Pugin, H.D. Schneider, F. Spindler, and A. Wegmann, *Chimia* **53** (1999) 275–280.

10. D. Forni, L. Gujer, S. Nyffenegger, U. Vogel, and U. Gantner, *Agrar Forschung* **6** (1999) 107–110.

11. H.R. Buser, T. Poiger, and M.D. Müller, *Environ. Sci. Technol.* **34** (2000) 2690–2696.

12. B. Aberg, In: R.L. Waine and F. Wightman (Eds.), *The Chemistry and Mode of Action of Plant Growth Substances*, Butterworth Scientific Publications, London, 1956, p. 102.

13. M.D. Müller and H.R. Buser, *Environ. Sci. Technol.* **31** (1997) 1953–1959.

14. A.W. Garrison, P. Schmitt, D. Martens, and A. Kettrup, *Environ. Sci. Technol.* **30** (1996) 2449–2455.

15. J.M. Schneiderheinze, D.W. Armstrong, and A. Berthod, *Chirality* **11** (1999) 330–337.

16. H.R. Buser and M.D. Müller, *Environ. Sci. Technol.* **31** (1997) 1960–1967.

17. E. Romero, M.B. Matallo, A. Pena, F. Sanchez-Rasero, P. Schmitt-Kopplin, and G. Dios, *Environ. Pollut.* **111** (2001) 209–215.

18. I.J. Buerge, T. Poiger, M.D. Müller, and H.R. Buser, *Environ. Sci. Technol.* **37** (2003) 2668–2674.

19. H.R. Buser and M.D. Müller, *Environ. Sci. Technol.* **32** (1998) 626–633.

20. T.D. Bucheli, S.R. Müller, A. Voegelin, and R.P. Schwarzenbach, *Environ. Sci. Technol.* **32** (1998) 3465–3471.

21. A.C. Gerecke, M. Scharer, H.P. Singer, S.R. Müller, R.P. Schwarzenbach, M. Sagesser, U. Ochsenbein, and G. Popow, *Chemosphere* **48** (2002) 307–315.

22. T. Poiger, M.D. Müller, and H.R. Buser, *Chimia* **56** (2002) 300–303.

23. A.J. Hutt and J. Caldwell, *J. Pharm. Pharmacol.* **35** (1983) 693–704.

24. Anonymous, *Scrip* (1988) April 21.

25. H.R. Buser, T. Poiger, and M.D. Müller, *Environ. Sci. Technol.* **33** (1999) 2529–2535.

26. I.J. Buerge, T. Poiger, M.D. Müller, and H.R. Buser, *Environ. Sci. Technol.* **37** (2003) 691–700.

Advances in Mass Spectrometry, Volume 16
A.E. Ashcroft, G. Brenton and J.J. Monaghan (Editors)

CHAPTER 6

Noncovalent Biomolecule Complexes Both Large and Small

Evan R. Williams*, John C. Jurchen, David E. Garcia,
Andrew S. Lemoff, and Matthew F. Bush

Department of Chemistry, University of California, Berkeley, CA 94720, USA

6.1 Introduction

Noncovalent interactions play a major role in the chemistry and structure of bio-molecules. These can be weak, nonspecific interactions, such as those between a molecule and the surrounding solvent, or they can be strong, specific interactions, such as those between an enzyme and substrate molecules or two complementary strands of deoxyribonucleic acid (DNA). With electrospray ionization (ESI), it is possible to produce a wide range of noncovalent complexes, ranging from weakly bound hydrated ions [2] to strongly bound double stranded DNA [3,4] and even intact ribosomes [5,6]. For specific complexes, information about the stoichiom-etry of the complex can be readily obtained from a simple mass measurement. To obtain structural information from gas-phase biomolecules, tandem mass spec-trometry can be used and can provide information about the sequence of peptides, proteins, DNA, and oligosaccharides, as well as information about sites and iden-tity of posttranslational modifications.

Recent evidence shows that multiply charged gas-phase ions produced by elec-trospray ionization are formed by the charge residue mechanism as originally pro-posed by Dole [7–9]. Further evidence for the charge residue mechanism comes from solvation studies [1,2] and a quantitative comparison between charge states of ions produced by these two mechanisms [9]. These studies suggest that the final structure of the ion is determined by the solvent evaporation process. As sol-

---

*Corresponding author. E-mail address: williams@cchem.berkeley.edu.

vent evaporates, energy is removed from the cluster and the cluster cools. For solvent to continue to evaporate, energy must be added to the cluster, either through gas-phase collisions with neutral atoms or molecules or through the absorption of blackbody photons generated by the surrounding surfaces of the mass spectrometer. Because energy is removed from the cluster each time a solvent molecule evaporates, the temperature of hydrated clusters can be very low particularly if the energy transfer process is minimized, either by reducing the number and energy of gas-phase collisions, or by reducing the temperature of the blackbody radiation field. This process can be made to be very gentle resulting in the potential for gas-phase ions to retain a "memory" of their solution-phase structures [10–12]. For example, the dissociation kinetics for the loss of a heme group from myoglobin is strongly dependent on the solvent conditions from which these ions are formed [10]. Similar results have been reported for very small molecules as well [12]. Thus, it is possible to form gas-phase ions that are not in their most stable gas-phase form and the potential exists to "lock-in" elements of solution-phase structure.

Here, we examine the effects of solvent on molecular structure in the gas phase and demonstrate that it is possible to convert the most favored gas-phase structure of a cationized amino acid from its preferred nonzwitterion form to its preferred zwitterion form with the addition of water molecules. We also examine the origin of asymmetric charge partitioning that occurs in the dissociation of noncovalent complexes of large biomolecules.

6.2 Experimental

Experiments investigating the structure of amino acids, hydration, and zwitterion formation were performed on a home-built 2.7 Tesla Fourier-transform ion cyclotron resonance (FTICR) mass spectrometer with an external electrospray ionization source. This instrument has been described in detail elsewhere. Blackbody infrared radiative dissociation (BIRD) experiments are performed in an ion cell that is surrounded by a copper jacket that can either be resistively heated or cooled with liquid nitrogen. This new cooled cell is described elsewhere [13]. To obtain the threshold dissociation energies from these BIRD experiments, it is necessary to model the experimental data by solving the master equation. The process for the master equation modeling is described in detail elsewhere [14]. Ions are formed by nanospray ionization using borosilicate capillaries that are pulled to a tip diameter of ~ 4 μm.

Experiments on peptide and protein clusters were performed on a 9.4 Tesla FTICR instrument that has unique capabilities for two-dimensional experiments. The clusters are dissociated using sustained off-resonance irradiation (SORI) collisionally activated dissociation (CAD) by applying a waveform slightly off-resonance to the natural cyclotron frequency of the ion while the pressure inside the ion cell is momentarily increased by introducing nitrogen gas through a pulsed valve. After the collision gas is introduced, the ion cell returns to a base pressure of $< 5 \times 10^{-9}$ Torr prior to product ion detection.

6.2.1 The role of water and basicity in zwitterion formation

Amino acids exist in their zwitterionic form in water within a wide range of pH. Yet in the gas phase, the nonzwitterion structure is more stable for all the naturally occurring amino acids. For example, the nonzwitterion form of glycine is ~ 20 kcal/mol more stable than the zwitterion form in the gas phase [15]. Arginine is the most basic amino acid, and the zwitterion form is stabilized because of the strongly basic guanidine side chain. But both experiment and theory indicate that the most stable form of this amino acid is nonzwitterionic although the relative energies of these two forms of arginine are similar [16–18].

The energetics related to gas-phase zwitterion formation are shown in Fig. 6.1 [19]. In order to produce a zwitterion in the gas phase, a proton must be removed from the acid site. This requires an energy corresponding to ΔH for deprotonation. For glycine, this value is ~ 340 kcal/mol. Energy corresponding to the proton affinity is gained by adding the proton to the basic site. For glycine, this value is 211.9 kcal/mol. Energy is also gained from the binding associated the with the attraction of the positive and negative charged groups. This corresponds to primarily the Coulomb attraction corresponding to bringing two opposite charges from infinite distance to their interaction distance. Finally, whether the zwitterion form is most stable depends on the sum of the above interactions compared to the stabilization associated with the noncharged form of the amino acid [19].

The zwitterion form of an amino acid can be stabilized by the addition of a charge, such as an alkali metal ion. For glycine, attachment of a singly-charged metal ion to form a salt-bridge structure can substantially stabilize the zwitterionic form of the amino acid, but this structure is still a few kcal/mol less stable than the nonzwitterion or charge-solvated structure [20,21]. Attachment of divalent metals ions can result in the salt-bridge structure, in which glycine is zwitterionic, being substantially more stable [22–24]. For arginine, attachment of a metal ion can also change the relative energetics of the nonzwitterion vs. zwitterion form of the amino acid, but there is a strong dependence on cation size [25]. Both experiment and theory indicate that the structure of cationized arginine changes as a function of increasing cation size from a charge-solvated structure in which arginine is nonzwitterionic to a salt-bridge structure in which arginine is zwitterionic [25]. For the smaller cations, arginine is effectively able to solvate the metal ion, but for the larger cations, arginine can only partially solvate the charge. This results in a destabilization of the charge-solvation structure with cation size.

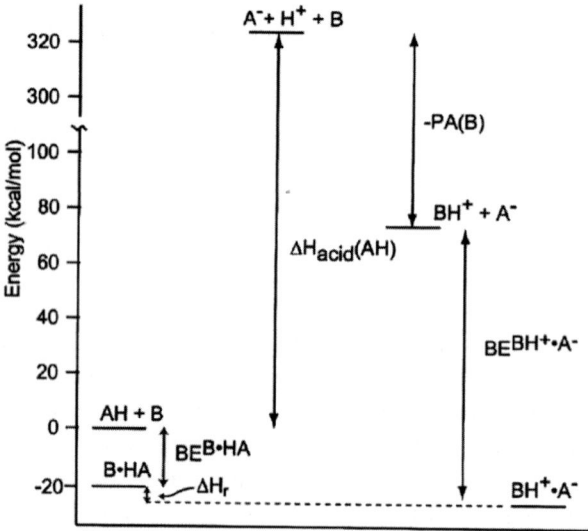

Figure 6.1 Energy diagram showing the relative energetics of producing a gas-phase ion pair or zwitterion vs. nonzwitterion structure (reprinted from Ref. [19]).

The basicity of the amino acid clearly plays an important role in the relative stability of the zwitterion and nonzwitterion forms (Fig. 6.1). Bowers and co-workers examined the effects of proton affinity of glycine, alanine and a related series of molecules and concluded that there was a nearly linear relationship between the proton affinity and the stability of the zwitterion [21]. Based on these calculations, the propensity of all 20 cationized amino acids to form charge-solvated vs. salt-bridge structures was predicted [21]. Experimental results on protonated trimers consisting of two identical basic molecules and trifluoroacetic acid showed that the structure of the trimers is directly related to the proton affinity of the bases [26]. In a gas phase "titration" experiment, it was found that trimers consisting of the least basic molecules exist as charge-solvated structures in which the proton is solvated by the neutral base and acid molecules. Trimers consisting of the most basic molecules exist as salt-bridge structures in which both bases are protonated and the trifluoroacetic acid is deprotonated.

6.2.1.1 Solvation

Investigation of the structure of hydrated ions can provide detailed information about how water influences the structure of an ion. Using electrospray ionization, it is possible to form extensively solvated ions consisting of complexes of an amino acid and an alkali metal ion. By investigating the dissociation of these complexes with BIRD, it is possible to obtain accurate values of the threshold dissociation of water from these complexes. Since there is not expected to be a significant reverse activation barrier, the threshold dissociation energy should be approximately equal to the water binding energy.

To deduce structural information from the measured water binding energies, reference molecules are chosen to model the nonzwitterionic and zwitterionic form of the amino acid. For valine, we choose the ethyl ester of alanine (AlaOEt) as a model for the nonzwitterionic form and betaine (Bet) as a model of the zwitterionic form. These compounds are isomers and absorb and emit radiation at similar rates [27,28].

For $AA \cdot Li^+(H_2O)_1$ (amino acid or amino acid analogue), we find that the binding energy of water to valine (0.88 ± 0.03 eV) is nearly the same as that for AlaOEt (0.85 ± 0.02 eV), but both are different from that of betaine (0.81 ± 0.02 eV). From these experimental data, we conclude that the structure of valine in these complexes is nonzwitterionic. Calculations indicate that the lowest energy structure of nonzwitterionic valine is one in which the lithium ion interacts with both the amino nitrogen and the carbonyl oxygen (NO coordination) and that water interacts directly with the lithium ion. This mode of water binding in the nonzwitterion form of valine is the same as that in AlaOEt. The lowest energy zwitterionic form of valine is one in which the lithium ion interacts directly with the carboxylate group (OO coordination) and the water molecule interacts with both the lithium ion and the carboxylate group. This mode of water binding in the zwitterionic form of valine is the same as for betaine. Thus, these two molecules are excellent references for these two forms of valine. Calculations indicate that the nonzwitterionic form is more stable, in excellent agreement with the experimental results.

For $AA \cdot Li^+(H_2O)_2$, the binding energy of water to valine and AlaOEt are indistinguishable (0.59 ± 0.01 eV), whereas that for betaine is higher (0.65 ± 0.02 eV). The experiment clearly shows that valine is nonzwitterionic in these clusters. The structures described above for the one water case are the same, except the second water molecule again interacts directly with the lithium. Calculations indicate that the zwitterionic and nonzwitterionic forms are nearly isoenergetic.

For $AA \cdot Li^+(H_2O)_3$, water is bound significantly more strongly to valine (0.50 ± 0.02 eV) than to either AlaOEt (0.26 ± 0.02 eV) or betaine (0.39 ± 0.02 eV). Calculations indicate that water binding to a NO coordinated structure should be the same for both nonzwitterionic valine and AlaOEt. The experiment clearly shows that the addition of the third water molecule changes the mode of the metal ion binding from NO coordination to OO coordination. Modeling indicates that the mode of water binding in the nonzwitterion valine with OO coordination is the same as that for betaine, but different for the zwitterionic form of valine with OO coordination. The significantly unique binding energy of water to valine provides strong evidence for the zwitterionic form of valine in this cluster.

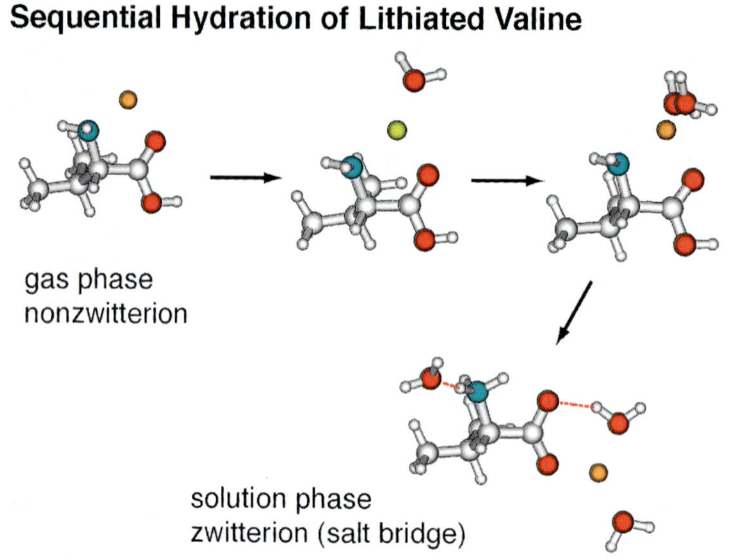

Sequential Hydration of Lithiated Valine

gas phase
nonzwitterion

solution phase
zwitterion (salt bridge)

Figure 6.2 Structure of lithiated valine as a function of the number of attached water molecules showing the transition from the gas-phase structure in which valine is nonzwitterionic with no water to the solution-phase structure in which valine is a zwitterion with three water molecules.

These results indicate that the addition of a singly water molecule to lithiated valine does not significantly change the structure or the relative energetics of the two forms of valine. Valine is nonzwitterionic with the metal ion NO coordinated (Fig. 6.2). Addition of a second water molecule does not change the structure significantly, but the relative energies of the two forms of valine are very similar. Addition of a third water molecule changes the mode of metal ion binding from NO coordination to OO coordination, and is the first water molecule to directly interact with valine (Fig. 6.2). Three water molecules are sufficient to produce the zwitterionic form of valine in these clusters [28].

6.2.1.2 Metal ion size and basicity

The effects of the proton affinity of the proton acceptor in the amino acid cluster was investigated by comparing the binding energy of water to sodiated clusters of valine, betaine, alanine ethyl ester, sarcosine ethyl ester, and methylaminoisobutyric acid (Maiba). The structures of these molecules are shown in Fig. 6.3. As before, alanine ethyl ester and betaine provide reference structures for the nonzwitterionic and zwitterionic form of valine, respectively. Sarcosine ethyl ester and betaine are references for the nonzwitterionic and zwitterionic form of Maiba. The calculated proton affinity of Maiba is 7.4 kcal/mol higher than that for valine,

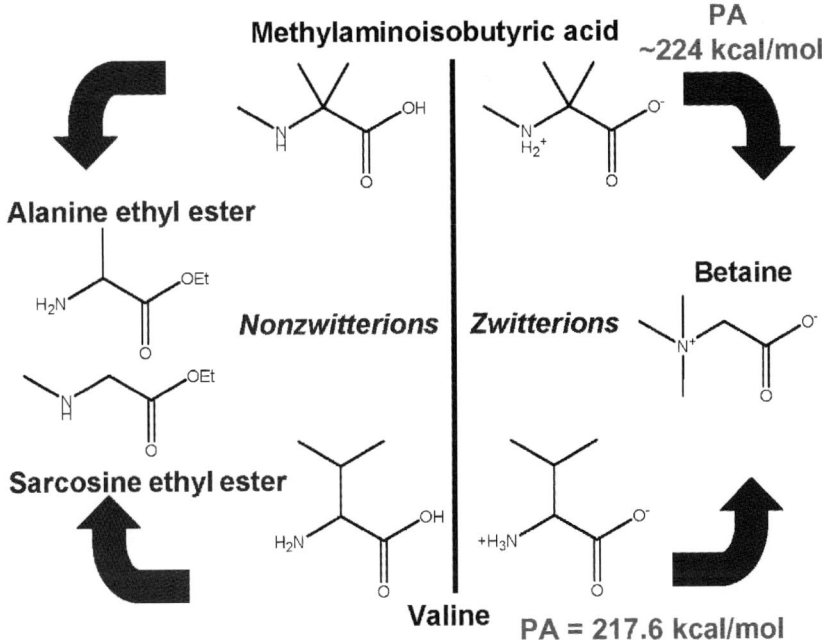

Figure 6.3 Structures of valine, methylaminoisobutyric acid (Maiba) and structural analogues of both the zwitterion and nonzwitterion forms of these isomers.

so this molecule should have a higher propensity to form a zwitterionic structure [29]. Predictions by Bowers and co-workers [21] based on the proton affinity of the amino acids suggest a zwitterionic form of sodiated valine, but kinetic experiments of Wesdemiotis and co-workers [30] indicate a nonzwitterion form.

BIRD Arrhenius plots for the loss of a single water molecule from each of these clusters is shown in Fig. 6.4. By modeling these data, we obtain the binding energy of water. These values are indicated in Fig. 6.4. We find that valine, alanine ethyl ester, and sarcosine ethyl ester all have the same water binding energies. Similarly, Maiba and betaine have the same water binding energies and these values are clearly distinguishable from the nonzwitterionic reference compounds. From these results, we conclude that valine is nonzwitterionic in these clusters whereas Maiba is a zwitterion [29].

Similar results for a model of proline strongly suggest that both lithiated and sodiated proline adopts a zwitterionic form with one water molecule attached. These results are consistent with theoretical predictions of others [30].

6.2.2 Asymmetric protein dimer dissociation

In contrast to nonspecific, weakly bound complexes, such as the hydrated amino acid complexes discussed earlier, specific complexes of biomolecules can be sub-

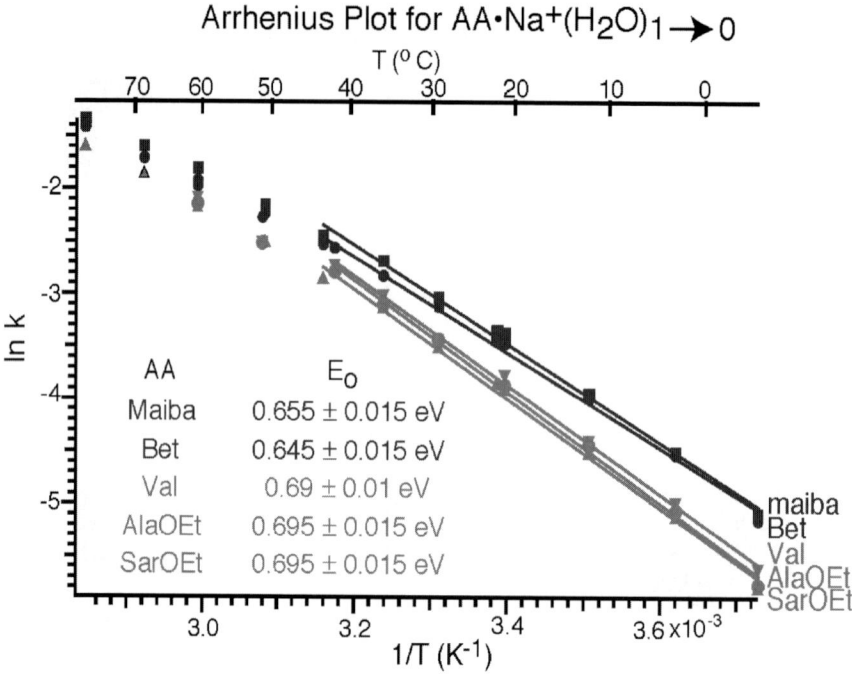

Figure 6.4 Zero-pressure Arrhenius plot for loss of a water molecule from lithiated valine, Maiba, and structural analogues using blackbody infrared radiative dissociation (BIRD) in a heated/cooled ion cell. Threshold dissociation energies obtained from master equation modeling of these data are inset.

stantially more stable in the gas phase. Evidence for the retention of specific interactions in the gas phase, such as the Watson–Crick base pairing in DNA duplexes has been reported [3,4], but other evidence suggests that this is not always the case [31]. Specific clusters of many proteins and other types of complexes can be readily produced using electrospray ionization. This has become a powerful method to determine the stoichiometry of a complex.

Dissociation of such complexes has led to the observation of asymmetric dissociation with respect to both charge and mass. For example, Smith and co-workers [32] showed that the tetramer of streptavidin, the form of this protein in solution, can be produced in the gas phase. Dissociation of the 14+ charge state of this complex results in primarily formation of the 7+ monomer and 7+ trimer. No dimer ions were observed. Thus, the monomer carries away half the charge of the complex yet has only half the mass. A liquid droplet model was proposed which could qualitatively account for the asymmetry but could not account for the extent of the asymmetry observed. Charge asymmetry has also been reported for protein dimers.

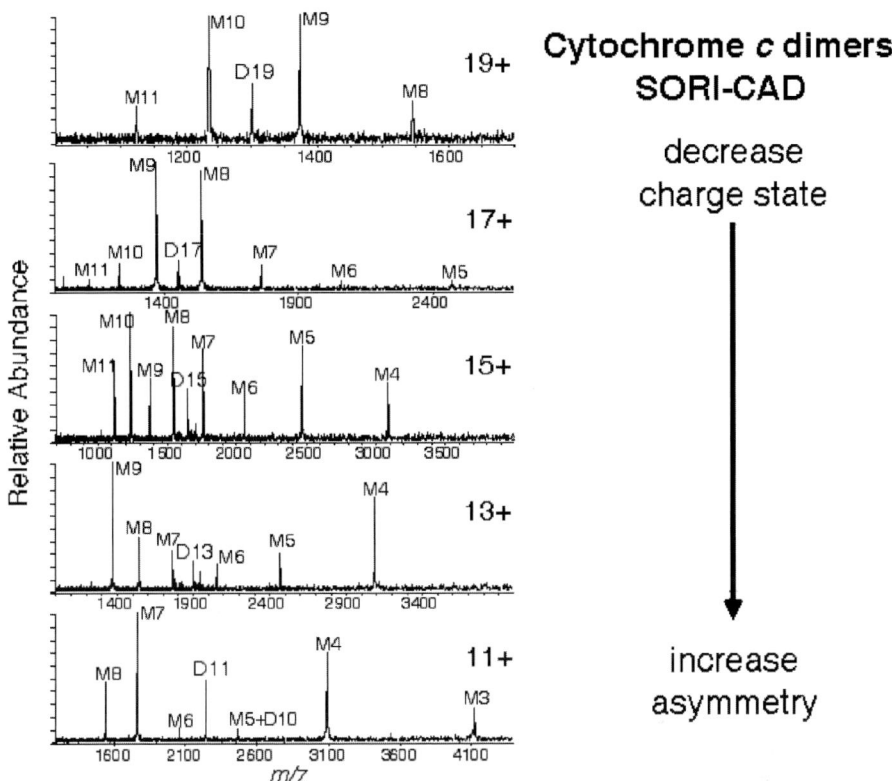

Figure 6.5 Sustained off-resonance irradiation collisionally activated dissociation (SORI-CAD) spectra of cytochrome *c* dimer ions as a function of charge state (19+ top to 11+ bottom). Dimers are indicated by "D," monomers are indicated by "M."

To understand the origin of the charge asymmetry in the dissociation of protein dimers, we investigated the effects of charge state, dissociation energy, solution-phase conformation, and conformational flexibility on the dissociation of nonspecific complexes of cytochrome *c* dimers, which can be readily formed at higher solution-phase concentrations ($\sim 10^{-4}$ M). Results for the dissociation of these dimer ions, formed from denaturing solution conditions (methanol/water/acetic acid) using sustained off-resonance irradiation (SORI) collisionally activated dissociation (CAD) as a function of charge state are shown in Fig. 6.5. For the dimer with 19 charges (D19), we see predominantly monomers with 10 (M10) and 9 (M9) charges. This corresponds to a symmetrical dissociation process. Dissociation of the D17 also results in a symmetric charge partitioning, although the distribution is somewhat broader than that for the D19. For D15, a distribution of both symmetric (M8/M7) and asymmetric (M10/M5) products are observed. The asymmetric charge partitioning is clearly dominant for both the D13 and D11.

D11 ions formed from solution-phase conditions in which cytochrome c is in a native conformation (water, 100 mM ammonium acetate) dissociate symmetrically, with M6 and M5 the major ions in this spectrum. Thus, these ions clearly "remember" the solution from which they are formed, and this solution-phase information can be detected in these gas-phase dissociation experiments! Increasing the collision energy results in the appearance of asymmetric product ions centered at M8 and M3. Thus, the internal energy deposited into these ions clearly plays an important role in the dissociation of these complexes.

The effects of the solution-phase conformation was further investigated by forming D17 ions out of denaturing solution, charge stripping these ions in the gas phase to the D13 via proton transfer with a neutral base molecule introduced into the ion cell, then dissociating this ion. The major product ions are M6/M7 corresponding to symmetrical dissociation. By comparison, D13 ions formed directly from the same solution conditions dissociate asymmetrically. This suggests that the different charge state ions formed from this same solution, have different conformations and that the solution-phase conformation is reflected in the asymmetric vs. symmetric dissociation products.

Conformational flexibility of the monomer subunits in these complexes also plays a key role in the appearance of the asymmetric dissociation. By chemically crosslinking lysine side chains within a molecule, it is possible to reduce the conformational flexibility of cytochrome c. After forming, on average, four cross links, dimers of cytochrome c dissociate symmetrically under all collision conditions. Thus, the ability of a molecule to unfold appears to be critical for asymmetric charge partitioning.

In a related experiment, the dissociation of α-lactalbumin, which has four disulfide linkages, was investigated. D13 ions of this protein dimer dissociate symmetrically at both low and high energy (Fig. 6.6, left). Upon reducing the disulfide bonds and acetamidated to prevent the disulfide bonds from reforming, D13 ions formed by charge stripping higher charge states in the gas phase were found to dissociate with an asymmetrical charge distribution (Fig. 6.6, right). These results clearly show that increasing the conformational flexibility of a monomer subunit within these clusters results in a change from symmetric to asymmetric dissociation.

A mixed cluster, consisting of one oxidized and one reduced molecule of α-lactalbumin with 15 charges can be formed. Dissociation of this D15 ion results in asymmetric dissociation, with the reduced species carrying away the majority of the charge. Thus, the molecule that is the least conformationally restricted is the molecule that carries away the majority of the charge.

In summary, these results show that the asymmetric dissociation is attributable to two separate factors. First, the solution-phase structure of the dimers is clearly playing a role in the appearance of symmetric vs. asymmetric dissociation. This is

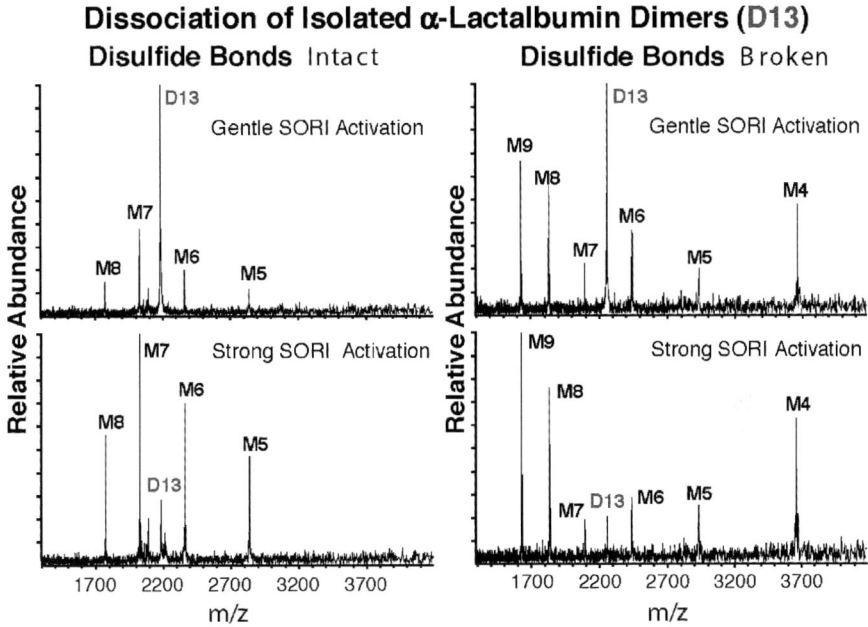

Figure 6.6 SORI-CAD spectra of dimers of α-lactalbumin with 13+ charges (D13) with the four disulfide bonds intact (left) and with the disulfide bonds reduced and acetamidated to prevent the disulfide bonds from reforming (right) at both low (top) and high (bottom) collision energy.

indicated by the experiments that clearly show that ions retaining a "memory" of the solution-phase conformation. Second, asymmetric dissociation can also come about from just the gas-phase chemistry, as clearly showed by the gas-phase energy deposition studies. Thus, some part of the asymmetric dissociation in these experiments is attributable to differences in the conformation of the monomer subunits in solution. But some fraction of the asymmetric dissociation can only be explained by the dynamics of the gas-phase dissociation process. The general model for the latter process that we have proposed [33] is one in which one of the protein monomers in the subunit unfolds, most likely at or near the transition state, and in the unfolding process, protons are transferred to the elongating species to minimize Coulomb repulsion. This results in the ejection of an elongated species that can carry away a greater fraction of the charge [33].

6.2.3 Peptide cluster dissociation

The dissociation pathways of clusters of the peptide leucine enkephalin were measured as a function of charge state and size. Dissociation of the 7-mer of leucine enkephalin with two charges (7LE2+) results in the formation of LE+, 2LE+,

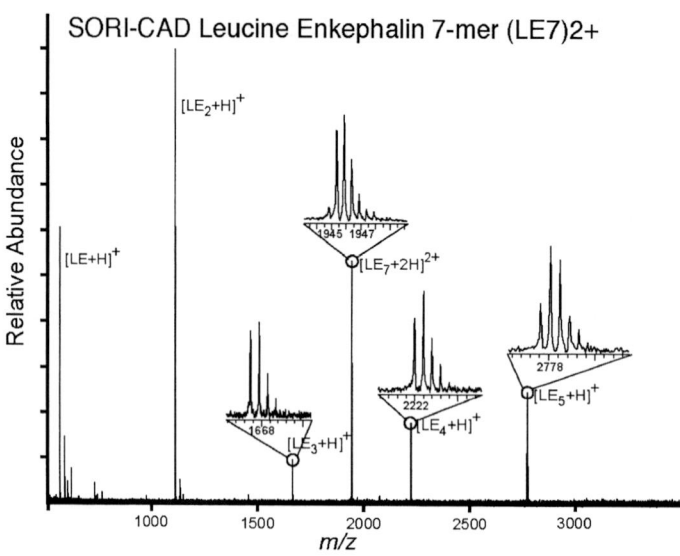

Figure 6.7 SORI-CAD spectrum of a cluster consisting of seven leucine enkephalin molecules with two protons (7LE2+).

3LE+, 4LE+, and 5LE+ (Fig. 6.7). Multiple dissociation products arising from a single multiply charged precursor ion are typical of a Coulomb explosion process which occurs when clusters are charged above the Rayleigh limit. In contrast, a fission process, which occurs when the cluster charge is below the Rayleigh limit, typically results in two or occasionally three products. To determine the origin of the product ions from 7LE2+, a series of double resonance experiments were performed. In a double resonance experiment, the precursor ion is dissociated and during this time, a RF waveform is applied to an ion of interest to eject the ion rapidly from the cell. If the abundances of ions at other m/z decrease, this indicates the fraction of these ions that are formed by subsequent dissociation of the ion of interest. For example, continuous ejection of 2LE+ under Fig. 6.7 conditions results in nearly the complete elimination of LE+. This demonstrates that the vast majority of LE+ is formed from subsequent dissociation of 2LE+ and not directly from the precursor (7LE2+). By performing these double resonance experiments for each of the fragment ions observed in Fig. 6.7, we conclude that the primary dissociation pathway for this ion (99+%) is formation of 2LE+ and 5LE+. The other ions are formed by subsequent dissociation of these ions. Thus, 7LE2+ dissociates by fission.

9LE2+ also undergoes fission to form 2LE+ and 7LE+, but the major dissociation process (87%) is loss of a neutral LE molecule to form 8LE2+. This ion can undergo fission to produce 2LE+ and 6LE+, and also neutral loss to form 7LE2+

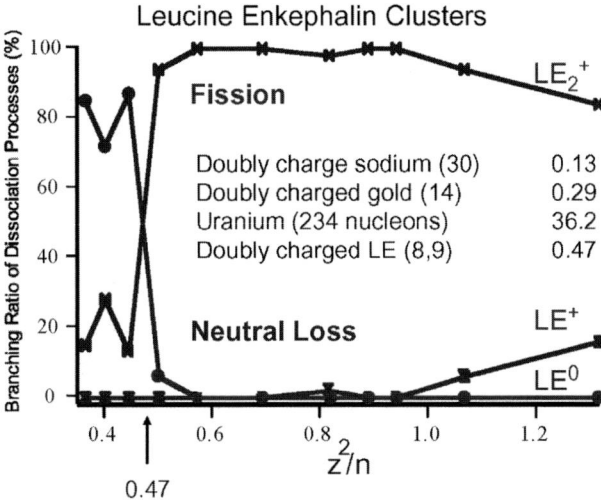

Figure 6.8 A plot of the branching ratio of dissociation to neutral loss as a function of z^2/n for leucine enkephalin clusters, where z is the number of protons on the cluster and n is the number of leucine enkephalin molecules in the cluster.

which then can undergo fission. With increasing cluster size and charge state, we find that the neutral loss pathway becomes less significant and fission becomes more dominant.

The data for all the nLEz+ clusters investigated to date are shown in Fig. 6.8 which presents the branching ratio for neutral loss vs. fission as a function of the z^2/n of the cluster, where z represents the overall charge on the cluster and n corresponds to the number of leucine enkephalin monomers in the cluster. Saunders [34] found that the log of the ratio of fission to neutral evaporation for metal clusters and nuclear decay varies linearly with z^2/n. The results of the peptide cluster dissociation resembles the dissociation of metal clusters and atomic nuclei. We find the branching ratio for neutral evaporation vs. fission changes dramatically at z^2/n of 0.47. For doubly-charged sodium clusters with 30 atoms, and for doubly-charged gold clusters with 14 atoms, this transition occurs at value of ~ 0.13 and 0.29, respectively. For uranium nuclei with 234 atoms, this transition occurs at 36.2. Thus, it appears that there is a qualitative similarity between the dissociation observed for the multiply charged peptide clusters and for multiply charged metal ion clusters and nuclear fission. However, important differences exist.

One key difference between the peptide cluster dissociation and other cluster dissociation observed previously is that conformational flexibility of the monomer subunits plays an important role. The dissociation pathways of pentamers of somatostatin with four charges were investigated. Somatostatin has one disulfide bond. Dissociation of the complexes consisting of somatostatin with the disulfide bonds intact, vs. reduced, results in the same primary dissociation pathways, but different branching ratios for fission vs. neutral evaporation. As with the protein dimers, the conformational flexibility plays a role in how these multimeric peptide clusters dissociate [33].

6.3 Conclusions

Electrospray ionization can generate abundant noncovalent complexes, both non-specific weakly bound complexes, such as hydrated biomolecule ions, and specific, more strongly bound complexes, such as protein–protein and DNA complexes. By investigating the dissociation energies and pathways, information about the structure of the complexes can be inferred. By studying hydrated ions, a detailed understanding of how water influences molecular structure can be obtained, one water molecule at a time. For lithiated valine, only three water molecules are necessary to change the structure of valine from the nonzwitterion form preferred in the gas phase, to the more stable zwitterionic form in solution. The binding energy of water can be a very sensitive probe of zwitterion vs. nonzwitterion structure.

For larger multiply charged complexes, the asymmetric charge and mass partitioning can be readily explained. In the case of protein homodimers, we show that asymmetric charge partitioning depends on the cluster charge state, the dissociation energy, the solution-phase structure, and the conformational flexibility of the monomers. The appearance of asymmetric charge partitioning can be due to different dimer structures in solution and it can be due to unfolding one of the monomer subunits in the gas phase. For larger nonspecific clusters of peptides, we find that the clusters dissociate by fission, not by Coulomb explosion, and that the fission process is in competition with neutral evaporation. There are qualitative similarities to the dissociation pathways observed for multiply charged metal ions and with atomic nuclei, but with one important difference. Conformational flexibility of the peptides influences the dissociation pathways.

Acknowledgments

Funding for this work has been generously provided by the National Science Foundation (Grant No. CHE-0098109) and the National Institutes of Health (Grant No. RO1-GM64712-01).

References

1.   X.F.D. Chillier, A. Monnier, H. Bill, F.O. Gulacar, A. Buchs, S.A. McLuckey, and G.J. Van Berkel, *Rapid Commun. Mass Spectrom.* **10** (1996) 299–304.

2.   S.E. Rodriguez-Cruz, J.S. Klassen, and E.R. Williams, *J. Amer. Soc. Mass Spectrom.* **10** (1999) 958–968.

3.   P.D. Schnier, J.S. Klassen, E.F. Strittmatter, and E.R. Williams, *J. Amer. Chem. Soc.* **120** (1998) 9605–9613.

4.   V. Gabelica and E. De Pauw, *J. Amer. Soc. Mass Spectrom.* **13** (2002) 91–98.

5.   A.A. Rostom, P. Fucini, D.R. Benjamin, R. Juenemann, K.H. Nierhaus, F.U. Hartl, D.M. Dobson, and C.V. Robinson, *Proc. Natl. Acad. Sci. USA* **97** (2000) 5185–5190.

6.   G. Siuzdak, B. Bothner, M. Yeager, C. Brugidou, C.M. Fauquet, K. Hoey, and C.M. Chang, *Chem. Biol.* **3** (1996) 45–48.

7.   M. Dole, L.L. Mach, R.L. Hine, R.C. Mobley, L.P. Ferguson, and M.P. Alice, *J. Chem. Phys.* **49** (1968) 2240–2249.

8.   J. Fernandez de la Mora, *Anal. Chim. Acta* **406** (2000) 93–104.

9.   A.T. Iavarone and E.R. Williams, *J. Amer. Chem. Soc.* **125** (2003) 2319–2327.

10.   D.S. Gross, Y.X. Zhao, and E.R. Williams, *J. Amer. Soc. Mass Spectrom.* **8** (1997) 519–524.

11.   C.L. Hunter, A.G. Mauk, and D.J. Douglas, *Biochemistry* **36** (1997) 1018–1025.

12.   F. Rogalewicz, Y. Hoppilliard, and G. Ohanessian, *Int. J. Mass Spectrom.* **227** (2003) 439–451.

13.   R.L. Wong, K. Paech, and E.R. Williams, *Int. J. Mass Spectrom.* **232** (2004) 59–66.

14.   R.A. Jockusch and E.R. Williams, *J. Phys. Chem. A* **102** (1998) 4543–4550.

15.   J.H. Jensen and M.S. Gordon, *J. Amer. Chem. Soc.* **117** (1995) 8159–8170.

16.   J. Rak, P. Skurski, J. Simons, and M. Gutowski, *J. Amer. Chem. Soc.* **123** (2001) 11695–11707.

17.   C.J. Chapo, J.B. Paul, R.A. Provencial, K. Roth, and R.J. Saykally, *J. Amer. Chem. Soc.* **120** (1998) 12956 12957.

18.   W.D. Price, R.A. Jockusch, and E.R. Williams, *J. Amer. Chem. Soc.* **119** (1997) 11988–11989.

19.   E.F. Strittmatter and E.R. Williams, *Int. J. Mass Spectrom.* **212** (2001) 287–300.

20.   S. Hoyau and G. Ohanessian, *Chem. Eur. J.* **4** (1998) 1561–1569.

21.   T. Wyttenbach, M. Witt, and M.T. Bowers, *Int. J. Mass Spectrom.* **183** (1999) 243–252.

22.   J. Bertran, L. Rodriguez-Santiago, and M. Sodupe, *J. Phys. Chem. B* **103** (1999) 2310–2317.

23.   T.R.N. Marino and M. Toscano, *J. Mass Spectrom.* **37** (2002) 786–791.

24.   E.F. Strittmatter, A.S. Lemoff, and E.R. Williams, *J. Phys. Chem. A* **104** (2000) 9793–9796.

25. R.A. Jockusch, W.D. Price, and E.R. Williams, *J. Phys. Chem. A* **103** (1999) 9266–9274.

26. E.F. Strittmatter, R.L. Wong, and E.R. Williams, *J. Phys. Chem. A* **104** (2000) 10271–10279.

27. R.A. Jockusch, A.S. Lemoff, and E.R. Williams, *J. Amer. Chem. Soc.* **123** (2001) 12255–12265.

28. R.A. Jockusch, A.S. Lemoff, and E.R. Williams, *J. Phys. Chem. A* **105** (2001) 10929–10942.

29. A.S. Lemoff, M.F. Bush, and E.R. Williams, *J. Amer. Chem. Soc.* **125** (2003) 13576–13584.

30. J.M. Talley, B.A. Cerda, G. Ohanessian, and C. Wesdemiotis, *Chem. Eur. J.* **8** (2002) 1377–1388.

31. E.N. Kitova, D.R. Bundle, and J.S. Klassen, *J. Amer. Chem. Soc.* **124** (2002) 5902–5913.

32. B.L. Schwartz, J.E. Bruce, G.A. Anderson, S.A. Hofstadler, A.L. Rockwood, R.D. Smith, A. Chilkoti, and P.S. Stayton, *J. Amer. Soc. Mass Spectrom.* **6** (1995) 459–465.

33. J.C. Jurchen and E.R. Williams, *J. Amer. Chem. Soc.* **125** (2003) 2817–2826.

34. W.A. Saunders, *Phys. Rev. Lett.* **66** (1991) 840.

Advances in Mass Spectrometry, Volume 16
A.E. Ashcroft, G. Brenton and J.J. Monaghan (Editors)

CHAPTER 7

AMS in Phytonutrition

Stephen R. Dueker [1,*] and Bruce A. Buchholz [2]

[1] Department of Nutrition, University of California, Davis, One Shields Ave,
Davis, CA 95616, USA

[2] Center for Accelerator Mass Spectrometry, Lawrence Livermore National Laboratory,
Livermore, CA 94551, USA

7.1 Introduction

"Let food be our medicine and medicine our food"—Hippocrate (4th century B.C.), the
father of medicine

Food as medicine—the notion is as relevant in our present age of phytonutrition, genetically engineered foods, and an increasing population of health-
conscious consumers, as it was in the age of Hippocrate several millennia back.
Popular conceptions of foods and plants, however, have transformed substantially
over the centuries. Presently, plants are esteemed for their reputed health benefits and we speak of them using the lexicon of modern chemistry. In medieval
times, though, plants were considered the residences of divinities or malevolent
spiritual forces. An example of one plant's conceptual evolution is illustrated by
the tomato plant. Today, a tomato is nearly synonymous with its bright colorant
lycopene, the hydrocarbon plant pigment responsible for its red color and putative enemy of cancer. In medieval Northern Europe, reference to what was then
called a tomato, would not imply any medicinal benefits, but more likely suggest
images of witches and werewolves (Fig. 7.1). This fearful perception arose from
the morphological resemblance of the tomato plant to poisonous members of the
Solanceae family (particularly Belladonna), which were common "hexing herbs"
of medieval witches. Old German folklore describes witches using plants of the
nightshade family to evoke werewolves, a practice known as lycanthropy. This

---

*Corresponding author.

Figure 7.1 Popular conceptions of foods have transformed substantially over the centuries. Now esteemed as a rich source of lycopene, to medieval Europeans the tomato had a baleful reputation.

lore was clearly noted by Carl Linnaeus, the father of modern biological classification, when he named the tomato *Lycopersicon esculentum*, which translates to 'edible wolf peach,' leading to the name of the phytonutrient—lycopene.

Plants, and their intrinsic properties, have been the subjects of increasing attention over the past several decades. Today, plants are considered complex chemical factories that produce an array of biologically active phytochemicals for the consumer. Some of the more commonly known and studied compounds include the antioxidant flavonoids in diverse products such as tea and apple peels, high molecular weight tannins in berries and wine, or the isoflavones in soybean-derived products. The interest of health-conscious consumers and the mass media has created a competition among the food and agricultural industries to create value-added food products enhanced or supplemented with phytochemicals. Creation of these products is possible through a combination of traditional plant breeding and genetic engineering [1–4]. The introduction of genetically engineered foods (GE foods or genfoods) into the food and agricultural industries has sparked heated debate. Many concerns have been raised regarding the safety of genfoods foods, as well as the social, political, and economic issues implicated with GE food production. As nutritionally enhanced products make the leap from laboratory to field, it will be the charge of the scientist to conclusively establish that any modifications in nutritional content have evident benefits.

Acute public interest in phytonutrition (the dietary use of phytochemicals found in plants for providing nutrients and optimal health) is propelled by the ongoing research of these microchemical food constituents as causative or preventative agents in disease. It is further supported by the findings in dietary research of people with higher fruit and vegetable intake demonstrating lower rates of cancer and cardiovascular disease [5]. These conclusions have prompted hypotheses centered on antioxidants, including the well-known redox vitamins such as vitamin E and C, as well as nonessential components such as selected carotenes and flavonoids. The positive association between fruits and vegetables and optimal health is popularly considered axiomatic. It is proven that dietary phytochemicals supply essential vitamins, minerals, and macronutrients to the consumer. It remains to be empirically established, however, if phytochemicals are indeed beneficial, or conversely, harmful, especially when given as high-dose supplements [6]. Indeed, the notion that plant products are 'harmless' was disproved when chemoprevention studies found that β-carotene can promote cancer in smokers [7,8]. These studies do not refute the beneficial role of phytochemicals in disease prevention, but they do favor a whole food-based approach to increasing photochemical intake and promoting good health.

A proven technique in pharmaceutical research, which is playing an integral role in phytonutrition, is mass spectrometry. Structural determination and identification are strengths of modern mass spectrometry. Moreover, hyphenated GC/MS [9] and LC/MS/MS [10–12] techniques are particularly well suited to establishing phytochemical content in plants, as well as the biological variance in human metabolism of phytochemicals, due to their high sensitivity, throughput, and ability to resolve overlapping components from complex matrices [13]. However, elaborating the full complement of biological metabolites of phytochemicals at trace levels is a strength of radioisotopes, typically ^{14}C or ^{3}H. In this regard, ^{14}C-accelerator mass spectrometry (AMS) is affording new opportunities in the study of human phytochemical metabolism [14]. AMS is a type of tandem isotope ratio mass spectrometer that measures the ratios of $^{14}C/C$ to parts per quadrillion or down to as few as 10^5 ^{14}C atoms. Accordingly, attomolar concentrations (10^{-18}) of ^{14}C labeled chemicals are quantified in milligram-sized biological specimens. Labeled phytochemicals and their metabolites can be traced in unprocessed plasma, tissue, and excreta, without regard to compound stability or the efficiency of the extraction procedure, thus enabling balance studies of excreta. Although AMS involves radioactivity, its high sensitivity reduces radiative exposure to negligible levels. Typical human doses range from 10–100 nCi (0.37–3.7 kBq) of activity. Even at levels such as these, labeled nutrients can be traced for months in plasma and excreta. For establishing metabolite profiles, small (microliter) aliquots of minimally processed plasma or unprocessed urine are separated by liquid chromatographic systems prior to analysis. These radiochromatograms can serve as detectors for assessing the range of biotransformations and identifying target analytes. The low background of ^{14}C in the environment provides an unambiguous baseline for clearly identifying ^{14}C enriched fractions collected from chromatographic systems. Metabolite identity is then finalized by applying traditional MS and other techniques to concentrated HPLC fractions.

AMS is poised to assume a leading role in phytonutrition. For effective expansion of AMS into phytochemical investigations, suitably labeled substrates must be available. Many natural products found in foods, though, are difficult to synthesize due to complex stereochemistries and a multiplicity of polymeric forms.

However, there are several options to overcome the aforementioned predicament. One option is the incorporation of a simple radiolabeled precursor, perhaps [14]C-*para*-aminobenzoic acid (folate) or phenylalanine (catechin), through the plant's roots, or via leaves and stems by surface application. If the label is efficiently incorporated into the target, the food can conceivably be ingested intact to examine the effects of the plant matrix on the nutrient digestibility. Alternatively, plants can be grown in [14]CO_2-enriched environments using atmospherically-isolated plant growth chambers. Photosynthetic labeling in the presence of [14]CO_2 randomly labels the entire phytochemical constitution of the plant, thus creating a radiopharmacy of sorts. When following the appropriate fractionation, this yields a spectrum of compounds that can be used individually or possibly in synergistic combinations for in vivo and in vitro studies of human metabolism and catabolism.

Throughout this article: *AMS in Phytonutrition*, a variety of MS tools and isotope labels will be discussed. As suggested by its title, however, the article will focus chiefly on exploring on recent developments in the area of accelerator-based techniques, substantiated by the author's knowledge and research conducted at the University of California, Davis. To establish an appropriate context for this exploration, the technique of AMS and its history will be surveyed. A selection of investigative studies will then illustrate the unique capabilities AMS brings to the realms of biological tracing and metabolite profiling, as well as noninvasive assessments of phytonutrient metabolism and biolabeling strategies. This article serves as an introduction to AMS and its developing role in phytochemical research. For further knowledge on the topic, several excellent reviews addressing the role of AMS in nutrition are available [15,16].

7.2 Overview of accelerator mass spectrometry

In the 1940s Willard Libby at the Institute for Nuclear Studies developed the method of carbon dating using a highly sensitive Geiger counter. In this form of radiometric dating, the concentration of ^{14}C, the longest-lived radioisotope of carbon, records the time since the last carbon exchange with the atmosphere. The specific activity of the material thus serves as a reverse chronometer, accurate—but not highly so—to about to 60,000 years into the past.

Upon its development, radiocarbon dating had immediate and widespread consequences on many fields. However, the method of decay counting, as described by Libby, is an inefficient means of quantifying the ^{14}C content of a sample since ^{14}C decays slowly: measuring 0.1% of the ^{14}C in a sample requires uninterrupted counting for 8.3 years {$0.1\% \times 5730$ years/$\ln 2$} (by contrast, 0.1% of the population of ^{32}P is measured in 29.7 min). To overcome this limitation, very large quantities of sample, anywhere from grams to hundreds of grams were required. More often than not, such quantities of material were either not available or could not be sacrificed to the measurement. In the late 1970s and throughout the 1980s a mass spectrometric method for direct detection of ^{14}C and other long-lived isotopes was developed in low-energy nuclear physics laboratories. The technique, accelerator mass spectrometry (AMS), although originally developed for the difficult task of radiocarbon dating, emerged in the 1990s as a useful bioanalytical tool for the quantification of ^{14}C and (other long-lived isotopes) in biochemical labeling and tracing.

AMS is a type of tandem isotope ratio mass spectrometer that measures the ratios of ^{14}C/C to parts per quadrillion or down to as few as 10^5 ^{14}C atoms. In AMS, negative ions are generated in a cesium ion source and accelerated by a potential of 0.5–10 MV. The accelerated ions are smashed through a thin carbon foil or inert gas that removes electrons and destroys all molecular species. After passing through a high-energy mass spectrometer and various filters, carbon nuclides are measured with current and particle detectors. A schematic of the compact bio-AMS instrument at Lawrence Livermore National Laboratories is shown in the adjoining figure (Fig. 7.2) with a discussion of its mode of operation [14].

AMS provides the lowest detection limits of any isotopic measurement technique for not only ^{14}C, but also for other long-lived radionuclides that include ^{3}H, ^{10}Be, ^{26}Al, ^{32}Si, ^{36}Cl, ^{41}Ca, ^{59}Ni, ^{99}Tc, ^{129}I, and several actinide isotopes. All isotopes for which AMS is developed have half-lives in the range of 10^3–10^7 years, with the exception of ^{3}H (half-life 12.25 years). On this basis, an AMS instrument could be described as an ultra sensitive scintillation counter. Such a description, however, is misleading for it suggests that the measurement is tied to the decay properties of the nuclide, which AMS is not. Radioactivity is not a requirement for AMS; rather, AMS profits from the extremely low natural abundance of radioisotopes. Stable isotopes occur at abundances of 0.1% to 50% of the natural element, whereas radioisotopes occur on Earth at parts per billion or less. For example, while ^{14}C is present at ~ 100 amol/mg of carbon in contemporary materials, the stable isotope of ^{13}C is present at levels around 1 mmol in the same amount of material. Quantifying on ^{14}C, therefore, extends the theoretical dynamic range of the measurement by a factor of 10^{-10}, translating into detection sensitivities of attomoles for ^{14}C in milligram-sized samples.

Most AMS systems operate at 3–10 MV and primarily measure natural radiocarbon. The larger systems analyze the elements important for geochronology, which typically require higher voltages. Conventional AMS systems are hardly "benchtop," or even laboratory-sized [15]. The LLNL conventional AMS system, for example, fills a 550 m^2 area. This is a general-purpose research AMS system, however, that is being used for analyses from mass 3 to mass 240 [14]. Radiocarbon dating with lower charge states has been demonstrated and several commercial spectrometers are becoming available, at varied size and cost, that will encourage widespread use in institutional and industrial research [17]. The Lawrence Livermore Laboratory also has a 1 MV BioAMS (described in Fig. 7.2) dedicated to

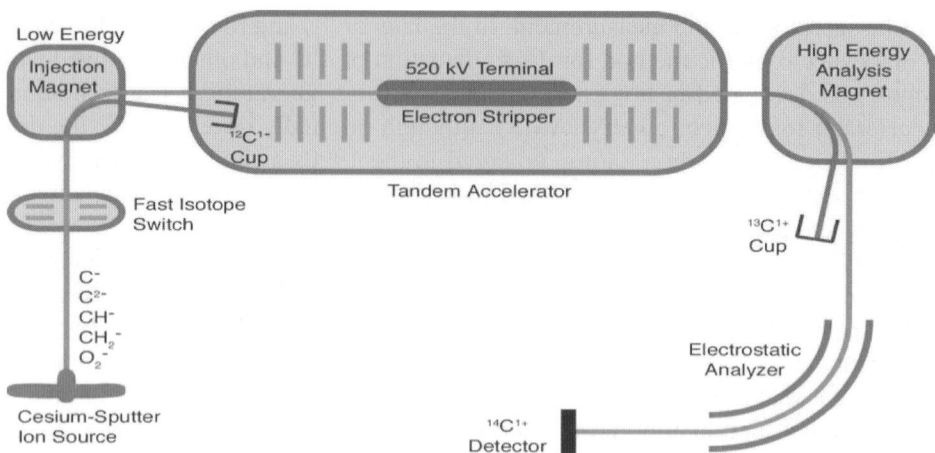

Schematic of compact bio-AMS system at LLNL (Courtesy of Bruce Buchholz, LLNL)

Figure 7.2 Biological samples for ^{14}C-AMS are typically combusted to CO_2 and then reduced to graphite. These samples are then bombarded by 3–10 keV Cs ions in a cesium-sputter ion source. This process physically knocks atoms and molecules out of the sample and contributes an electron to a fraction of the ejected particles, forming negative elemental or molecular ions. The production of negative ions removes the primary isobaric interference for radiocarbon, ^{14}N, since nitrogen does not form a stable negative ion. Singly ionized negative ions are accelerated to 40 keV and then filtered by a low energy mass spectrometer that alternately switches between ions of mass about ^{14}C, i.e., ^{14}C, ^{13}CH, and ^{12}CH$_2$ and mass 13, for the stable isotope measurements, into the accelerator. The field in the injection magnet is held constant and the isotopes are changed by quickly switching the energy (voltage) of the incoming ion beam. Negative ions are accelerated through 520 kV in the first stage of a tandem accelerator. At the end of this first acceleration stage these ions pass through an electron stripper, a small volume of ~ 45 mT gas constrained by differential pumping. Here, particles undergo collisions, losing valence electrons. Depending on the accelerator potential, charge states of $+1$ to $+4$ are created. These positive ions then accelerate away from the positive potential to ground potential in the second half of the accelerator. The loss of electrons in the collision cells destroys all molecules, leaving only nuclear ions at relatively high energies (1.04 MeV for C). Positive ions exiting the accelerator are analyzed with a magnet and an electrostatic analyzer. An off-axis Faraday cup measures the ^{13}C^{1+} current after the analyzing magnet. A solid state particle detector counts individual ^{14}C^{1+} ions. The footprint of the compact Bio-AMS system is approximately 50 m^2.

biochemical tracing of ^{3}H [18,19] and ^{14}C analyses. A Single Stage AMS System (SSAMS) consisting of an open air 250 kV deck (not a tandem accelerator) is presently being marketed at a cost below $1 million.

7.3 Biological tracing

The approach of phytochemical research can be considered comparable with that of pharmacokinetic studies. Pharmacokinetic data—absorption, distribution, metabolism, and excretion—is central to drug development and is attainable from the ability to measure drugs, at the concentration that they are found, in the human body during the course of therapy. Without the ability to execute such concentration measurements, pharmacokinetics could only be a theoretical exercise. This same statement holds true for phytochemical research. A bioactive phytochemical, in order to be an effective public health prophylactic, nutraceutical, or functional food component, must either be absorbed intact or as some active metabolite by most individuals and attain effective concentrations at the receptors or tissues. To draw conclusion, these concentration measurements must then be associated to epidemiological evidence, modulation of biomarkers, or some clinically defined endpoint. It merits noting, however, that phytochemical kinetic studies are technologically more challenging than similar drug studies in several ways. First, drugs are most often tested using large milligram to gram-sized doses, whereas phytochemicals are generally consumed in microgram to milligram concentrations. Secondly, preexisting concentrations in tissues and fluids limit the ability to follow the fate of dietary constituents when given as a 'dose,' either with food or as an isolate. These challenges can be surmounted through application of sensitive detection methods towards isotope labeled compounds that facilitate discrimination of exogenous intake from endogenous pools.

The application of AMS in phytochemical research is relatively new; however, isotopic labels have been a primary tool in tracing chemicals in natural systems for over 60 years [15,20]. The value of isotopes as elemental or molecular tracers depends on the detection methods: isotope ratio mass spectrometry (IRMS), nuclear magnetic resonance (NMR), decay production counting, AMS, etc. Short-lived radioactive isotopes have high sensitivity in decay counting above their very low backgrounds, but expose the participants to possible radiative harm. Tritium and radiocarbon, because of their availability, ease of detection by decay counting, and appropriateness to biologic questions, have been incorporated into human nutrition research. For example, radiotracer studies with β-carotene conducted in the 1960s established that β-carotene is first cleaved within the intestinal enterocyte to vitamin A active retinyl esters, then transported, via the lymphatics, into circulation [21]. The full potential of radioisotopes has not been realized in regard to human nutrient metabolism due to health concerns over the use of ionizing radiation. In addition, practical issues related to disposal cost has restricted widespread acceptance and application. In order to achieve adequate counting statistics with permissible blood sample, the use of radiotracers in many studies is not practical.

[14]C-AMS was originally developed for, and remains most applied to, the area of carbon dating at levels of ratios of 14C/C at or below the contemporary levels of [14]C. The importance of ^{14}C as a biological tracer at concentrations above contemporary levels was not lost on the early developers of AMS for carbon dating. The possibility for the measurement of ^{14}C-glucose in humans was explored in the late 1970s at the University of Rochester (New York). This was one of two laboratories barely 100 miles apart that first published its methods of detecting natural ^{14}C at part-per-trillion levels using tandem Van de Graaff accelerators [22]. For all intents and purposes, AMS did not establish a foothold in bioanalytical tracing until AMS operations at Lawrence Livermore National Laboratory began in late 1988 [14]. In the first study of its type, AMS was used to quantify the amount of meat carcinogen ^{14}C-MeIQX, covalently bound to mouse liver DNA (DNA adduct) following a very low-level dose [23–25]. The first nutrients followed with the long-term biological tracing of folic acid in 1997 [26] and β-carotene in 2000 [27] at the University of California, Davis, in collaboration with LLNL. An important demonstration of AMS in human breath analysis came from Lund University in 1996 [28]. The first use of AMS analysis in human mass balance and HPLC metabolite profiling of a pharmaceutical appeared in 2002 at the CBAMS Institute (now Xcerleron) at York, United Kingdom [29].

7.4 Natural and synthetic α-tocopherols

Public interest in nutritional supplements has escalated dramatically over recent decades. To gain a market advantage, certain formulations are labeled as being more efficacious (biopotent) than competing products for reasons that are rarely scientifically based. For example, a consistently presented idea is that 'natural,' meaning a plant derived concentrate, is superior to its 'synthetic' alternative. Ultimately, the biopotency of one form over another needs to be determined on the basis of clinical and biochemical endpoints [30]. However, in lieu of these data, relative availabilities and retention are assessed using tracer studies that can record the biological fate of the compound.

Natural and synthetic forms of α-tocopherol are available for use as vitamin E supplements and fortified foods. When discussing the biopotency of either form, it is necessary to carefully examine its correspondent data. Natural α-tocopherol is RRR-α-tocopherol, a pure stereoisomer. The conventional synthetic form, all rac-α-tocopherol (all rac), however, consists of equal amounts of 8 different stereoisomers (RRR-, RRS-, RSR-, RSS-, SRR-, SSR-, SRS-, SSS-) with only 1/8 the quantity represented as the natural RRR form. The long accepted ratio of biopotency of natural to synthetic α-tocopherol is 1.36 on the basis of an assay in pregnant rats. Such bioassays are not practical for humans, however, as they require condition of vitamin E deficiency. To assess the relative biopotency in humans, several authors have measured the relative plasma exposures of *natural* and *synthetic* α-tocopherol as a surrogate measure for biopotency using deuterium labels and GC/MS and LC/MS detectors [31–33]. These data suggested a biopotency ratio closer to 2 than to 1.36, and consequently, an adjustment in the official value has been proposed. The validity of this data has been questioned, though, for reasons that are both methodological and biological [30,34]. A primary concern centers around the use of pharmacological doses (15–150 mg) that do not mimic normal dietary intake patterns and overwhelm physiological transport systems [34]. Indeed, capacity limitation of plasma vitamin E concentrations is likely mediated by the hepatic α-Tocopherol Transfer Protein (α-TTP), which is critical for biodiscrimination of vitamin E homologues and stereoisomers, as they exhibit differing affinities to the protein [35]. Since binding to TTP is saturable and stereoselective, the pharmacological dosing approach may not be applicable for the assessment of vitamin E biopotency.

AMS sensitivity allows tracing nutrients at dietary relevant levels that do not saturate or perturb metabolic pathways. This model was perfectly suited, therefore, for a study of natural and synthetic forms of α-tocopherol recently conducted at the University of California, Davis. The results are detailed in this presentation to highlight the unique capabilities that AMS brings to biochemical tracing rather than attempting to resolve the current dispute over the biopotency of α-tocopherols.

7.5 Plasma biokinetics

[14]C-labeled tocopherols (natural and synthetic) were individually tested in a longitudinal design in a single male participant: the first dose consisted of *natural* α-tocopherol ([14]C-*RRR* α-tocopherol 3.7 kBq or 100 nCi, 0.85 ug). Three months later, a similarly sized dose of the *synthetic* mixture was administered ([14]C-*all rac* 3.7 kBq, 0.79 ug or 100 nCi). Both doses were given with a cup of whole milk (10 g fat) to facilitate absorption. Frequent blood samples (5 mL) were taken via an indwelling venous catheter over the first two days after dosing. Less frequent samples were obtained via venipuncture over the course of 2 months post dosing. 25 uL aliquots of plasma were graphitized and carbon isotope ratios were measured to ±3% precision by AMS. Data are expressed as fmol of ^{14}C per ml of plasma; these values can easily be converted to labeled tocopherol equivalents using the specific activity of the labeled compounds.

The plasma concentration time course of the ^{14}C is shown in the adjoining figure for 63 days (Fig. 7.3), with the 0–300 h region expanded in the bottom panel. The doses were equal in terms of their ^{14}C content, and, therefore, similar metabolism would yield highly equivalent biokinetic patterns. For both compounds, there was a 2 h lag in the appearance of activity, consistent with absorption with chylomicrons via the lymphatic system. An early appearance (30 min) could

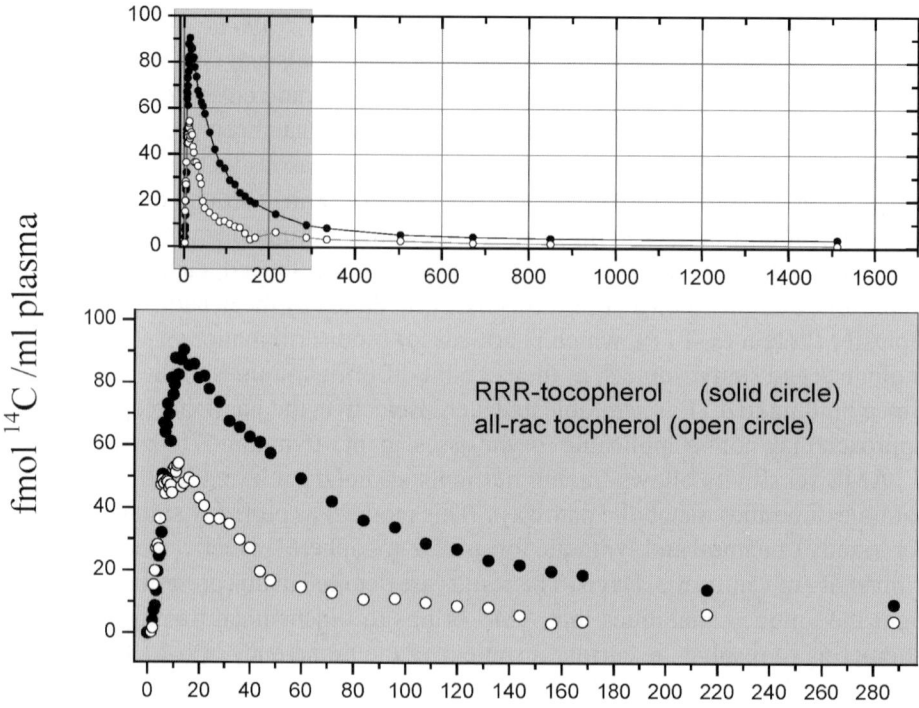

Figure 7.3 Two ^{14}C doses were given in a longitudinal design in a single male volunteer: the first dose consisted of the single RRR isomer (RRR 3.7 kBq (100 nCi), 0.85 ug). Three months later a similarly sized dose of the *all-rac* mixture was administered. Both doses were given with a cup of whole milk (10 g fat) to facilitate absorption. Dose masses were 10,000 to 125,000 times less than similar studies using deuterium labeled isotopes with GC/MS or LC/MS detection.

suggest alternative absorption in the portal blood, but this was not indicated. Concentrations of both forms rose equivalently until ~ 6 h, suggesting an absence of biodiscrimination in the absorptive phase. Thereafter, the rate of rise slowed and concentrations vacillated. The increased complexity of this phase is attributed to continual inputs from the intestine, as well as resecretion of the tocopherols with hepatic lipoproteins (VLDL) and secretions from other tissues. Peak concentration occurred for both compounds between 12 and 14 h, although concentrations close to the maximum were maintained over a time span from 10 to 24 h. Concentrations did not begin to consistently decay until after the 24 h time point. Post 24 h, the decay of the *natural* α-tocopherol form was biphasic whereas the *synthetic* form displayed a complex decay pattern characterized by several resuspensions in concentration until settling into a terminal decay pattern after ~ 220 h post dose administration.

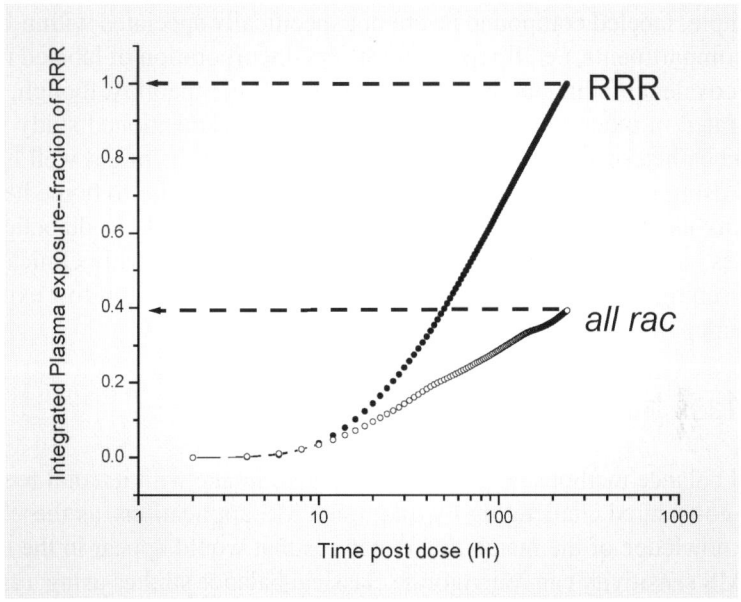

Figure 7.4 Total plasma exposure to ^{14}C following administration of equivalent amount of natural (RRR) and synthetic (*all rac*)-α-tocopherol out to 250 h post dose. A value of 1 is assigned to the RRR form. After 250 h, the relative exposure to the synthetic form was about 40% that of the natural α-tocopherol.

Visual inspection of the concentration profiles reveals a muted plasma response for the *synthetic* relative to the *natural* α-tocopherol experiment: peak concentration in the 10–24 h period for the *synthetic* were roughly 50% those of the *natural* form. This response cannot be attributed to differences in gut absorption since fecal measurements indicated similar digestibility (presented below). The magnitude of the differential response was confirmed by plotting cumulative total plasma exposure (Fig. 7.4) for 10 days post dose. This plot clearly suggests the total availability of the *synthetic* dose relative to the *natural* dose, after 10 days is about 40%. Based on this information, the setting of the biopotency of *natural* compounds to twice that of the *synthetic* compounds would be supported (by this measure only) when using small, nonsaturating doses.

As demonstrated in this study, the small sample requirements for an AMS measurement (minimum requirements of one mg or less of carbon or about 25 uL of plasma) provided for frequent blood sampling and a subsequent detailed temporal record of the kinetic behavior of the labeled tocopherols. The observed differences in the fine structure of the plasma concentrations can then be related to underlying mechanism, as well as provide a basis for validation of existing theories—or development of new theories—for the biologic handling of these compounds. In this example, labeled compounds were not specifically speciated within biochemical subcompartments, i.e., lipoprotein carriers, incorporation of labeled red blood cells, or covalent binding to circulating albumin. This capability, though, has been demonstrated in other studies. The results of the aforementioned study were obtained via catheterization. It merits noting that such results might well have been obtained using microliter sized capillary finger sticks, similar to home health care testing kits used for the monitoring of blood glucose levels in diabetics. These techniques have dual advantages; they prevent the need for costly clinical support while lessening participant resistance associated with the discomfort experienced from venipuncture or catheterization.

7.6 Mass balance: urine and feces

Classical balance methods, comparing compound intake with its total fecal excretion, are considered challenging by traditional MS applications, as these methods require knowledge of the metabolite structures that would appear in the urine and stool. AMS sensitivity can reinvigorate classical balance studies using a radiolabel tracer at doses that will not saturate the binding capacity of the gastrointestinal system. Since AMS is a combustive process, performing balance measurements is relatively straightforward. Fecal samples are dispersed in alkaline solutions and subjected to a series of mechanical mixing and heating cycles before removing small aliquots for ^{14}C determinations. This process leads to a homogenous suspension from which representative samples can be removed for AMS analysis. Urine samples are analyzed either neat or subject to isotope dilution prior to graphitization using tributyrin as a carbon diluent.

Cumulative urine and feces were collected for 8 days. The loss of label in the stool was biphasic, the transition between the phases occurring between 2 and 2.5 days (Fig. 7.5). Early stool losses (< 3 days) are attributed to unabsorbed compound whereas later losses are attributed to gastrointestinal tract epithelial cell loss and inefficient recovery of tocopherols in the enterohepatic pool with bile loss. Losses were virtually equivalent for both compounds and accounted for 25% of the dose in this phase, suggesting at subphysiological doses the gut does not discriminate between the forms [32]. Accordingly, the compounds were estimated to be 75% bioavailable. This approximation, however, does not take into account potential losses in the gut and other contribution of the rapidly reexcreted tocopherol and, therefore, may underestimate the true absorption.

Biodiscrimination between the two forms was not evident in the recovery of the label in the stool. The urine patterns, however, showed the loss of the *synthetic* form to be $\sim$ 2.5-times that of the *natural* α-tocopherol form at 8 days post dose. This result is consistent with the observation that synthetic stereoisomers are less retained than the natural form, presumably due to differential affinities for hepatic

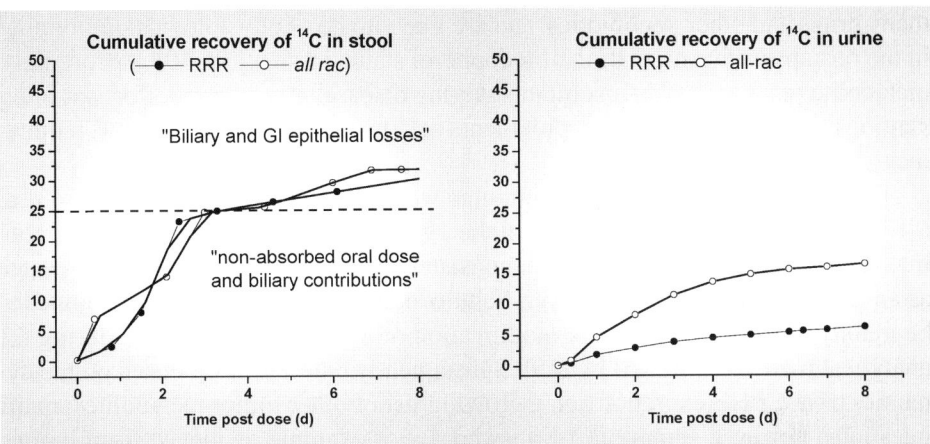

Figure 7.5 Recovery of ^{14}C in urine (left) and stool following the two oral doses of labeled tocopherol. Intestinal absorption was equivalent for the two forms. Postabsorptively, the synthetic *all-rac* form underwent more rapid elimination via the urinary route. After 8 days, approximately 50% of the *all rac*, and 65% of the RRR were still present in the body.

binding proteins. Cumulative urine and fecal ^{14}C losses indicated after 8 days, that approximately 50% of the *synthetic* and 65% of the *natural*, were still present in the body. The value shown here is considerably longer than the 2–3 day half-lives of α-tocopherols, which are typically reported. In fact, such a disparity is a recurrent observation in AMS-based studies. In other words, where dilution of the label into the background can occur only after several days in other protocols, AMS sensitivity affords the recording of true terminal elimination rates arising from the low seepage of retained compounds from tissues and organs.

7.7 Urinary metabolites: metabolite profiling

Isotope-labeled compounds reveal all of their metabolites in HPLC separation, but sensitivity when using LSC detection can be impacted by the need to keep chemical and radiative exposures low when using decay counting [36,37]. The small sample requirements and high sensitivity make AMS an excellent quantification tool for biochemical separatory processes that result in low concentration samples, such as urinary profiles.

Urine contains a myriad of information that can be sampled noninvasively. Vitamin E is extensively metabolized by side chain degradation initiated by cytochrome P450 (CYP)-mediated ω-hydroxylation of the phytyl side chain, followed by β-oxidation. These P450 enzymes are responsible for the metabolism of many drugs and other xenobiotics, and are likely to be of physiological importance in the postabsorptive regulation of tocopherol status in vivo [38–40]. Accordingly, metabolite profiles may be useful in studying disease progression, effects of interventions (such as antioxidant supplements or drugs, or defining vitamin E requirements) [41,42].

The first described urinary metabolites were identified by Simon et al. [43] as α-tocopheronic acid and its lactone in the 1956. The eponymous 'Simon' metabolites had an open chroman structure, consistent with α-tocopherol that had reacted as an antioxidant. In 1995, however, Schultz et al. found the Simon metabolites to be mainly products of oxidative modification of a precursor, α-CEHC, during the analytical workup (Fig. 7.6) [44,45]. Subsequently, interest has focused on the dynamics of α-CEHE excretion and a growing list of other minor metabolites; many questions, however, remain to be answered. For example, evidence from limited studies have reported that only a small fraction—as low as 5% of administered α-tocopherol—is recovered as α-CEHC in the urine [44,46]. Therefore, methods that can account for the full complement of biological metabolites are sought currently.

Aspects of this issue were addressed in the aforementioned study of the individual who had consumed true tracer doses of natural and synthetic α-tocopherols. Within this study, the urinary metabolites were speciated. The procedure employed was that described by Lodge et al. [46], involving deconjugation by glucuronidase treatment, and partitioning of the metabolites into ether. Urine samples were 'cold spiked' with the α-CEHC and 'Simon' metabolites at concentrations that would be easily viewed in the column effluent by UV/vis absorbance. Similar to standard HPLC means of quantification by spectrometry, compound identification comes from the characteristic retention times of the known metabolites. In the described example, α-CEHC and the 'Simon' metabolites were available and added to the urine, prior to processing. An injection corresponding to ~ 80 uL of urine was then separated by reversed-phase HPLC, with the 20–30 s eluent fraction collected and analyzed for ^{14}C by AMS.

In the adjoining radiochromatogram (Fig. 7.7), peak concentrations of ^{14}C coincided with the α-CEHC standard. In contrast, little or no ^{14}C appeared at the characteristic elution time for the 'Simon' metabolites. This later result affirmed that the integrity of at least some of the metabolites, namely α-CEHC, was preserved in the extraction procedure. The activity of ^{14}C was not exclusive to the α-CEHC metabolites. Observing 'peak tops' in the radiochromatograms, a minimum of 9 chemically distinct metabolites carry the ^{14}C tag in both the natural and *synthetic* urine samples. Speculation as to the identity of these other peaks is beyond the scope of this presentation, but educated guesses can be made based upon literature information, in vitro experiments, and understanding of the chromatographic system. The *natural* and *synthetic* profiles present some observable

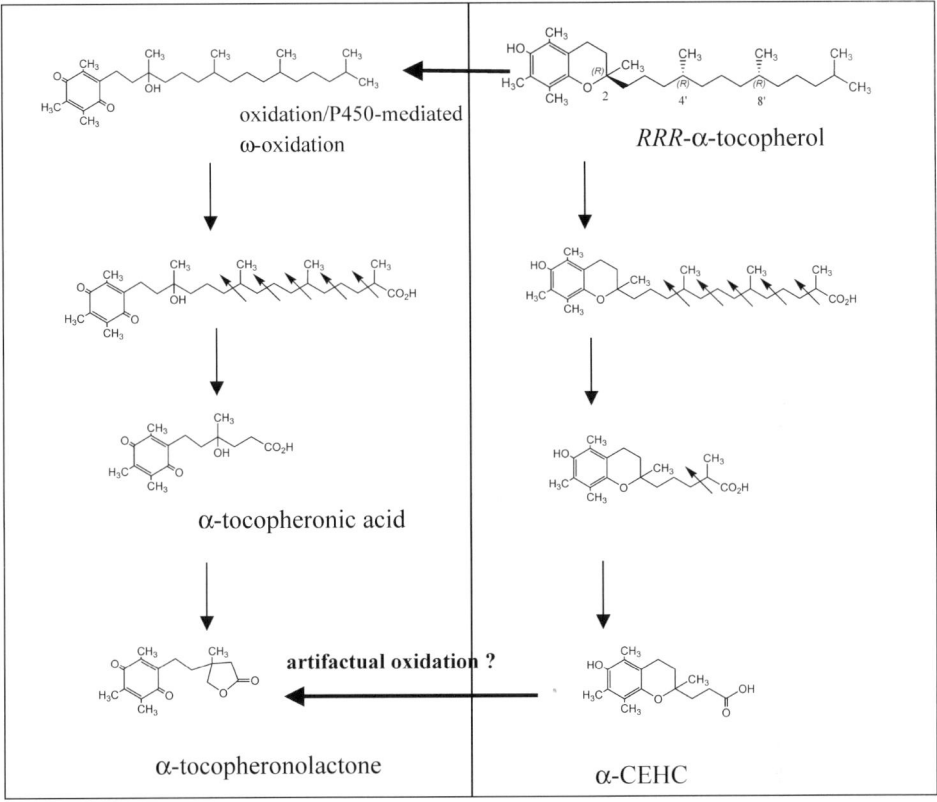

Figure 7.6 Pathways of tocopherol metabolism. Current understanding states that participation as an antioxidant leads to the opening of the chroman ring and lactone products (left side), whereas an intact chroman ring (right side) is hypothesized to represent excretion of excess tocopherol. The Simon metabolite, α-tocopheronolactone, appears to be a produced by artifactual oxidation of α-CEHC during the extraction procedure (scheme adapted from Pope et al., 2001).

differences: the peak after 8 min in the *synthetic* profile was absent in the *natural* profile, indicating differential pathways of catabolism for the natural and synthetic forms.

The limit of quantification (LOQ) for HPLC-AMS is determined by the average of the baseline fractions plus two standard deviations and is typically in the vicinity of 2–20 amol [14]C [36]. Even lower levels of detection can be achieved through application of cleanup methods for removing other sources of biogenic carbon (lipid, proteins, etc.) and higher specific activity compounds (but not higher radiative doses). HPLC fractions such as these have too little carbon for direct processing

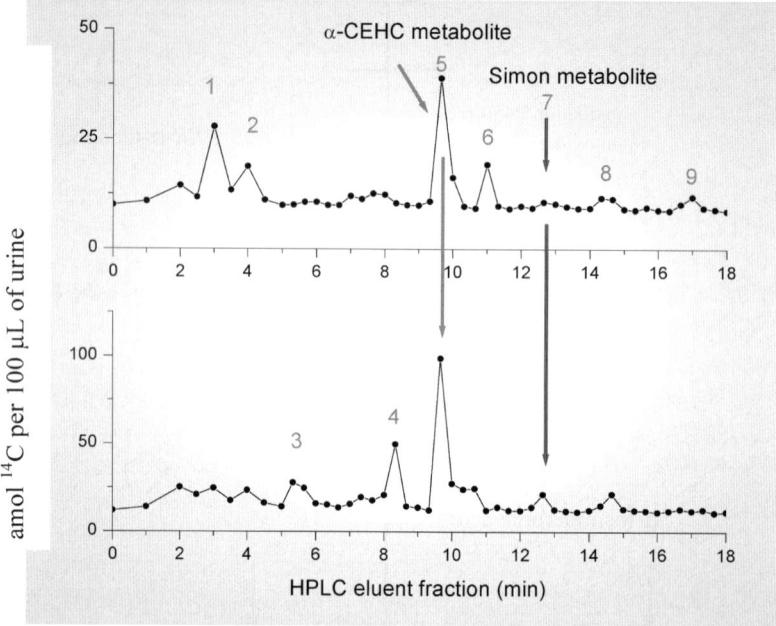

Figure 7.7 ^{14}C contents in HPLC eluents from human urine after consumption of RRR (top panel) and *all-rac* (bottom panel) α-tocopherol. Baseline is 9.0 amol $\pm$0.55 using a blank injection run. The limit of quantification is 1.6 amol per eluent fraction. α-CHEC is the single major metabolite in both tests, although at least 9 metabolites are indicated in the combined radiochromatograms.

graphite; therefore, 1 mg or more carbon is added to the dried eluents in the form of tributyrin.

Many factors can affect recovery from biological samples; in the case of reactive metabolites, the formation of artifacts can lead to spurious theories of metabolism. The ideal approach to sample preparation is to exclude the step altogether or "*dilute and shoot*" [47]. It was hoped that multiple sector MS/MS techniques might achieve this possibility. However, this ideal remains to be realized, primarily due to problems and limitations associated with matrix-suppression of ionization or increased incidence of ion-molecule interactions. Similar concerns do not pertain to AMS-based method for the obvious reason that the sample is converted to graphite prior to analysis and quantitation is based upon the ratio of the carbon isotopes. Therefore, minimal, or even no sample preparation affords an opportunity to view the metabolite profile in their native states. Metabolite profiling by direct injection of unprocessed urine has been demonstrated in other applications [29,48].

7.8 β-Carotene: cooperative behavior

The pigments found in plants exercise vital functions of plant metabolism. Furthermore, they provide visual attraction in nature and, for humans, provide us with nutrients. Indeed, many plant pigments perform essential roles in human and animal health. An early record of the positive association between plant pigments and human health is found in 1919. In this record, Steenbock noted that yellow corn (*Zea mays* Lat.) and "yellow" vegetables, such as carrots (*Daucus carota* Lat.) and sweet potato (*Ipomoea batatas* Lat.) eliminated the symptoms of vitamin A deficiency in rats. Conversely, white corn and "white" vegetables, such as parsnip (*Pastinoca sativa* Lat.), potato (*Solanum tuberosum* Lat.), and beets (*Beta vulgaris* Lat.) did not. An explanation of these observations was found when it was discovered that enzymes in the intestine and other parts of the human body enzymatically convert certain yellow colored carotene to vitamin A active compounds [49]. Of these carotenes, β-carotene is one of the most commonly consumed and easily metabolized forms of provitamin A pigments. In fact, it provides about 80% of the vitamin A needs globally. However, despite its abundance, vitamin A deficiency is presently plaguing many developing nations, afflicting millions of children each year with xerophthalmia, blindness, or death.

β-Carotene's relationship to health may extend well beyond its function as a vitamin A source. Numerous studies have observed inverse correlations between plasma carotene concentrations and certain cancers. Based on these and other data, several large chemoprevention trials have been conducted over the last several decades. These trials found supplemental β-carotene to provide no benefit or, in fact, to cause harm in smokers. In the aftermath of these studies, it has become clear that many fundamental aspects of β-carotene metabolism were poorly understood in humans. The observed toxicity of carotene has called into question the use of isolated phytochemicals at supraphysiological levels, as well as strengthened the case for food-based approaches to the prevention of chronic disease.

There are several reasons why food could succeed where supplements fail. One hypothesis states that nutrients act cooperatively and the body is optimized towards utilizing these nutrients when given as complex mixtures and at dietary levels. Marked deviations from this balance may then lead to altered distributive kinetics, affecting the chemical and anatomical disposition of the nutrients. An example relevant to β-carotene and AMS capabilities involves the effect of vitamin A status on β-carotene metabolism. Although marginally tested, it is popularly suspected that vitamin A status inversely relates to the utilization of β-carotene as a vitamin A source in humans [50]. To test this hypothesis, at risk population of marginal vitamin A status, typically children, are selected and then fed β-carotene. Individuals of low status will supposedly derive greater vitamin A benefit from the administered β-carotene than those that who are vitamin A adequate [51]. These studies, however, require the participation of experimental subjects of marginal nutrient status and such subjects are difficult to locate in industrialized countries. Moreover, to the extent that phytonutrition is often focused on optimal health in nutritionally replete population, rather than correcting deficiencies, it is of interest to have methods that are sensitive to subtle physiological changes in replete population.

A recent study conducted at the University of California, Davis and Lawrence Livermore National Laboratory, highlights the unique opportunities that AMS provides in the study of replete populations where other biomarkers might respond to subtle changes in dietary intake. The effect of vitamin A supplements on the absorption and metabolic behavior of a physiological dose of ^{14}C-β-carotene was investigated in test-retest format. Specific details of the dose time line are shown in Fig. 7.8. Complete fecal and urine samples were collected for 14 and 30 days, respectively. After the test, both subjects consumed daily supplements of 10,000 IU retinyl palmitate for 21 days to raise their vitamin A nutritional status. Both subjects then took a second and similar dose of ^{14}C-β-carotene.

In the vitamin A adequate individual, the effect of supplemental dose in moderate excess of normal dietary levels (2 X RDA) might not be expected to substantially impact the utilization of β-carotene. The results suggest otherwise and are

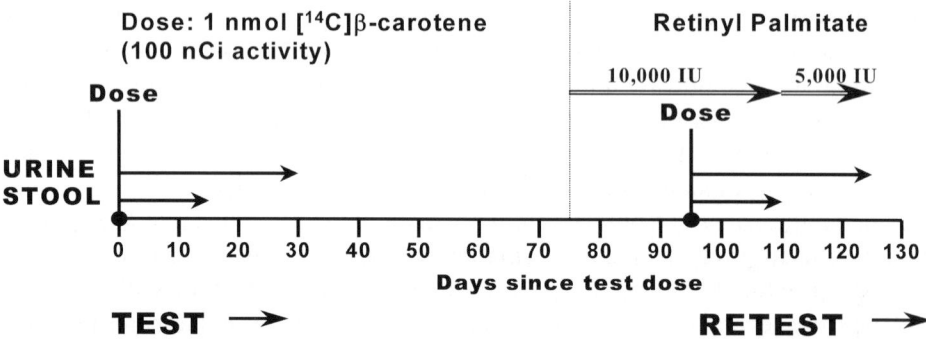

Figure 7.8 Experimental design and time line. The experiment was designed to incorporate a test and retest period. In the test period, subjects began complete fecal and urine collection 24 h in advance of the dose and continued complete 24-h collections until days 16 and 30, respectively. The subjects were then given a 1 nmol dose of [^{14}C]β-carotene in an emulsified drink. Cumulative urine was collected for 30 days and cumulative stool for 17 days. This was followed by a seven-week wash-out period. After that time, the administration of a second [^{14}C]β-carotene dose marked the beginning of the retest period. Three weeks prior to the start of the retest, subjects began consuming 10,000 IU (3000 µg RE) of vitamin A supplement daily and continued at that level until two-weeks after the retest dose administration. Then the supplement was continued at 5000 IU (1500 RE) until the completion of the testing.

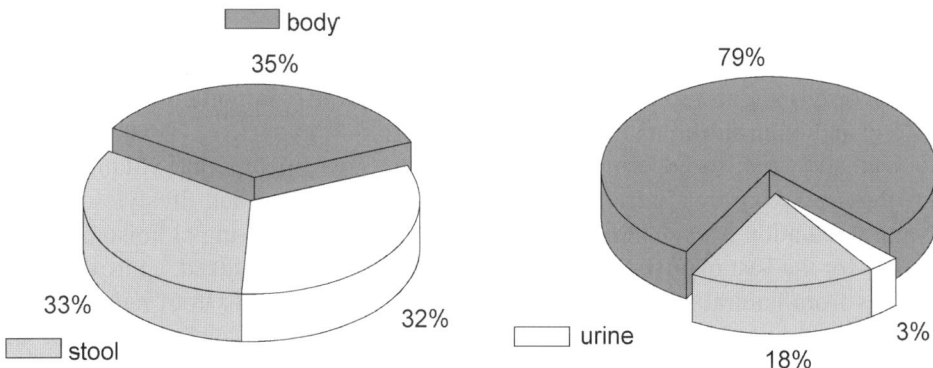

Figure 7.9 Effects of vitamin A supplementation upon the disposition of a tracer dose of ^{14}C-β-carotene in a longitudinal test in a female participant. The charts illustrate the dose disposition in excreta and subsequent body burden prior to (left) and following supplementation with vitamin A (right). Supplemental vitamin A resulted in increased absorption (stool) and retention (urine and body burden) of a tracer dose of β-carotene in vitamin A replete participants.

illustrated in the adjoining chart (Fig. 7.9) where the dose disposition 72 h post-dose for both the test and retest periods is exhibited. Absorption, as assessed by the cumulative recovery of the ^{14}C label in the stools, was raised by about 50% in both subjects. Conversely, the recovery of ^{14}C in the urine was reduced by 10-fold in both subjects. As a result, 72 h after dosing, about 80% of the consumed vitamin A was retained in the body of the supplemented individuals, whereas only about 35% was retained in the untreated and same subjects.

These data suggest that raising the vitamin A status in already replete subjects can markedly raise absorption and subsequent retention of a physiological dose of β-carotene. Such data has important implications for recommendations regarding the consumption of β-carotene either in food-based or supplemental forms. Minor changes in the consumption of one nutrient can have additive, or even synergistic, effects on the utilization of a second nutrient, even in nutritionally replete populations that, by all definitions, are not lacking in either nutrient.

Additionally, these data illustrate the value of *noninvasive* collections, i.e., urine and stool for the assessment of phytonutrient metabolism. Because the dose input is known, a full accounting of the ^{14}C is possible without the collection of blood specimens. Accordingly, the total of the body burden of the labeled compound is obtained. Such experimental models clearly represent an attractive method for assessing the biological variance in the absorption and retention characteristics of phytonutrients with specific analyte identification methods. This noninvasive feature, causing minimal discomfort in participants, would greatly facilitate population surveys into the utilization of selected phytonutrients.

7.9 Radiation risk

When discussing the role of radiation in AMS, it is first important to discuss the role of radiation in the natural world. Our world is radioactive—when assessing the radiation risk in experimental subjects, one must incorporate voluntary and involuntary exposures presented by our environment, diet, and lifestyle choices. Natural radiation from cosmic rays and from naturally occurring radioisotopes accounts for the vast majority of the total average effect dose. Indeed, approximately 81% is from natural sources of radiation and, of that, most is from radon. Every food has radionuclides, which lead to the accumulation of radioisotopes within the body. The common radionuclides in food are radiocarbon (^{14}C), as well as potassium 40 (^{40}K), radium (^{226}Ra), and uranium (^{238}U) and their associated progeny. An average human body experiences about 500,000 radioactive disintegrations per minute due to these and other radionuclides. The integrated average exposure is about 3600 µSv, a Seivert being the deposited energy equivalent to a joule per kilogram.

It is important to consider the amount of ^{14}C that naturally occurs in the body. The human body is about 23% carbon. The natural background levels of ^{14}C are due to formation in the upper atmosphere (cosmogenic production) as well as contribution from anthropomorphic sources (the burning of fossil fuels and atmospheric testing of nuclear weapons) is 6.11 pCi/g carbon (0.23 Bq). Accordingly, approximately 100 nCi (37 kBq) is in a 70 kg person. An average adult might, in a 24-h period, consume 200 g of carbon, which would contain 2.1 nCi of ^{14}C. By contrast, a 'large' administered radio-dose in AMS experiments is 100 nCi. In the case of vitamins, this quantity was determined to obtain long-term kinetic information associated with the longer turnover times inherent to fat soluble vitamins (β-carotene $t_{1/2} \sim 40$ days) and even retained water soluble forms (folate $t_{1/2} \sim 100$ days). Experiments with β-carotene and folic acid were conducted for greater than 200 days [26,27]. A majority of studies will be primarily interested in those events occurring within the period of absorption and distribution of a dose; therefore, studies on the limited duration of several days can be conducted. Where long-term tracing is not necessary, doses of 10 nCi or less suffice. In such cases, the amount of administered radiocarbon would be equal to that consumed in several meals. By this analogy, it is hard to imagine that low radiation dose AMS studies pose any quantifiable risk to the participant.

7.10 Labeling strategies

Stable and radioisotopes of hydrogen and carbon are often incorporated into organic molecular structures to serve as isotopic tracers. However, despite the ingenuity of modern organic chemists, many natural products found in common foods or medicinal plants are not amenable to isotopic synthesis due to their complex stereochemistry and distribution of unusual structures. For example, within the flavonoid family, proanthocyanidins (condensed tannins) are polymeric entities, which constitute one of the most abundant polyphenols in the human diet. These compounds are responsible for the astringency and, to some extent the bitterness, of many fruits such as grapes, apples, and persimmons. They bring these same qualities to fruit-derived beverages, most notably wine. The molecular weight of proanthocyanidins can vary between 600–3000 (dimers), although molecular weights as great as 30,000 Da have been recorded. Such a variance clearly poses a daunting synthetic challenge.

As a more practical alternative to isotopic synthesis, labeled compounds can be obtained by the introduction of isotopically labeled precursors into various portals. Labels can be incorporated into intact plants, as well as excised tissues, cultured plant cells, algae or isolated enzymatic homogenates. Harvesting the natural capabilities of biosystems has several advantages. First, detailed knowledge of synthetic chemistry is unessential. Second, the appropriately selected organism will replicate the natural complexity of the desired analytes. Finally, there is minimal production of hazardous waste products pertinent to organic synthesis [52].

The value of isotopes (stable or radioactive) as molecular tracers depends on the detection methods: isotope ratio mass spectrometry (IRMS), nuclear magnetic resonance (NMR), decay production counting, or AMS. Furthermore, the final detection method will often dictate the level of sophistication employed in the labeling protocol. It is established that methods developed around the detection of stable isotopes generally require much higher levels of isotope incorporation into the molecule—to facilitate detection—than corresponding radiometric methods (particularly AMS).

One demonstration of biolabeling is that of carotene in the photosynthetic alga, *Dunaliella Salina* (Fig. 7.10). This organism accumulates large amounts of

Figure 7.10 Photo of *Dunaliella Salina*, a salt tolerant photosynthetic alga that makes copious quantities of *all-trans* and 9-*cis* β-carotene.

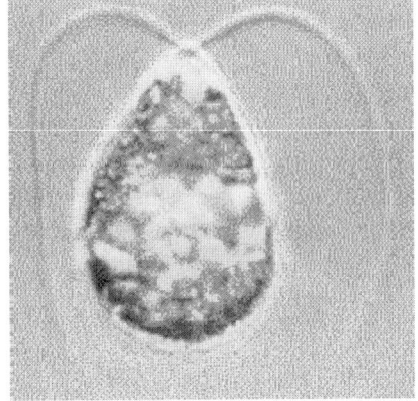

β-carotene in response to the stress of nutrient depletion and increased light intensity. As it is photosynthetic, it offers the opportunity of incorporating labeled carbon into its metabolic product by the feeding of labeled CO_2. Wilson and co-workers detailed the construction of a 4.5 L photosynthetic bioreactor for the production of ^{13}C-labeled carotenes using a $^{13}CO_2$ precursor [53]. The design of the reactor was optimized for the conservation of the label. This was achieved by using a gas recycling protocol during the feeding of the label into the system and a high level of incorporation of label into the β-carotene target. Between 0.7 and 1.4 g/day of $NaH^{13}CO_3$ were fed for $\sim$ 25 days in several experiments. Mass spectral data demonstrated high levels of incorporation of ^{13}C into the β-carotene. In one experiment, the average molecular formula of the β-carotene showed 30 of the possible 40 carbons in the molecule possessed the ^{13}C label. These high levels of label were deemed necessary for the efficient detection of β-carotene in human tracer studies using traditional mass spectrometry. Parker demonstrated the value of highly labeled β-carotene in human kinetic studies using GC-IRMS detection [54].

Unlike the aforementioned example, the high sensitivity of AMS obviates the need for high levels of isotope incorporation (specific activity) or sophisticated labeled chambers. In one experiment (unpublished results), 3 mCi of $NaH^{14}CO_3$ were added to 50 mL culture of *D. Salina* grown in a 50 mL screw cap bottle. A light source was provided and the contents of the closed vessel were allowed to grow for 1 week. The final specific activity of the β-carotene measured 30 Ci/mol, which corresponded to approximately $1\text{-}^{14}C$ atom per two molecules of β-carotene (by contrast, a perlabeled ^{14}C-β-carotene, as used in Parker's study, would possess a specific activity of 2500 Ci/mol if ^{14}C were substituted for the ^{13}C nuclide). Nonetheless, even at these low levels of incorporation, a nCurie-sized dose could be traced in human plasma for $>$ 200 days by AMS [27].

The incorporation of labeled CO_2 as a bicarbonate into *D. Salina* and other organisms represents a generic labeling strategy. A similar approach can be applied for the photosynthetic labeling of intact plants as well using atmospherically sealed plant chambers. An example of a chamber appropriate for radiolabeling with $^{14}CO_2$ is illustrated in Fig. 7.11. This chamber ensures controlled cycles of irradiance, temperature and relative humidity while quantitatively containing the administered radiolabel. The system is primarily comprised of: a stainless steel chamber, a refrigerated water chiller, a water reservoir with pump, a 1000 W metal halide lamp suspended from an adjustable support arm, a control panel, a pressure relief system, and an on-time totalizer for administering CO_2.

An application of the described isolation chamber is illustrated in a biolabeling study of common spinach (*Spinacia oleracea*). In this study, mature plants (30 days of age) were placed into the enclosed chamber. Exposure was initiated by adding 10 mCi/day as a solution of $NaH^{14}CO_3$ to excess acid, and this was repeated for

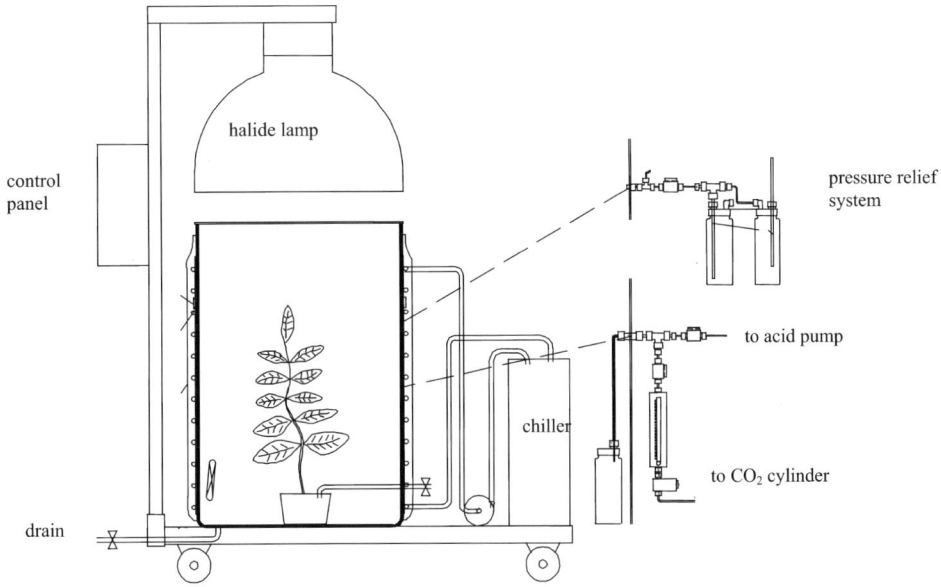

Figure 7.11 Photosynthetic labeling chamber.

an additional 4 days. Following the final exposure, the plants were maintained for 72 h in the chamber prior to harvesting to remove residual $^{14}CO_2$. The aerial parts of the plant were harvested and were extracted for β-carotene and lutein: specific activities were 1.45 and 0.35 Ci/mol, respectively, corresponding to 0.023 and 0.0056 ^{14}C atoms per molecule, respectively.

Within the interest of phytonutrition, there lies an increasing attention on the assessment of available nutrients from the intact plant source (as opposed to chemically isolated concentrates). Photosynthetic radiolabeling cannot be applied to address this interest, as it would expose the participant to high, ethically unacceptable, levels of radiation. Stable isotope approaches circumvent this concern, however. These approaches are applicable when sufficiently high levels of isotope incorporation can be achieved to overcome sensitivity limitations. An important example of this concept is demonstrated in the biological tracing of deuterium labeled vitamin K (phylloquinone). In this example, broccoli grown hydroponically using 31 at% deuterium hydroxide was fed to a single participant. The plasma concentration of the labeled vitamin K in the plasma could be traced for $\sim$ 12 h using GC/MS detection operated in the negative chemical ionization mode [55].

The concept of intrinsic labeling has been applied to the radiolabeling of folates in the pea plant [56]. In these experiments, pea seeds were allowed to imbibe ^{14}C-*para*-aminobenzoic acid (pABA) prior to germination and vegetative growth [56] (Fig. 7.12). Significant portions of the administered labeled appeared in the

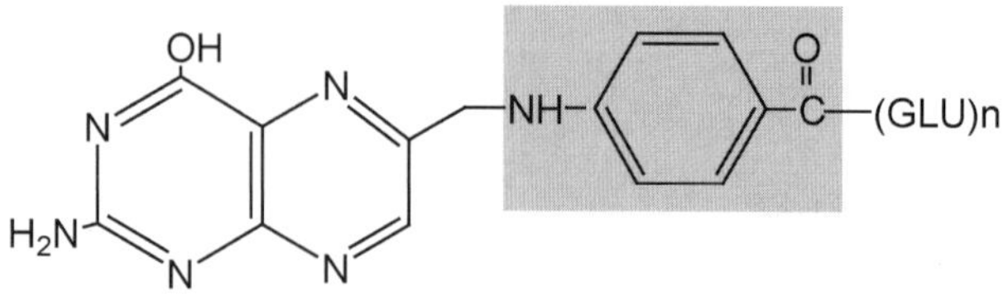

Figure 7.12 Stylized representation of folic acid. Grey area represents moiety derived from pABA.

plant folate pools. While human testing was not the aim of this investigation, it does validate the general concept. Our group also applied the concept for the specific purpose of assessing the availability of folates when consumed by an intact plant. In this study, ^{14}C-pABA was administered to a mature kale plant (*Brassica oleracea acephala*) using the cotton-wick method. After uptake of the label, the plant was then allowed to grow for one week to promote translocation of the dose to the actively growing leaf tips, the area of most active respiration. This experiment was unsuccessful, however, as quantities of the administered label were not assimilated into the folate pools in the degree necessary to facilitate testing. Most of the activity, rather, resided with unmodified ^{14}C-pABA. Nonetheless, further experimentation with growth cycles and label deliver techniques should eventually prove successful. Upon surveying various labeling strategies, it is clear that AMS sensitivity will provide a most amenable approach for difficult to synthesize phytochemicals.

7.11 Conclusion

As public interest in phytonutrition continues to increase, the result will be an augmented demand for extensive phytochemical research. The fact that foods are inherently phytochemically complex dictates a need to apply scientific techniques, which can detect synergistic interaction among the many active principles and adjuvant substances in the plant, and, furthermore, modify the activities of these components. As illustrated by the experiments discussed in this presentation, the advantages of AMS are unique and extensive. These advantages are best summarized by Dr John Vogel, an originator of biological AMS experimentation: "AMS brings (at least) three advantages to biochemical tracing: high sensitivity for finding low probability events or for use of physiologic-sized doses; small sample sizes for painless biopsies or highly specific biochemical separations; and reduction of overall radioisotope exposures, inventories, and waste streams." AMS opens the door to increased phytochemical tracing in humans to obtain biochemical data concerning human health at dietary relevant levels of exposure. AMS, thus, obviates the need for uncertain extrapolations from animal models, which express marginal relevance to human metabolism. The unparalleled capabilities and benefits of AMS will undoubtedly establish this particular MS technique as an important analytical tool in phytochemical research.

Acknowledgments

Yumei Lin, Jennifer Follett, Andrew Clifford, Colleen Carkeet, Charlene Ho, Bruce Buchholz, and John Vogel all provided original data of published work or data in preparation for the work presented here. The author also *sincerely* thanks Megan Patterson for significant contributions to the editing and refinement of the text. Part of this work was performed under the auspices of the U.S. Department of Energy by the University of California, Lawrence Livermore National Laboratory under Contract No. W-7405-Eng-48, NIH NCRR P41 RR 13461, CSREES#99-35200-7584 and NIDDK DK48307.

References

1. M.A. Grusak, *Curr. Opin. Biotechnol.* **13** (2002) 508–511.

2. I. Potrykus, *Plant Physiol.* **125** (2001) 1157–1161.

3. M.W. Farnham, P.W. Simon, and J.R. Stommel, *Nutr. Rev.* **57** (1999) S19–S26.

4. P.M. Mullineaux and G.P. Creissen, *Biochem. Soc. Trans.* **24** (1996) 829–835.

5. G. Block, E. Norkus, M. Hudes, S. Mandel, and K. Helzlsouer, *Amer. J. Epidemiol.* **154** (2001) 1113–1118.

6. C.H. Halsted, *Amer. J. Clin. Nutr.* **77** (2003) 1001S–1007S.

7. D. Albanes, *Amer. J. Clin. Nutr.* **69** (1999) 1345S–1350S.

8. G.S. Omenn, *Ann. Rev. Public Health* **19** (1998) 73–99.

9. C.K. Van Pelt, P. Haggarty, and J.T. Brenna, *Anal. Chem.* **70** (1998) 4369–4375.

10. F.A. Mellon, R.N. Bennett, B. Holst, and G. Williamson, *Anal. Biochem.* **306** (2002) 83–91.

11. L.W. Sumner, P. Mendes, and R.A. Dixon, *Phytochemistry* **62** (2003) 817–836.

12. R. Goodacre, E.V. York, J.K. Heald, and I.M. Scott, *Phytochemistry* **62** (2003) 859–863.

13. G. Hopfgartner and E. Bourgogne, *Mass Spectrom. Rev.* **22** (2003) 195–214.

14. J.S. Vogel, K.W. Turteltaub, R. Finkel, and D.E. Nelson, *Anal. Chem.* **67** (1995) 353A–359A.

15. J.S. Vogel and K.W. Turteltaub, *Adv. Exp. Med. Biol.* **445** (1998) 397–410.

16. G.S. Jackson, C.M. Weaver, and D. Elmore, *Nutr. Res. Rev.* **14** (2001) 317–334.

17. D.J.W. Mous, K.H. Purser, W. Fokker, R. Van den Broek, and R.B. Koopmans, *Nucl. Instr. Meth. Phys. Res. B* **123** (1997) 159–162.

18. M.L. Chiarappa-Zucca, K.H. Dingley, M.L. Roberts, C.A. Velsko, and A.H. Love, *Anal. Chem.* **74** (2002) 6285–6290.

19. K.H. Dingley, M.L. Roberts, C.A. Velsko, and K.W. Turteltaub, *Chem. Res. Toxicol.* **11** (1998) 1217–1222.

20. F.A. Mellon and B. Sandstrom, *Stable Isotopes in Human Nutrition*, Academic Press, London–San Diego, 1996.

21. D.S. Goodman, R. Blomstrand, B. Werner, H.S. Huang, and T. Shiratori, *J. Clin. Invest.* **45** (1966) 1615–1623.

22. H.E. Gove, *From Hiroshima to the Iceman*, Institute of Physics Publishing, Bristol, UK, 1999.

23. K.W. Turteltaub, J.S. Felton, B.L. Gledhill, J.S. Vogel, J.R. Southon, M.W. Caffee, R.C. Finkel, D.E. Nelson, I.D. Proctor, and J.C. Davis, *Proc. Natl. Acad. Sci. USA* **87** (1990) 5288–5292.

24. K.W. Turteltaub, J.S. Vogel, C. Frantz, M.H. Buonarati, and J.S. Felton, *Environ. Health Perspect.* **99** (1993) 183–186.

25. K.W. Turteltaub, J.S. Vogel, C.E. Frantz, and E. Fultz, *IARC Sci. Publ.* (1993) 293–301.

26. A.J. Clifford, A. Arjomand, S.R. Dueker, P.D. Schneider, B.A. Buchholz, and J.S. Vogel, *Adv. Exp. Med. Biol.* **445** (1998) 239–251.

27. S.R. Dueker, Y. Lin, B.A. Buchholz, P.D. Schneider, M.W. Lame, H.J. Segall, J.S. Vogel, and A.J. Clifford, *J. Lipid Res.* **41** (2000) 1790–1800.

28. K. Stenstrom, S. Leide-Svegborn, B. Erlandsson, R. Hellborg, S. Mattsson, L.E. Nilsson, B. Nosslin, G. Skog, and A. Wiebert, *Appl. Radiat. Isot.* **47** (1996) 417–422.

29. R.C. Garner, I. Goris, A.A. Laenen, E. Vanhoutte, W. Meuldermans, S. Gregory, J.V. Garner, D. Leong, M. Whattam, A. Calam, and C.A. Snel, *Drug Metab. Dispos.* **30** (2002) 823–830.

30. P.P. Hoppe and G. Krennrich, *Eur. J. Nutr.* **39** (2000) 183–193.

31. G.W. Burton, M.G. Traber, R.V. Acuff, D.N. Walters, H. Kayden, L. Hughes, and K.U. Ingold, *Amer. J. Clin. Nutr.* **67** (1998) 669–684.

32. M.G. Traber, G.W. Burton, K.U. Ingold, and H.J. Kayden, *J. Lipid Res.* **31** (1990) 675–685.

33. C. Lauridsen, S.W. Leonard, D.A. Griffin, D.C. Liebler, T.D. McClure, and M.G. Traber, *Anal. Biochem.* **289** (2001) 89–95.

34. W. Cohn, *Eur. J. Clin. Nutr.* **51** (1997) S80–S85.

35. Y. Sato, K. Hagiwara, H. Arai, and K. Inoue, *FEBS Lett.* **288** (1991) 41–45.

36. B.A. Buchholz, S.R. Dueker, Y. Lin, A.J. Clifford, and J.S. Vogel, *Nucl. Instr. Meth. Phys. Res. B* **172** (2000) 910–914.

37. S.D. Gilman, S.J. Gee, B.D. Hammock, J.S. Vogel, K. Haack, B.A. Buchholz, S.P. Freeman, R.C. Wester, X. Hui, and H.I. Maibach, *Anal. Chem.* **70** (1998) 3463–3469.

38. R. Brigelius-Flohe, *Biochem. Biophys. Res. Commun.* **305** (2003) 737–740.

39. T.J. Sontag and R.S. Parker, *J. Biol. Chem.* **277** (2002) 25290–25296.

40. R.S. Parker, T.J. Sontag, and J.E. Swanson, *Biochem. Biophys. Res. Commun.* **277** (2000) 531–534.

41. S.A. Pope, G.E. Burtin, P.T. Clayton, D.J. Madge, and D.P. Muller, *Free Radic. Biol. Med.* **33** (2002) 807–817.

42. R. Brigelius-Flohe, F.J. Kelly, J.T. Salonen, J. Neuzil, J.M. Zingg, and A. Azzi, *Amer. J. Clin. Nutr.* **76** (2002) 703–716.

43. E.E. Simon, A.L. Sundheim, and A. Milhorat II, *J. Biol. Chem.* **221** (1956) 807–817.

44. M. Schultz, M. Leist, M. Petrzika, B. Gassmann, and R. Brigelius-Flohe, *Amer. J. Clin. Nutr.* **62** (1995) 1527S–1534S.

45. M. Schultz, M. Leist, A. Elsner, and R. Brigelius-Flohe, *Methods Enzymol.* **282** (1997) 297–310.

46. J.K. Lodge, M.G. Traber, A. Elsner, and R. Brigelius-Flohe, *J. Lipid Res.* **41** (2000) 148–154.

47. J. Henion, E. Brewer, and G. Rule, *Anal. Chem.* **70** (1998) 650A–656A.

48. B.A. Buchholz, E. Fultz, K.W. Haack, J.S. Vogel, S.D. Gilman, S.J. Gee, B.D. Hammock, X. Hui, R.C. Wester, and H.I. Maibach, *Anal. Chem.* **71** (1999) 3519–3525.

49. J.A. Olson and O. Hayaishi, *Proc. Natl. Acad. Sci. USA* **54** (1965) 1364–1370.

50. N.W. Solomons and J. Bulux, *Ann. NY Acad. Sci.* **691** (1993) 96–109.

51. M. Van Lieshout, C.E. West, and R.B. Van Breemen, *Amer. J. Clin. Nutr.* **77** (2003) 12–28.

52. R.J. Molyneux, *J. Agric. Food Chem.* **50** (2002) 6939–6942.

53. P.D.G.H. Wilson, M.G., C.R. Waspe, D.C. Steer, and D.R. Wilson, *Biotechnol. Lett.* **19** (1997) 401–405.

54. R.S. Parker, J.T. Brenna, J.E. Swanson, K.J. Goodman, and B. Marmor, *Methods Enzymol.* **282** (1997) 130–140.

55. G.G. Dolnikowski, Z. Sun, M.A. Grusak, J.W. Peterson, and S.L. Booth, *J. Nutr. Biochem.* **13** (2002) 168–174.

56. E.A. Cossins and L. Chen, *Phytochemistry* **45** (1997) 437–452.

Advances in Mass Spectrometry, Volume 16
A.E. Ashcroft, G. Brenton and J.J. Monaghan (Editors)

CHAPTER 8

Genotyping Single Nucleotide Polymorphisms by MALDI Mass Spectrometry

Jörg Tost and Ivo Glynne Gut*

Centre National de Génotypage, Bâtiment G2, 2, Rue Gaston Crémieux, 91057 Evry Cedex, France

Abbreviations

A	adenosine
bp	base pairs
C	cytosine
cSNP	coding SNP
ddNTP	dideoxynucleotide triphosphate
DNA	deoxyribose nucleic acid
dNTP	deoxynucleotide triphosphate
FEN	flap endonuclease
G	guanosine
HPA	3-hydroxypicolinic acid
MALDI	matrix-assisted laser desorption/ionization
PCR	polymerase chain reaction
PNA	peptide nucleic acid
PROBE	primer oligo base extension

---

*Corresponding author. E-mail address: ivo.gut@cng.fr.

RNA	ribose nucleic acid
SAP	shrimp alkaline phosphatase
SNP	single nucleotide polymorphism
STR	short tandem repeat
T	thymidine
UTP	uracil triphosphate
VSET	very short extension

Abstract

The last decade has seen an increased demand for high-throughput DNA analysis. This is mainly due to the human genome sequencing project that was recently announced finished. Even though mass spectrometry did not contribute to this project, it is clear that it will have an important role in the post-genome sequencing era, in genomics and proteomics. In genomics, mainly MALDI mass spectrometry will contribute to large-scale single nucleotide polymorphism (SNP) genotyping projects. Several strategies for allele-discrimination (hybridization, cleavage, ligation, and primer extension) have been combined with MALDI mass spectrometric detection. In practice only primer extension has been integrated for SNP genotyping at high throughput. The methods using primer extension use quite diverse strategies in order to get to grips with the intricacies of MALDI and the time-of-flight detector. Adduct formation with ubiquitous sodium and potassium is a major problem and is circumnavigated by stringent purification procedures or charge-neutralization of the DNA backbone. Resolution of the MALDI in the mass range of 5–8 kDa is not sufficient to distinguish A/T SNPs (9 Da mass difference). Thus, either a strategy where two alleles are separated by one or more bases is implemented, or a significant portion of the primer is removed prior to mass spectrometric detection. Problems surrounding the integration of SNP genotyping by MALDI mass spectrometry at high throughput are discussed, as well as applications that go beyond standard SNP genotyping and are more demanding, such as molecular haplotyping, DNA methylation analysis, and mutation detection. Examples are shown that illustrate the specific advantages of mass spectrometric analysis of DNA compared to other DNA analysis techniques.

8.1 Introduction

It is the 50th anniversary of Watson and Crick's elucidation of the molecular structure of DNA [1]. Their discovery has had a most profound impact on biology. One of the most recent examples is the human genome sequencing project that was spearheaded again by Watson [2]. The emergence of the automated DNA sequencer at that time made the human genome sequencing project conceivable. With the human genome sequencing project now completed [3], attention is turning to the elucidation of gene function [4,5]. Variation analysis—genotyping—is an important tool to determine correlation of genes with phenotypes. Association studies preferably make use of very densely spaced polymorphisms. In the human genome single nucleotide polymorphisms (SNPs) are spaced, on average, one every 500–1000 bases. A large systematic effort was launched by the SNP consortium to identify several hundred thousand SNPs [6]. Currently there are about 4 million SNPs in the public domain [3,7]. This emergence of wealth of polymorphisms goes hand-in-hand with a frenzy of development of methods to genotype them efficiently. Until today, many methods have been introduced but no single one is dominating. A few have actually been applied at very high throughput, and new ones are still emerging [8]. High-throughput SNP genotyping methods that use MALDI mass spectrometry for detection are among the most powerful and reliable [9].

8.1.1 A brief history of DNA analysis by mass spectrometry

Concurrent with the initiation of the human genome sequencing project, Tanaka, Karas, and Hillenkamp discovered matrix-assisted laser desorption/ionization time-of-flight mass spectrometry (MALDI) [10,11]. Initially, this method predominantly found its application in the analysis of proteins and peptides. The uptake for DNA analysis was significantly more sluggish. Initial work centered around the investigation of desorption conditions and the resolution that could be achieved with different sequence compositions. Thymidine containing sequences were easiest to analyze in negative ion-mode, and limitations for the size of oligonucleotide around 20 were encountered. However, it was not until Becker's group published that 3-hydroxypicolinic acid (HPA) was a good matrix for DNA analysis in the positive ion-mode that DNA analysis by MALDI started to gain pace [12]. With this matrix, two problems became blatantly apparent—first, that larger DNA fragments showed depurination, and, second, that sodium and potassium formed adducts with DNA. As a result, peaks were broad and cross-talk made it inconceivable to use MALDI for DNA analysis beyond 50 bases. A number of strategies to deal with these problems were devised. Solutions included transcribing DNA into RNA, because RNA tended to show less depurination [13], and using DNA with deazapurine bases [14,15]. A key observation made by Pieles et al. [16] was that adding ammonium ions during matrix preparation significantly reduced problems with sodium and potassium adducts. This publication also marks the first enzymatic sequencing of oligonucleotides by partial exonuclease digestion from the $5'$- and the $3'$-end with mass spectrometric detection. The initial problems surrounding the mass spectrometry of nucleic acids are described in great detail in a review by Nordhoff et al. [17].

In 1995, MALDI instrumentation took a quantum leap with Vestal's rediscovery and integration of delayed extraction into the MALDI process [18]. This dramatically improved the resolution of spectra. MALDI instruments became accessible to non-mass spectrometrists because many instrument manufacturers started to offer "turnkey instruments." Research groups with more of a background in molecular biology entered the field. As it became apparent that the problems of limited size range of detection, resolution, fragmentation, and adducts were very difficult to improve on, DNA analysis by MALDI started shifting to an emerging problem—that of analyzing mutations, part of which were later termed SNPs, using, for example, single-base primer extension [19]. However, with single-base primer extension at a mass of 5500–7000 Da, it was impossible to resolve the smallest base change (T to A, with a 9 Da mass difference). Thus, another strategy with multiple-base extension and specific termination was used to space alleles of a mutation further apart [20]. With this technology, Sequenom initiated a campaign for mutation analysis by MALDI mass spectrometry. This campaign coincided perfectly with the discovery that SNPs are very frequently found throughout the human and other genomes, and that they might be well suited for association studies. Thus, MALDI turned into one of the major players in the field of SNP genotyping.

8.1.2 What are SNPs?

A single nucleotide polymorphism (SNP) is a single-base change that occurs at a specific position in a genome (Fig. 8.1). Although, single-base insertions and deletions are strictly not SNPs, they are often included in this definition. For these positions different sequence alternatives (alleles) exist in individuals in some populations [21]. By definition, the less-frequent allele has an abundance of 1% or greater; otherwise, it is referred to as a mutation. In diploid species, like humans, SNPs are usually biallelic (a SNP has two alleles). Rarely are SNPs found with three or four alleles. On average, one SNP is found in every 500 to 1000 bases in humans. Only a small portion of them lie within coding regions (cSNPs), and an even smaller percentage is responsible for amino acid changes in expressed proteins. However, SNPs in noncoding regions can still affect gene regulation. To geneticists, the fact that a SNP marker, in contrast to microsatellites (also called

Single Nucleotide Polymorphism (SNP)

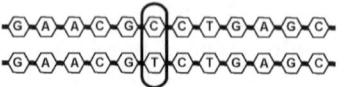

Insertion / Deletion Polymorphism

Figure 8.1 (a) The two alleles of a single nucleotide polymorphism; (b) is a single nucleotide deletion/insertion.

short tandem repeat, STR) can be lodged directly in the genomic region of interest, for example a candidate gene, has great value. An allele of a SNP can constitute a genetic risk factor. In contrast to microsatellites, SNPs are significantly more stable over generations, and are less liable to spontaneous change [22]. Due to their binary nature, SNPs are fairly easy to genotype, and the interpretation of the readout can be automated [8].

The current effort in the development of high-throughput SNP genotyping methods is based on the assumption that, with a dense map of markers, it will be possible to localize and characterize genes involved in complex human diseases such as cancer, cardiovascular diseases, diabetes, and psychiatric disorders [23]. Association studies, in which the genotypes of a cohort of cases and controls are compared, have been proposed as a tool to detect even moderate effects of the individuals' genotypes in relation to their phenotype [24]. The quality of results must be very high. MALDI mass spectrometry-based SNP genotyping methods with primer extension for allele-distinction deliver data of highest quality if controlled properly. Errors are comparatively easy to pinpoint. Pharmacogenomics will be another application, where the promise is that patients will be medicated as a function of their genetic profile, thus reducing the risk of adverse drug effects and loss of valuable time in therapy. Currently, some of the highest volume applications of SNP genotyping are found in agricultural applications such as generating genetic fingerprints of animals for identification and traceability purposes. The field of SNP genotyping methods and high-throughput versions thereof has been reviewed extensively in the past few years [8,25,26].

8.1.3 Why use MALDI for DNA analysis?

MALDI is a very rapid process, and—combined with time-of-flight analysis— is well suited for the analysis of complex mixtures. The current generation of MALDI instruments is capable of recording a single spectrum within less than one second. Because a mass spectrum provides many detection channels, a MALDI time-of-flight analysis can be used to record many data points in a single experiment. In terms of SNP genotyping, this capability means that multiple SNPs can be genotyped in a single experiment (multiplexing). Most other SNP genotyping methods cannot provide this degree of multiplex-detection. Further, in contrast to other DNA analysis methods, mass spectrometry relies on the direct mass determination of products rather than on an indirect analysis, where a fluorescent or radioactive report tag is interpreted. Fluorescent labels for DNA analysis are expensive, whereas radioactivity is expensive and undesirable. Thus, also a marked price advantage goes with the direct mass determination of DNA fragments. Troubleshooting is far easier for methods that use direct observation than for indirect tag methods.

8.2 Methods for allele-discrimination (Fig. 8.2)

All methods for SNP genotyping combine two elements: first, the generation of an allele-specific product, and, second, the analysis thereof. Currently, four principles exist for the generation of the allele-specific products: hybridization, ligation, cleavage, and primer extension.

For allele-specific hybridization two probes, each representing one allele of a polymorphism, are hybridized to query a target sequence. The fully complementary probe hybridizes to the target sequence better than a probe with one or more mismatches. In the case of mass spectrometric detection, the mass of the hybridized probe is used for its identification, whereas other analysis methods require fluorescent labeling of the probe. Hybridized and nonhybridized probes must be separated for mass spectrometric detection. Hybridization is a physically controlled process and lacks the discrimination power of methods including an enzymatic step. The efficiency of discrimination depends on the position of the mismatch, the length of the probe, the sequence context, and reaction conditions. Using a DNA analog—peptide nucleic acid (PNA) [27,28]—that contains the four nucleobases of DNA attached to an amide backbone (N-(2-aminoethyl)glycine units) and which perfectly Watson–Crick base pairs with DNA [29] has some advantages for mass spectrometric analysis. PNA has a charge neutral backbone and is thus insensitive to adduct formation and less prone to fragmentation [30]. Hybridization can be carried out under low salt conditions. PNAs behave more like peptides and are best analyzed with peptide matrices. Two similar approaches for the analysis of SNPs, using hybridization with PNAs, were shown independently [31,32]. PCR is performed with one biotinylated primer. The biotinylated strand is captured on streptavidin-coated magnetic beads. Two PNA probes are allowed to hybridize with the target sequence and after stringent washing the bead solution is transferred to a MALDI target for analysis. Alleles are assigned by measuring the presence of a PNA probe at a certain mass.

Hybridization is also the underlying strategy used in the monitored nuclease-selection assay [33], demonstrated on two SNPs in exon 10 of the cystic fibrosis transmembrane conductance-regulator gene. An asymmetrically amplified PCR product (the amplification of the strand of interest is favoured) is purified by magnetic beads, and short oligonucleotides (9 bases) that represent alleles of the queried SNPs are allowed to hybridize with the PCR product. All probe oligonucleotides not bound tightly to the target sequence are degraded with phosphodiesterase I. Finally residual, not degraded probes are detected by mass spectrometry for allele assignment.

For oligonucleotide ligation (OLA), two specific oligonucleotides with a length of 15–30 nucleotides are allowed to anneal adjacent to each other on a single-stranded target sequence. In the case of complete complementarity they are linked

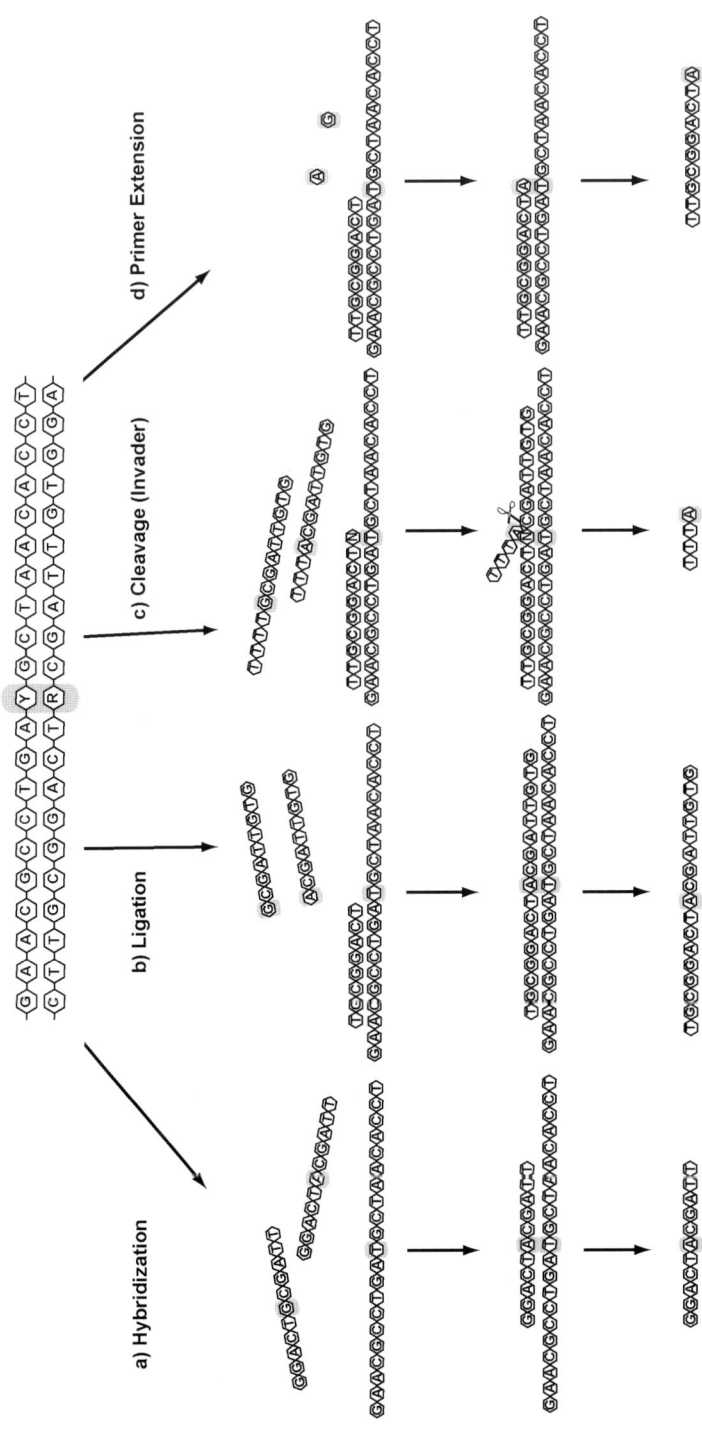

Products analyzed in the MALDI mass spectrometer

Figure 8.2 The four principles for allele-discrimination, and the products that are subsequently analyzed in the MALDI mass spectrometer: (a) hybridization, (b) ligation, (c) cleavage (Invader assay), and (d) primer extension. An allele that carries an A of an A/G polymorphic site was chosen for demonstration. The length of the utilized probes was randomly chosen, and does not represent the actual size of probes used in the assays. R and Y are IUPAC polymorphism codes (R = A or G, Y = C or T).

to each other by a thermostable DNA ligase. The presence or absence of a ligated product identifies alleles of point mutations and SNPs. Analysis of purified ligation products by MALDI mass spectrometry has been shown on cloned DNA and synthetic templates [34]. However, a demonstration of this method on genomic DNA is still awaited.

Detection of SNPs by creating or abolishing recognition sites for restriction enzymes, similar to conventional restriction fragment-length polymorphism (RFLP) analysis, has been used in combination with MALDI mass spectrometric detection [35]. Only a small number of polymorphisms alter known restriction sites, and the design of amplification primers to create restriction sites in connection with one allele is not straightforward in most cases.

The Invader assay [36] is based on the cleavage of specific DNA or RNA structures by the $5'$-nucleases domain of eubacterial Pol A DNA polymerases and structurally homologous FLAP endonucleases (FENs) that are involved in the DNA repair process [37,38]. These enzymes or genetically engineered thermostable versions thereof called cleavases specifically recognize and cleave an overlapping structure of two oligonucleotides that are hybridized adjacent to each other on a DNA template. Mainly, fluorescence resonance energy transfer is used for the detection of either solution-based [36] or solid-phase Invader assays [39]. Mass spectrometry was used in connection with a squared Invader format [40]. In this method [41], the released flap of the probe acts as an invader oligonucleotide in a secondary reaction, leading to an exponential amplification of the signal. Two allele-specific probes for the target sequence are used for the first invader reaction, and two probes for the second invader reaction. The latter differ in size of the flap (three, respectively four nucleotides), thus separating the masses of the two alleles by 300 Da. Biotin labels on the $5'$-end of the second flap allow sample purification on streptavidin-coated magnetic beads prior to MALDI analysis. Most notably the Invader assay does not rely on PCR amplification [36].

Primer extension is the most widely applied method for allele-distinction for SNP genotyping and point mutation analysis [42]. Incorporation of a complementary nucleotide by a DNA polymerase distinguishes more accurately between the two alleles than the different thermal stability of hybridizing allele-specific probes [43]. Primer extension assays are flexible, robust, and well-suited for high-throughput applications. An important feature for the analysis by MALDI is the generation of small products (3 to 25 bases). Duplex structures of extension primers and templates dissociate under MALDI conditions so that only the short extension primer and primer extension products are detected.

Several primer extension assays that use MALDI as the detection platform have been developed, and some of them have been commercialized. In all of these assays, the DNA target sequence is amplified with PCR. Remaining dNTPs and

Table 8.1: Overview: methods for allele-discrimination and assays with MALDI detection

Method for allele distinction	Method	Purification	Mass range m/z	Main advantage	Main disadvantage	Ref.
Hybridization	PNA	Magnetic beads	2000–4000	Simple principle	Prize of PNAs	[31,32]
	Monitored nuclease selection assay	None	5000–8000	Cheap, robust enzyme	Time-consuming analysis	[33]
Oligonucleotide ligation	Ligase chain reaction	Magnetic beads	> 9000	Exponential product amplification	Low enzymatic specificity	[34]
Cleavage	Restriction enzymes	Desalting columns	10,000–100,000	Simple principle	Not universally applicable, expensive enzymes	[35]
	Invader assay	Magnetic beads	1000–2000	No PCR, isothermal reaction	Large amount of DNA consumed	[40]
Primer extension	Pin Point assay	ZipTips	4000–9000	Accurate, rapid data accumulation, multiplexing	Mass difference between A/T may not be resolved	[19,61,62]
	PROBE assay	Magnetic beads	4000–8000	Accurate, rapid data accumulation	Solid-phase purification	[46]
	Homogenous MassEXTEND assay	Ion-exchange resin	4000–8000	High-throughput, rapid, accurate	Cost of instrumentation set-up	[57]
	VSET assay	Ethanol precipitation	4000–8000	Accurate, rapid data accumulation	Purification	[47]
	GenoLINK assay	Magnetic beads	4000–8000	Accurate, rapid data accumulation	Solid-phase purification	Bruker Saxonia GmbH
	GenoSNIP assay	Magnetic beads	1000–2000	Increased sensitivity, small products	Solid-phase purification	Bruker Saxonia GmbH
	GOOD assay	None	1000–2000	No purification	Modification chemistry	[44,45]

primers are removed from the reaction by various methods, because these compounds would be undesirable and interfere in the following primer extension reaction. A primer for primer extension anneals with its 3′-terminal base immediately upstream of the SNP on the target sequence. Discrimination of the alleles is achieved by either extending the primer with a single nucleotide, using terminating dideoxynucleotides (ddNTPs) as applied in the PinPoint assay [19], the GOOD assay [44,45], and the GenoSNIP assay (Bruker Saxonia Analytik GmbH, Leipzig, Germany), or by using a mixture of deoxynucleotides (dNTPs) and ddNTPs. Alleles are spaced further apart as termination occurs at the first nucleobase in the template complementary to a ddNTP. This concept is used in the PROBE assay [46], the MassEXTEND assay [9], the VSET assay [47], and the GenoLink assay (Bruker Saxonia Analytik GmbH, Leipzig, Germany) (see Table 8.1).

8.2.1 Purification procedures and sample preparation

One has to keep in mind that SNP genotyping might be carried out at very high throughput. Studies can easily require the accumulation in excess of a million data points. Consequently, all handling and processing has to be optimized. Mass spectrometric analysis of DNA is very sensitive to impurities. In many instances, rigorous sample clean-up is necessary. Components of enzyme buffers such as detergents, stabilizers, and glycerol, as well as salts can severely interfere with MALDI analysis [48]. In principle, ethanol precipitation yields sufficient purity for sample analysis by MALDI. However, because this procedure is difficult to automate it is suitable only for limited sample throughput. The same arguments hold for approaches based on hybridization to complementary DNA sequences on solid surfaces [49]. Streptavidin-coated magnetic beads have been widely applied for this purification [20,50]. Alternatively, float dialysis can be used for purification [51,52]. Low molecular weight contaminants are removed by diffusion through a membrane. Although this procedure is a cheap alternative and biotin-modified primers become unnecessary, its application in automated sample processing might be technically demanding. Pipette tips loaded with POROS 50 (Perspective Biosystems) [19] are efficient for purification and desalting. Reversed-phase purification with ZipTips (Millipore) was demonstrated [53,54]. Their use contributes significantly to cost. Recently a 96-well format with reversed-phase material was presented [55]. Homogenous purification procedures facilitate the process, but dedicated robotic systems are necessary to implement them at high throughput. The application of an ion-exchange resin is one possibility [56], which is now used in a similar way in Sequenom's homogenous MassEXTEND™ assay (Spectro-CLEANER™) [57]. Affinity binding with paramagnetic beads is another purification method (Genopure, Bruker). A recent approach uses glass slides that are coated with a polymer (polyvinylpyrrolidone) film [58]. Oligonucleotides diffuse into the film, the supernatant is removed, and spots are extensively washed. After application of matrix the glass slide with the purified oligonucleotides is introduced into the mass spectrometer. Signal intensities are improved compared to conventional purification procedures. In summary, purification is cumbersome, tricky in automation, and can be expensive. The GOOD assay employs a chemical modification procedure to circumvent purification [44]. The use of charge-tag chemistry increases the sensitivity ca. 100-fold [59]. A defined charge is introduced into the extension primer and backbone charges are neutralized by either alkylation of thiophosphate bridges [44] or the use of charge-neutral methylphosphonates [60] to render the DNA significantly less prone to adduct formation.

8.2.2 The PinPoint assay

The PinPoint assay, commercialized by Applied Biosystems (Foster City, CA, USA) as the Sequazyme™ PinPoint SNP Typing kit, was first published by Haff and Smirnov in 1997 [19]. Residual dNTPs and primers of the PCR amplification are degraded by incubation of the reaction mixture with shrimp alkaline phosphatase (SAP) or a combination of SAP and exonuclease I for the degradation of PCR primers. Primer extension is carried out in the presence of all four dideoxynucleotides to extend the primers by a single nucleotide. For homozygous SNPs, one complementary ddNTP is incorporated, and for heterozygous SNPs the two complementary ddNTPs are incorporated. Primer extension products were purified and concentrated with pipette tips filled with reversed-phase chromatography media (POROS 50 R1) or gel filtration on Sephadex G-25 spin columns. Mixtures of HPA and sample were further treated with ion-exchange beads prior to analysis. Products are in the mass range of 4000 to 9000 Da. Several assay formats for multiplex genotyping reactions were subsequently described [61]. For multiplex analysis of a maximum of six SNPs, extension primers contained a template complementary core-sequence of constant length and mass-tags of 5′-oligothymidylic acid residues (dT) [62]. A high degree of multiplexing was demonstrated by applying multiplex PCR (several amplicons are generated simultaneously) and parallel genotyping of twelve SNPs [63]. The practical limit of multiplexing was determined to be at 12- to 15-plex. However, it must be kept in mind that such high levels of multiplexing are only possible with the time-consuming optimization of multiplex PCR and primer extension reactions, because the choice of combined SNPs strongly affects the quality of the enzymatic reactions. An inherent problem of the PinPoint assay is that the smallest mass difference between two nucleotides (A and T) is only 9 Da, which is demanding to resolve in the 4000–9000 Da mass range. The use of mass-tagged ddNTPs with dyes used for fluorescent applications such as fluorescein can overcome this problem [52,64]. These tags shift the mass of one allele by 300 to 800 Da. Cleavable extension primers have also been integrated in the PinPoint assay [65]. The cleavage reduces the masses of the products to a range where the mass spectrometer has sufficient resolution to differentiate between the smallest nucleotide mass difference (A/T) without any problems.

8.2.3 The PROBE assay

The primer oligo base extension (PROBE) assay relies on a slightly different strategy for allele-distinction [20,46]. The DNA target sequence that contains the mutation(s) or SNP(s) to be analyzed is amplified by PCR. One of the PCR primers carries a 5'-biotin label. PCR products are purified and captured with streptavidin-coated magnetic beads. Subsequent denaturation of the double-stranded DNA releases the nonbiotinylated strand. The primer extension reaction is carried out on the immobilized strand in the presence of three deoxynucleotides (dNTPs). The fourth nucleotide is added in the form of a ddNTP. In this fashion, the extension of the primer is terminated at the first position complementary to the ddNTP, and different numbers of bases added to the primer distinguish the different alleles of heterozygous polymorphisms. Typically, one allele is extended by only one base, and the second by two or more, to produce a mass difference of at least 300 Da that facilitates allele-discrimination. Prior to analysis, the extended primers are denatured, purified, and concentrated by ethanol precipitation. For the PROBE assay, a novel principle for sample preparation was developed. Piezoelectric pipetting of nanoliter quantities of purified sample onto silicon chips facilitated the analysis of the extension products in the MALDI mass spectrometer because spot size is similar to the laser irradiation area [66]. With this method, because the entire preparation is volatized with a few laser pulses, the need to search for 'sweet' spots on the HPA matrix is avoided, and the spot-to-spot reproducibility increases. The procedure of the PROBE assay has been continuously improved, and is now referred to as the MassEXTEND™ assay running on a highly-automated, high-throughput platform, the MassARRAY™ system (Sequenom) [9,67]. Dedicated software allows the automatic design of amplification and extension primers (SpectroDE-SIGNER™). Primer extension is performed on a double-stranded PCR template. Because the homogenous MassEXTEND™ assay uses an ion-exchange resin for sample conditioning prior to analysis, biotinylated primers and immobilization are avoided, and sample clean-up is facilitated [57]. The analyte is robotically transferred to silicon chips that are precharged with matrix (SpectroCHIPS™). The MassEXTEND™ assay has been applied to a large-scale study of over 9000 gene-based SNPs in pooled DNA samples [9], to the identification of differences between bacterial strains [68] and to numerous others studies [69,70].

8.2.4 The VSET—assay

A variation of the assays described above is the VSET (Very Short Extension) assay [47], which addresses some of the problems associated with the first demonstrations of the PinPoint and the PROBE assay. The primer extension reaction is performed in the presence of three ddNTPs and one dNTPs so that for one allele extension is one base and for the other two bases, which separates the masses of alleles by 300 Da. However, because ethanol precipitation, and in some cases, further desalting with ion-exchange beads was required for purification, this assay is less suitable for automation.

8.2.5 GenoLINK and GenoSNIP

Bruker (Bruker Saxonia Analytik GmbH, Germany) proposes a similar assay format. PCR products are purified with the genopure *ds*™ purification kit. Primer extension reactions are carried out in the presence of a mixture of dNTPs and ddNTPs, and are purified by the genopure *oligo*™ purification kit prior to analysis in the mass spectrometer.

Recently, Kostrzewa et al. (50th ASMS Conference on Mass Spectrometry and Allied Topics, Orlando, FL, USA, 2002) presented a method for SNP genotyping based on the extension of UV photocleavable oligonucleotides in two variations (GenoSNIP™). The DNA fragment of interest is amplified by PCR. In the first variation the PCR product is purified with genopure magnetic beads. Single-base primer extension is carried out with oligonucleotides that contain a special UV-cleavable base [71] and the four ddNTPs. Products are desalted with again the genopure beads, and transferred onto a MALDI target. The linker is cleaved by UV irradiation ($\lambda = 366$ nm) and the resulting small oligonucleotides (5–10 bases) are measured by MALDI. In the second variation, a SAP digest to remove dNTPs follows PCR. Extension primers have an additional 5′-biotin link for purification in streptavidin-coated 384-well plates (genostrep) after single-base primer extension. Oligonucleotides are cleaved in the plate, and supernatant that contains the 3′-end of the cleaved oligonucleotides is spotted onto the MALDI target. By photocleaving the extension primers, products are shifted into a mass range of 1200 to 1650 Da that reduces depurination and salt adduct formation, and yields more uniform ionization efficiency.

8.2.6 The GOOD assay

In contrast to the above-described assay formats, the GOOD assay [44,45] requires no purification step due to a chemical modification strategy—charge tagging [59]. DNA products are conditioned to carry either a single positive or a single negative charge rather than being polyanionic like native DNA. This chemistry enhances sensitivity ca. 100-fold compared to conventional DNA analysis by MALDI. The modified oligonucleotides are less susceptible to adduct formation. As a consequence, conventional buffers are used for the molecular biology, and are not removed prior to analysis. After PCR, remaining dNTPs are degraded by the addition of shrimp alkaline phosphatase. Single-base primer extension is carried out with α-S-ddNTPs. Part of the primer is then digested with 5′-phosphodiesterase II.

Phosphorothioate bridges resist this digestion, and a core-sequence of four or five nucleotides that contains the allele information is obtained. Recently, a version of the GOOD assay, where a photocleavable base in the primer replaces the 5′-phosphodiesterase digestion was shown [72]. Remaining charges of the oligonucleotide backbone are neutralized by selective and quantitative alkylation of the phosphorothioates with methyliodide [73]. α-Cyano-4-hydroxy-cinnamic acid methyl ester—a weak protonating matrix—is used for desorption. The GOOD assay has been developed further to reduce the cost of an assay and to reduce the amount of handling required [44]. Recently, the simplified GOOD assay was shown to reduce the five-step procedure to three steps [60].

8.3 High-throughput SNP genotyping

Genotyping hundreds of thousands or even millions of SNPs efficiently and reliably puts great scrutiny on a technology. Typically the demands are accuracy of allele-calling, ease of operation, decrease of risk of mix-ups, possibilities of pinpointing errors, and cost. Automation of sample preparation can greatly reduce problems related to this but requires capital investment and a great effort in integration. In general, SNP genotyping by MALDI mass spectrometry is best suited to the analysis of a limited number of SNPs in a large number of DNA samples. Sequenom offers an off-the-shelf integrated solution for SNP genotyping by MALDI mass spectrometry. It is constituted of liquid and plate handling robotics, dedicated PCR machines, robotics to transfer samples onto the MALDI target plates, a dedicated mass spectrometer, software for data accumulation and interpretation.

8.4 Other applications

8.4.1 Quantitation

Recently, several MALDI approaches were presented for the quantitative analysis of SNPs in pools of DNA. Pooling of DNA has been proposed as a means to reduce the number of genotypes necessary for large-scale association studies. Numbers of 500,000 SNPs in 50,000 individuals have been proposed to detect subtle effects of genetic variations in multifactorial diseases [74]. The use of pooled DNA samples that are analyzed separately for cases and controls significantly decrease the effort, cost, and time required for such ambitious projects, and only SNPs that show significant differences in allele-frequencies between the pools must be further examined. A crucial point consists of adjusting well-defined mixtures of DNA of different individuals. Ross et al. [75] showed the accurate quantification of pooled

PCR products at different loci. The chip-based DNA MassARRAY™ system (Sequenom) was used in a study to characterize allele-frequencies of $\sim$ 9000 gene-based SNPs in DNA pools of 94 individuals [9]. Equimolar pools of 94 DNAs were used for PCR amplification and primer extension reactions. Assays were automatically designed by the MassEXTEND™ software to generate a single-base extension product for one allele and a double-base extension product, respectively, thereby separating the two signals by ca. 300 Da. Nanoliter quantities of the products were transferred onto SpectroCHIPS™ with piezoelectric pipetting. Four replicates of each sample were measured, recorded, and analyzed (by calculation of peak areas) in fully automated mode. High reproducibility of $\pm 1.6\%$ for alleles with a frequency of at least 10% was demonstrated. The deviation of the measured allele-frequencies ($\sim$ 0.7%) was determined by comparison of genotypes obtained in pooled DNA and for individual DNAs at several loci. The same assay format was recently used for the determination of allele-frequencies in pools of 94 to 280 individuals [76]. By combining the accuracy of primer extension assays with the flexible and high-throughput of MALDI mass spectrometry, the analysis of pooled DNA samples is a cost- and time-effective method—if common alleles are of interest. However, rare alleles may be missed and individual information like haplotypes is lost.

8.4.2 Haplotyping

Haplotypes, the phase of several SNP alleles on one parental allele, reflect the sequence and structure of a transcribed protein. They are more informative than individual SNPs because the interaction of several single-base changes can have stronger impact on the phenotype. Therefore, they may be more appropriate for pharmacogenetic assessment [77]. For the analysis of isolated SNPs a variety of high-throughput techniques is available, whereas most procedures for molecular haplotyping are tedious, labor, and cost-intensive. Inferring haplotypes by mathematical algorithms is an often-used alternative, but is unacceptable in clinical practice due to their inherent statistical uncertainty. Recently, the first high-throughput method for the physical determination of haplotypes across genomic regions of substantial size that combines a highly allele-specific PCR with the fast and simultaneous analysis of several SNP positions by MALDI mass spectrometry was demonstrated [78]. Samples that are heterozygous for at least two SNP positions in the DNA fragment are interesting for haplotyping. Normal PCR is replaced by allele-specific PCR as the first step to amplify only one of the haploid strands. Thereafter, the same steps as for regular SNP genotyping are carried out and the phase of SNP alleles is determined by the observed SNP alleles that are correlated with the allele-specific reaction.

8.4.3 Epigenotyping

With the increasing amount of sequence of the human genome available, research is trying to understand gene regulation and function. DNA methylation is one of the key mechanisms in activation and silencing of genes [79]. In mammals, DNA methylation almost exclusively takes place on the $5'$-position of cytosines in the sequence context of CpG dinucleotides. Because methylation patterns are lost during PCR amplification, and nearly all methods for analysis depend on the amplification of the target sequence, the methylation status of samples has to be 'frozen' before amplification. The most widely used method is bisulphite conversion of the genomic DNA [80]. This method relies on the hydrolytic deamination of cytosines to uracil when exposed to sodium bisulphite, whereas methylated cytosines are not converted under the reaction conditions. After PCR amplification, all initially nonmethylated cytosines are displayed as thymines, whereas formerly methylated cytosines remain cytosines. The DNA methylation status is analyzed as virtual C/T 'SNPs.' Bisulphite conversion of genomic DNA samples has been combined by us with the GOOD assay [44] to accurately analyze and quantitate the methylation status of CpG dinucleotides [81]. The resolution of quantitation of methylation is important because even minor changes in the methylation profile can have severe consequences. The limitation of MALDI for quantitation is described above.

8.4.4 Mutation detection

The detection of previously unidentified polymorphisms is gaining great importance. A method for this with mass spectrometric detection was recently presented by Rodi et al. [57] and Hartmer et al. [82]. These methods use fragment analysis of RNA, transcribed from DNA and digested with sequence specific RNases.

8.5 Conclusions

Numerous methods for SNP genotyping with mass spectrometric detection have been devised, and still new ones are being developed. Most of them were demonstrated as proof-of-principle, and only a handful has been integrated at high-throughput with all the necessary automation and software to deal with the large-scale experiments that geneticists are interested in performing. Compared to SNP genotyping methods with other detection devices, mass spectrometers, and, in particular, MALDI mass spectrometers, are serious contenders and very competitive. Key advantages are the quality of data, data accumulation, speed, and that no fluorescent or radioactive labeling is required. Clearly, two promises that SNP genotyping by mass spectrometry has not yet delivered on are the degree of multiplexing and miniaturization. Fulfilling either or both of these promises would result in a dramatic decrease of cost and increase of throughput. Some of the methods presented so far would be suitable for further development in this direction. However, there are limits to decreasing volumes for generating allele-specific products. Mainly, these limits are associated with putting reagents into, and taking products out of, miniaturized devices. Usually, genomic DNA is used as starting material for an assay. At least 1 ng of genomic DNA, which corresponds to 300 copies of the genome, is currently required for established methods. An individual heterozygous at a given position carries 150 copies of each allele. Reducing a system 10-fold could result in a statistical imbalance of alleles, and would complicate automated allele-calling.

Mass spectrometers are mature and robust devices for the analysis of biomolecules and, as shown here, they are perfectly suited to the task of SNP genotyping. Over the past ten years, instrument development has advanced at similar rates as methods for SNP genotyping have evolved. The current generation of instruments is beyond what is required for SNP genotyping in terms of resolution, and future efforts will have to concentrate on ways to increase the multiplex preparation of allele-specific products and the speed of data acquisition.

MALDI mass spectrometers have been started to be used for more demanding DNA applications such as molecular haplotyping, epigenotyping, and mutation detection with great success.

Acknowledgments

We thank the French Ministry of Research and the European Community for financial support.

References

1. J.D. Watson and F.H.C. Crick, *Nature* **171** (1953) 737.

2. J.D. Watson and E. Jordan, *Genomics* **5** (1989) 654.

3. F.S. Collins, M. Morgan, and A. Patrinos, *Science* **300** (2003) 286.

4. E.S. Lander, L.M. Linton, B. Birren, C. Nusbaum, M.C. Zody, et al., *Nature* **409** (2001) 860.

5. J.C. Venter, M.D. Adams, E.W. Myers, P.W. Li, R.J. Mural, et al., *Science* **291** (2001) 1304.

6. D. Altshuler, V.J. Pollara, C.R. Cowles, W.J. Van Etten, J. Baldwin, et al., *Nature* **407** (2000) 513.

7. R. Sachidanandam, D. Weissman, S.C. Schmidt, J.M. Kakol, L.D. Stein, et al., *Nature* **409** (2001) 928.

8. I.G. Gut, *Hum. Mutat.* **17** (2001) 475.

9.   K.H. Buetow, M. Edmonson, R. MacDonald, R. Clifford, P. Yip, et al., *Proc. Natl. Acad. Sci. USA* **98** (2001) 581.

10.   K. Tanaka, H. Waki, Y. Ido, S. Akita, Y. Yoshida, et al., *Rapid Commun. Mass Spectrom.* **2** (1988) 151.

11.   M. Karas, F. Hillenkamp, *Anal. Chem.* **60** (1988) 2299.

12.   K.J. Wu, A. Steding, and C.H. Becker, *Rapid Commun. Mass Spectrom.* **7** (1993) 142.

13.   F. Kirpekar, E. Nordhoff, K. Kristiansen, P. Roepstorff, A. Lezius, et al., *Nucleic Acids Res.* **22** (1994) 3866.

14.   F. Kirpekar, E. Nordhoff, K. Kristiansen, P. Roepstorff, S. Hahner, et al., *Rapid Commun. Mass Spectrom.* **9** (1995) 525.

15.   K. Schneider, B.T. Chait, *Nucleic Acids Res.* **23** (1995) 1570.

16.   U. Pieles, W. Zurcher, M. Schar, and H.E. Moser, *Nucleic Acids Res.* **21** (1993) 3191.

17.   E. Nordhoff, F. Kirpekar, and P. Roepstorff, *Mass Spectrom. Rev.* **15** (1996) 67.

18.   M.L. Vestal, P. Juhasz, and S.A. Martin, *Rapid Commun. Mass Spectrom.* **9** (1995) 1044.

19.   L.A. Haff and I.P. Smirnov, *Genome Res.* **7** (1997) 378.

20.   D.P. Little, A. Braun, B. Darnhofer-Demar, and H. Köster, *Eur. J. Clin. Chem. Clin. Biochem.* **35** (1997) 545.

21.   A.J. Brookes, *Gene* **234** (1999) 177.

22.   J.L. Weber and C. Wong, *Hum. Mol. Genet.* **2** (1993) 1123.

23.   N. Risch and K. Merikangas, *Science* **273** (1996) 1516.

24.   L.R. Cardon and J.I. Bell, *Nat. Rev. Genet.* **2** (2001) 91.

25.   P.Y. Kwok, *Pharmacogenomics* **1** (2000) 95.

26.   A.C. Syvänen, *Nat. Rev. Genet.* **2** (2001) 930.

27.   P.E. Nielsen, M. Egholm, R.H. Berg, and O. Buchardt, *Science* **254** (1991) 1497.

28.   P.E. Nielsen, *Curr. Opin. Biotechnol.* **12** (2001) 16.

29.   M. Egholm, O. Buchardt, L. Christensen, C. Behrens, S.M. Freier, et al., *Nature* **365** (1993) 566.

30.   J.M. Butler, P. Jiang-Baucom, M. Huang, P. Belgrader, and J. Girard, *Anal. Chem.* **68** (1996) 3283.

31.   T.J. Griffin, W. Tang, and L.M. Smith, *Nat. Biotechnol.* **15** (1997) 1368.

32.   P.L. Ross, K. Lee, and P. Belgrader, *Anal. Chem.* **69** (1997) 4197.

33.   J. Stoerker, J.D. Mayo, C.N. Tetzlaff, D.A. Sarracino, I. Schwope, et al., *Nat. Biotechnol.* **18** (2000) 1213.

34. C. Jurinke, D. van den Boom, A. Jacob, K. Tang, R. Wörl, et al., *Anal. Biochem.* **237** (1996) 174.

35. Y.-H. Liu, J. Bai, Y. Zhu, X. Liang, D. Siemieniak, et al., *Rapid Commun. Mass Spectrom.* **9** (1995) 735.

36. V. Lyamichev, A.L. Mast, J.G. Hall, J.R. Prudent, M.W. Kaiser, et al., *Nat. Biotechnol.* **17** (1999) 292.

37. D.J. Hosfield, G. Frank, Y. Weng, J.A. Tainer, and B. Shen, *J. Biol. Chem.* **273** (1998) 27154.

38. M.W. Kaiser, N. Lyamicheva, W. Ma, C. Miller, B. Neri, et al., *J. Biol. Chem.* **274** (1999) 21387.

39. M. Lu, M.R. Shortreed, J.G. Hall, L. Wang, T. Berggren, et al., *Hum. Mutat.* **19** (2002) 416.

40. T.J. Griffin, J.G. Hall, J.R. Prudent, and L.M. Smith, *Proc. Natl. Acad. Sci. USA* **96** (1999) 6301.

41. J.G. Hall, P.S. Eis, S.M. Law, L.P. Reynaldo, J.R. Prudent, et al., *Proc. Natl. Acad. Sci. USA* **97** (2000) 8272.

42. A.C. Syvänen, K. Aalto-Setala, L. Harju, K. Kontula, and H. Soderlund, *Genomics* **8** (1990) 684.

43. T. Pastinen, A. Kurg, A. Metspalu, L. Peltonen, and A.C. Syvänen, *Genome Res.* **7** (1997) 606.

44. S. Sauer, D. Lechner, K. Berlin, H. Lehrach, J.L. Escary, et al., *Nucleic Acids Res.* **28** (2000) e13.

45. S. Sauer, D. Lechner, K. Berlin, C. Plancon, A. Heuermann, et al., *Nucleic Acids Res.* **28** (2000) e100.

46. A. Braun, D.P. Little, and H. Koster, *Clin. Chem.* **43** (1997) 1151.

47. X. Sun, H. Ding, K. Hung, and B. Guo, *Nucleic Acids Res.* **28** (2000) e68.

48. B. Guo, *Anal. Chem.* **71** (1999) 333R.

49. C.-W. Chou, S.E. Bingham, and P. Williams, *Rapid Commun. Mass Spectrom.* **10** (1996) 1410.

50. C. Jurinke, D. van den Boom, V. Collazo, A. Lüchow, A. Jacob, et al., *Anal. Chem.* **69** (1997) 904.

51. U. Kallweit, K.O. Bornsen, G.M. Kresbach, and H. Widmer, *Rapid Commun. Mass Spectrom.* **10** (1996) 845.

52. Z. Fei and L.M. Smith, *Rapid Commun. Mass Spectrom.* **14** (2000) 950.

53. E. Nordhoff, C. Luebbert, G. Thiele, V. Heiser, and H. Lehrach, *Nucleic Acids Res.* **28** (2000).

54. M.G. Pluskal, *Nat. Biotechnol.* **18** (2000) 104.

55. M. Gilar, A. Belenky, and B.H. Wang, *J. Chromatogr. A* **921** (2001) 3.

56. E. Nordhoff, A. Ingendoh, R. Cramer, A. Overberg, B. Stahl, et al., *Rapid Commun. Mass Spectrom.* **6** (1992) 771.

57. C.P. Rodi, B. Darnhofer-Patel, P. Stanssens, M. Zabeau, and D. van den Boom, *Biotechniques Suppl.* (2002) 62.

58. I.P. Smirnov, L.R. Hall, P.L. Ross, and L.A. Haff, *Rapid Commun. Mass Spectrom.* **15** (2001) 1427.

59. I.G. Gut, W.A. Jeffery, D.J.C. Pappin, and S. Beck, *Rapid Commun. Mass Spectrom.* **11** (1997) 43.

60. S. Sauer, D.H. Gelfand, F. Boussicault, K. Bauer, F. Reichert, et al., *Nucleic Acids Res.* **30** (2002) e22.

61. L.A. Haff, A.C. Belden, L.R. Hall, P.L. Ross, and I.P. Smirnov, In: J.N. Housby (Ed.), *Mass Spectrometry and Genomic Analysis*, Kluwer Academic, Dordrecht, 2001, p. 16.

62. L.A. Haff and I.P. Smirnov, *Nucleic Acids Res.* **25** (1997) 3749.

63. P. Ross, L. Hall, I. Smirnov, and L. Haff, *Nat. Biotechnol.* **16** (1998) 1347.

64. Z. Fei, T. Ono, and L.M. Smith, *Nucleic Acids Res.* **26** (1998) 2827.

65. J. Li, J.M. Butler, Y. Tan, H. Lin, S. Royer, et al., *Electrophoresis* **20** (1999) 1258.

66. D.P. Little, T.J. Cornish, M.J. O'Donnell, A. Braun, R.J. Cotter, et al., *Anal. Chem.* **69** (1997) 4540.

67. J. Leushner and N.H. Chiu, *Mol. Diagn.* **5** (2000) 341.

68. G. Amexis, P. Oeth, K. Abel, A. Ivshina, F. Pelloquin, et al., *Proc. Natl. Acad. Sci. USA* **98** (2001) 12097.

69. J.R. Cashman, J. Zhang, J. Leushner, and A. Braun, *Drug Metab. Dispos.* **29** (2001) 1629.

70. K. Nakai, W. Habano, T. Fujita, J. Schnackenberg, K. Kawazoe, et al., *Hum. Mutat.* **20** (2002) 133.

71. P. Ordoukhanian and J.-S. Taylor, *J. Amer. Chem. Soc.* **117** (1995) 9570.

72. S. Sauer, H. Lehrach, and R. Reinhardt, *Nucleic Acids Res.* **31** (2003) e63.

73. I.G. Gut and S. Beck, *Nucleic Acids Res.* **23** (1995) 1367.

74. L. Kruglyak, *Nat. Genet.* **22** (1999) 139.

75. P. Ross, L. Hall, and L.A. Haff, *Biotechniques* **29** (2000) 620.

76. M. Werner, M. Sych, N. Herbon, T. Illig, I.R. König, et al., *Hum. Mutat.* **20** (2002) 57.

77. R. Judson and J.C. Stephens, *Pharmacogenomics* **2** (2001) 7.

78. J. Tost, O. Brandt, F. Boussicault, D. Derbala, C. Caloustian, et al., *Nucleic Acids Res.* **30** (2002) e96.

79. J.T. Attwood, R.L. Yung, and B.C. Richardson, *Cell. Mol. Life Sci.* **59** (2002) 241.

80. M. Frommer, L.E. McDonald, D.S. Millar, C.M. Collis, F. Watt, et al., *Proc. Natl. Acad. Sci. USA* **89** (1992) 1827.

81. J. Tost, P. Schatz, M. Schuster, K. Berlin, and I.G. Gut, *Nucleic Acids Res.* **31** (2003) e50.

82. R. Hartmer, N. Storm, S. Boecker, C.P. Rodi, F. Hillenkamp, et al., *Nucleic Acids Res.* **31** (2003) e43.

Advances in Mass Spectrometry, Volume 16
A.E. Ashcroft, G. Brenton and J.J. Monaghan (Editors)

CHAPTER 9

The Role of Mass Spectrometry in Systems Biology: Data Processing and Identification Strategies in Metabolomics

Jan van der Greef [1,2,3], Rob van der Heijden [3], and Elwin R. Verheij [1,3]

[1] TNO Pharma, P.O. Box 360, 3700 AJ Zeist, The Netherlands

[2] Beyond Genomics Inc., 40 Bear Hill road, Waltham, MA, USA

[3] Leiden/Amsterdam Center for Drug Research, P.O. Box 9502, 2300 RA Leiden University, The Netherlands

9.1 Introduction

The knowledge of biological systems has increased enormously in the last decade and the completion of the Human Genome project is certainly a major milestone in this development. The understanding that the genome-based information is not sufficient to correlate genotype and phenotype has resulted in increased attention to other levels of biological information, especially the mapping of the proteome and the metabolome. Although, of great importance in Life Sciences in its broad sense, pharmaceutical perspectives have mainly driven this research and emphasis has been put on the differential analyses of healthy and diseased systems to reveal a detailed insight into the disease etiology and at the same time possibly detect new targets and biomarkers. This approach certainly encompasses enormous analytical challenges and especially demands the development of novel strategies in biostatistics and bioinformatics.

Understanding biological systems in terms of the mechanisms, requires in-depth studies on pathways/networks at the cellular, tissue/organ level and the whole multicompartment systems level especially related to the communication/control aspects. This has evolved into *Systems Biology* [1–4] defined as *study-*

ing biology as an integrated system of genetic, protein, metabolite, cellular, and pathway events that are in a state of flux and interdependent.

The application of Systems Biology in pharmaceutical sciences has become particularly focused on biomarker discovery [5] and the understanding of the biology involved to enable:

- Elucidation of complex disease pathways, target systems, and drug response;

- Predictive safety and early validation of disease models (animal ⇔ human);

- Identification of next-generation biomarkers with broad applicability in preclinical and early clinical drug development enabling reduction of clinical trial size and improved patient inclusion criteria;

- Discovery of common disease mechanisms to provide new indications and leverage existing products.

In this context the biomarker approach is believed to deliver a better understanding of the drug discovery and especially the drug development trajectory to make knowledge-based decisions early on in the process thereby improving the attrition rates. Biomarkers can, in addition, help to reduce the cost for the clinical phase considerably by improvement of clinical trial design, reducing the size of the trial and improving the selection criteria. The concept from a single biomarker has emerged into the understanding that a biomarker fingerprint is necessary to describe a multifactorial disease and the drug response profile of a complex organism [5]. Translated into the analytical and biostatistics/bioinformatics issues it necessitates the profiling of biological systems at all biological levels in a quantitative way and preferably in a longitudinal way as biological systems are very dynamic. Hence, the biostatistics needs to be of a multivariate nature preferably focused on discovery of patterns linked to clinical observations or of an unsupervised manner. The bioinformatics needs to take into account the nonlinearity and should provide the connectivity between the elements.

In the analytical domain mass spectrometry plays a dominant role as technology within Systems Biology [5], in particular, in the analysis of the proteome and the metabolome. For both approaches enormous progress has been made but still a great challenge exists both for the comprehensive coverage and, for providing quantitative data. In relation to biomarker discovery especially, the analyses of body fluids such as blood (plasma, serum) and urine as the most important fluids for clinical trials are of major importance but with many other options for more specialised experiments using for instance saliva, synovial fluid, lymphe, etc. The analytical challenge for serum, for instance, is tremendous as the dynamic range to be covered within proteomics as well as metabolomics is estimated to be of 10 orders of magnitude. Using an upfront digesting step and multidimensional chromatography—mass spectrometry has become the premier biomarker approach for broad quantitative profiling of the serum proteome, but still major improvements in terms of coverage need to be achieved [6–12]. Routinely 500–1000 proteins are monitored in the current approaches but this is still less than by the more qualitative but still powerful 2D-gel technology [13]. For metabolomics the diversity of chemical structures is another challenge and the coverage of the metabolome is best achieved by using a combination of techniques such as LC-MS and GC-MS in combination with different sample preparation or separation tools and for the high concentration NMR [14–16]. The latter is a very quantitative technique and allows quantitation simultaneously over a broad diversity of components but it is unfortunately restricted to $>$ mid-ng/mL and higher concentration ranges. The field of mass spectrometry has been leading in the field of body fluid profiling and originates from the strong GC-MS work of groups in the clinical chemistry domain in the seventies, especially in the area of inborn errors of disease. In the early eighties this was extended to body fluid profiling in combination with pattern recognition techniques. The development of mass spectrometry to include soft ionisation techniques and especially LC-MS technologies has resulted in the development of metabolomics platforms to cover as much of the diverse metabolome as possible. From these studies it already appeared from the early days of this strategy that robust profiles could be created but that minor alignments still need to be made to allow profile comparisons to be made. For NMR it was shown that a partial-linear-fit algorithm [17] allowed correction for minor chemical shifts without the use of a binning approach which results in a loss of resolution. In mass spectrometry the alignment of GC-MS and LC-MS data was described by Gaspari et al. [18]. In this chapter, we described an improved method for data alignment to be used for hyphenated techniques applicable to both proteomics and metabolomics. Also, we addressed the novel opportunities to approach the identification procedure of low concentration components found in biomarker fingerprints obtained by a metabolomic platform, by applying a new linear trap FTMS instrument capable of delivering high resolution and high accuracy mass measurements on a routine basis.

9.2 Scale and complexity of LC-MS data in Systems Biology

The application of a typical analytical platform consisting of LC-MS based proteomics, XC-MS (X = L or G) metabolomics and NMR metabolomics in Systems Biology research produces vast amounts of data. An example is presented in Fig. 9.1 for a small study of 50 subjects, three matrices at a single time point. Already this relatively small study produces some 5500 raw data files with a total size of about 100 GB. Standards and quality control samples are not included

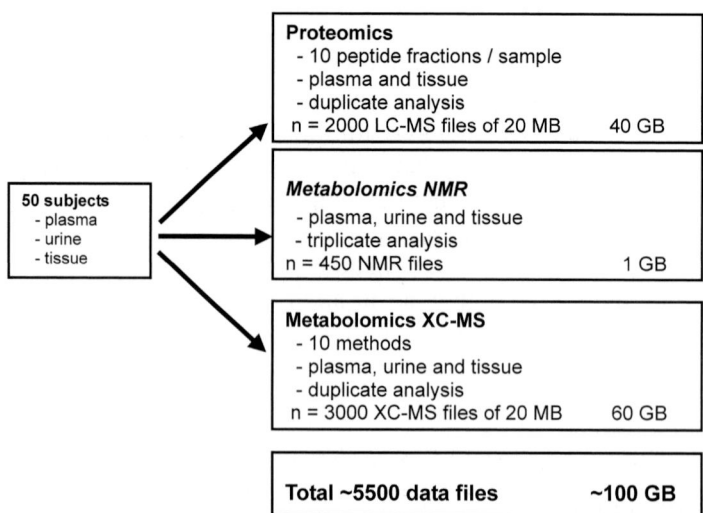

Figure 9.1 The number of LC-MS, GC-MS, and NMR files and the total amount of data generated in a Systems Biology study comprising LC-MS proteomics and LC-MS, GC-MS, and NMR metabolomics.

in this example. Studies with more subjects and/or time points are quite common.

Apart from key issues such as analytical technology, methods development, statistical evaluation, and biological interpretation the processing of these amounts of data is key. Manual data processing combined with the suboptimal vendor software (for this task) is not an option. It is too time consuming and results in many errors. Automated high quality data processing tools justify the use of high-throughput, brute-force analysis and enable sensible statistical analysis and biological interpretation. The sole purpose of data processing is to transfer raw data into clean data.

This section describes the general strategy and some of the tools for processing of LC-MS and GC-MS data used in our Systems Biology research. All these tools were developed in house for the simple reason that software from instrument manufacturers was never designed to perform the specific tasks in non-target analysis. Hardware manufacturers mainly focus on quantitative target compound analysis, qualitative biotransformation of xenobiotics, and qualitative peptide/protein analysis for which black-box software is provided with the instruments.

The philosophy behind in-house software development is to generate a generic, but flexible, toolbox that reduces manual data handling to an absolute minimum. Speed is nice, but not essential because analysis is by far more time consuming.

9.3 Information

Irrespective of the application, all data processing software has the same fundamental objective, namely to distil the real information from raw data and to remove
non-information (background, noise, etc.). The first step in designing data processing tools is to understand the nature of the raw data and to define information. The
dimensionality of LC-MS and GC-MS data is three (a data cube). The 2D-map plot
with m/z axis, retention time axis and greyscale/colour coded signal intensity is an
excellent means of representing the raw data. Fig. 9.2 shows a 2D map of mouse
plasma lipids analysed with LC-MS containing a wealth of information. Actually,
this resembles the 2D images of 2D-PAGE proteomics data except for the shape of
the 'spots.' In principle, one could use (modified) 2D-PAGE proteomics software
for processing LC-MS data and further statistical analysis. As a matter of fact, our
approach has much in common as will be shown later.

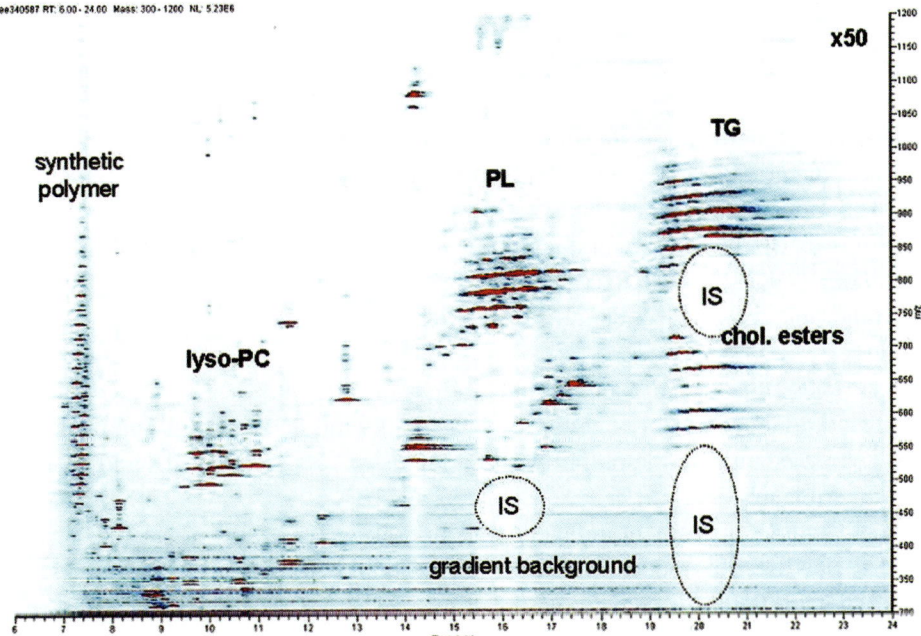

Figure 9.2 A typical result obtained from the analysis of 10 μL human plasma with the
plasma lipid RP LC-MS method. The data is shown as a 2D map: time on the x-axis, m/z
on the y-axis and greyscale or colours to represent signal intensity. To show minor peaks
the intensity scale is blown up 50 times. *Labels*: TG: triglycerides, PL: phospholipids,
lyso-PC: lyso-phosphatidylcholine, chol. esters: cholesterol esters, IS: region of ion
suppression.

The definition of information is quite obvious from this 2D map, namely the position and intensity of the spots. The rest is not relevant. In this specific example the spots only occupy at most 5% of the total plot area (rough estimate, probably less). The information/data ratio is very unfavourable. This loops back nicely to the common understanding and fundamental that the objective of method development should always be to improve this ratio by maximising the utilisation of the available peak capacity (either put more peaks in or reduce the peak capacity).

9.4 Peak picking and integration

With this knowledge it is obvious that processing of LC-MS data is nothing else than picking the peaks and characterising these with relevant parameters. The software package IMPRESS^{TM} was developed in-house for this. The input of IMPRESS^{TM} is LC-MS, GC-MS, or CE-MS datafiles (1 to n) as either NetCDF or ThermoFinnigan Xcalibur XRaw files. Peaks are found by analysing all m/z traces to obtain position (m/z and retention time) and quantitative information (baseline corrected area, height, and s/n ratio). The output is a simple list of the important descriptors for all the peaks found in each XC-MS file. A very nice feature of the IMPRESS^{TM} algorithm is its stability towards user defined settings for peak picking and integration. The same settings are used for LC-MS metabolomics data, GC-MS metabolomics data, and LC-MS proteomics data acquired on a variety of instruments from several manufacturers. Actually, the only relevant user definable parameter is the s/n ratio to exclude noisy peaks.

Important design specifications of IMPRESS^{TM} were:

(1) 100% rejection of background artefacts, spikes, etc.:
Experience tells it is better to occasionally miss a small but real peak than to contaminate the output with hundreds of very noisy peaks and baseline artefacts;

(2) Generic, instrument independent solution:
NetCDF file import (NetCDF toolbox, public domain);

(3) Fast:
Coded in C++ and VB. NET, example 100 LC-MS files, 1500 m/z traces, 45 min acquisition at 1 scan/s are processed in < 10 min on a 2.4 GHz P4, 512 MB computer;

(4) Accurate integration:
Performance was checked by evaluating LC-MS/MS data from validated drug bioanalysis methods;

(5) Avoid time consuming manual steps (there is no time for this);

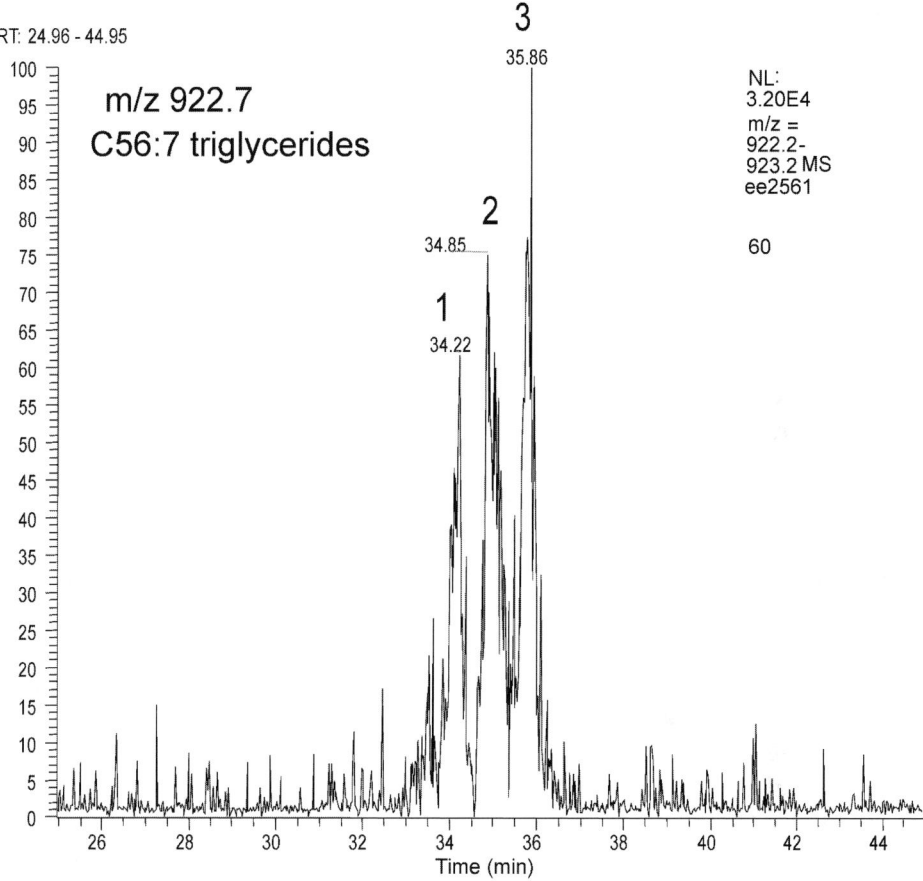

RT: 24.96 - 44.95

m/z 922.7
C56:7 triglycerides

NL:
3.20E4
m/z =
922.2-
923.2 MS
ee2561

60

Time (min)

Figure 9.3 Chromatogram of C56:7 triglycerides (M + NH$_4^+$, m/z 922.7) showing 3 main peaks corresponding to groups of isoforms. The small peak just before the peak 1 is the M + 2 isotope of C56.8 triglycerides. The chromatogram was obtained by the analysis of the lipid fraction from 5 μL of mouse plasma (injection corresponds to 0.5 μL of plasma) using LC-MS lipid method (C4 column, ESI, ThermoFinnigan TSQ700).

(6) Detect poorly resolved peaks:
IMPRESS^{TM} will not detect a small shoulder, there should be a valley between to incompletely resolved peaks.

The current version meets all these design specifications. IMPRESS^{TM} has proven to be a crucial and also very reliable tool for processing huge amounts of LC-MS and GC-MS data. The only feature lacking at this moment is the ability to handle very poorly resolved peaks with similar mass spectra; for example, deconvolution based on minute but significant spectral differences (e.g., GC-MS in EI mode). This feature will be added to IMPRESS^{TM} in the very near future to improve the performance, especially for GC-MS data.

It is important to realise that IMPRESS^{TM} does not introduce significant errors, even though a small change in a single data point might have a marginal effect on retention time (± 1 scan or ± 1 s). The main problem is variation in retention time due to the chromatography, e.g., minute changes in temperature, gradient reproducibility, column ageing, and small changes in organic by evaporation, etc. Typical analysis sequences will take 1 to 3 days of continuous acquisition meaning all these factors will contribute whatever precautions are taken.

Statistical analysis requires that all the peaks are aligned with respect to retention time. For example, assume the same metabolite is observed at 11.10 min in some of the samples (group A) and at 11.15 min in the rest (group B). This leads to the situation that the intensity of the peak at 11.10 min in group B is 0 (the peak is at 11.15 min) and the intensity of the peak at 11.15 min in group A is 0 (the peak is at 11.10 min). Whatever, statistical analysis method is used, this situation will always lead to a significant difference between the groups. Only after checking the raw data will it be concluded that the difference between the groups is a chromatographic artefact. Therefore, the next step in data processing is peak alignment.

9.5 Peak alignment

It is obvious that manual alignment of several hundreds to thousands of peaks in 100 or more data files is simply impossible (about 2 man weeks of work per dataset). Software (EQUEST^{TM}) was developed to do this automatically. The peak alignment is achieved using a 4-step procedure:

(1) Define a set of time references in the dataset, e.g., internal standards or common, abundant metabolites;
(2) Find these peaks in the peak lists from IMPRESS^{TM};
(3) Correct retention times of all peaks using linear interpolation (reference peaks);
(4) The actual alignment, classification of each peak using a probabilistic model.

The process is visualised in Figs. 9.4 and 9.5 using an atypical dataset with severe retention time variation and drift (see legends for explanation).

The peak alignment software only requires user input for defining the reference peaks, and occasionally when a reference peak is found outside the indicated retention time $\pm$ tolerance. When this is the case the user should either go back to original raw data to check if this is indeed the reference peak and/or use a larger

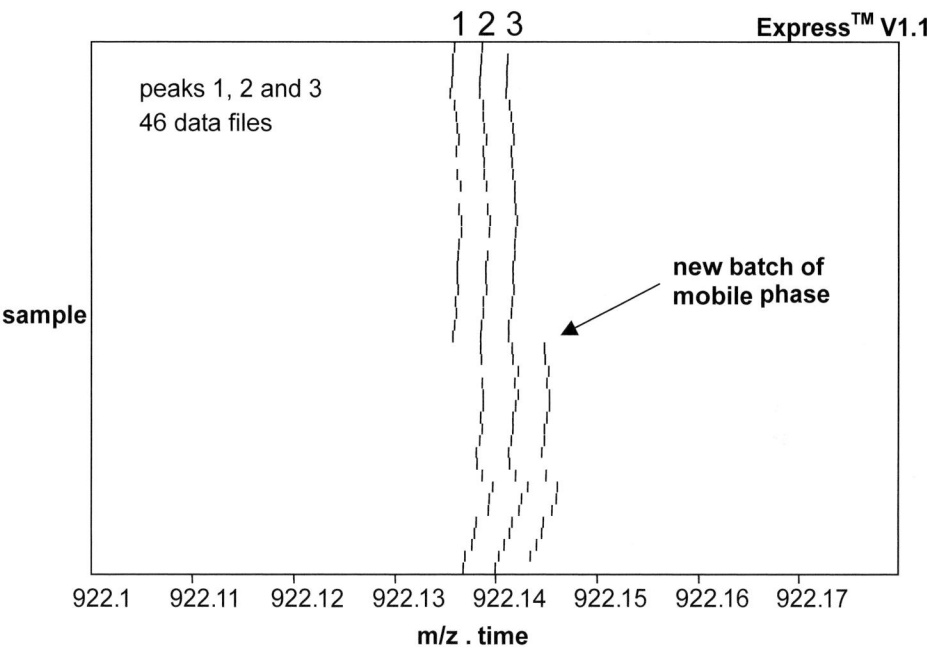

Figure 9.4 Graphic representation of the 3 triglyceride peaks (see Fig. 9.1) in 46 mouse plasma samples in EXPRESS^{TM} data processing and multivariate statistics software. The horizontal axis is a combination of nominal m/z and the retention time (or scan number). For example, 922.1400 means m/z 922 and a peak top at scan 1400. Each vertical bar is a peak. The EXPRESS^{TM} software uses colours to represent the peak area or height. This plot only shows m/z 922 (the total dataset covers a mass range of m/z 300 to 1500).

tolerance. Definition of the reference peaks is only required once if the method is not changed and retention times are reproducible on the long run.

The alignment of the dataset above (example used for IMPRESS^{TM}) is achieved within a couple of minutes. A report containing tables and figures is produced automatically by EQUEST^{TM} to check the quality of the unaligned dataset and the actual alignment process.

The IMPRESS^{TM} and EQUEST^{TM} software are key tools to produce clean data. The EQUEST^{TM} output is ASCII text and is compatible with virtually all statistical analysis software. Data scaling, internal standard scaling, normalisation, and statistical analysis is done using EXPRESS^{TM} (developed in house) and MATLAB because both can be adapted with relative ease to changing needs, new algorithms, or to link the clean data with metabolite identification databases, protein identification tools, data warehouse, LIMS system, etc. This is more difficult or not possible at all with many of the commercial statistical software.

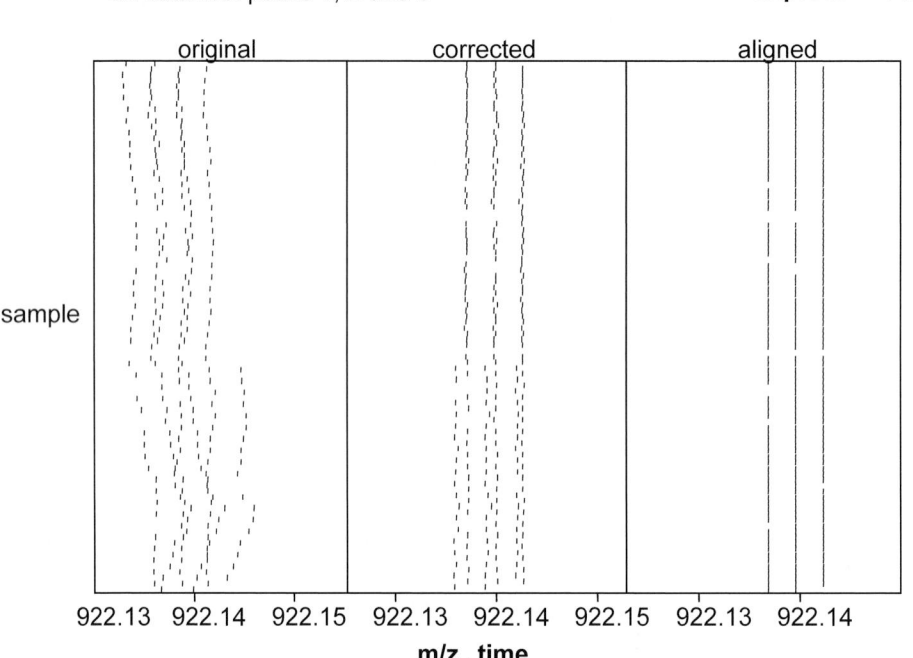

92 datafiles: peaks 1, 2 and 3   **Express™ V1.1**

Figure 9.5 The EQUEST™ peak alignment is a 2-step process, i.e., retention time correction and the actual alignment. The 3 boxes show the same data as in Fig. 9.2 and in addition also the duplicate measurements (92 results in total, corresponding to a total runtime of approx. 100 h). The left section shows the original raw data. Combined gradient formation errors, column ageing, minor temperature fluctuation, and the effect of mobile phase change make it nearly impossible to distinguish 3 peaks. The middle section shows the result after the first step, i.e., correction by linear interpolation using retention time markers (internal standards and a few abundant plasma lipids). The results indicate that the correction improves the data dramatically. The presence of 3 peaks at this m/z is obvious. On the other hand, the mobile phase change causes a systematic and fundamental change in the retention characteristics that cannot be corrected completely. The right section shows the results after alignment. The 3 peaks are fully aligned which makes the data ready for statistical analysis.

The process of producing clean data with IMPRESS™ and EQUEST™ is truly generic. It is a 100% data driven process, and information about the nature of the analytes or the analytical application is not required. Therefore, the application range is very broad: metabolomics, proteomics, biotransformation, drug impurities, drug stability, complex mixture analysis (e.g., petrochemical applications), protein modification, and many more. Application specific issues are only relevant and necessary for the analysis of the clean data and the scientific interpretation of its outcome. The described processing tools reduce the amount of data. A typical metabolomics LC-MS file of 20 MB is reduced to a peak list of just 10 to 20 KB. The compression factor is a function of the number of peaks, but a value of 1000 is quite typical.

9.6 The benefits of high-resolution accurate mass data

The use of high-resolution accurate mass instruments will have an enormous impact on data processing. Peak picking and integration is much simpler because for the vast majority of peaks in an LC-MS run the high resolution ion-chromatograms will be almost free of background. This is demonstrated in Fig. 9.6 showing results from the analysis of human plasma metabolites after butylation of the carboxylic acid groups using the ThermoFinnigan LTQ-FT. The top panel shows the base peak chromatogram. The middle panel shows the chromatogram of m/z 188 with a window of 1 mass unit. This m/z corresponds to the $[M + H]^+$ ion of amongst others butylated leucine, iso-leucine, and hydroxyproline. The resolution of approximately 100.000 results in a peak width of about 4 mmu at the baseline. The bottom panel shows the HR chromatogram of m/z 188.1645 corresponding to the element composition of butylated Leu and Ile with a window of ± 0.002 mass units (2 mmu). The only peaks observed are indeed leucine and a small peak of iso-leucine just before leucine. The hydroxyproline peak is no longer observed because it has a different elemental composition. With respect to peak picking and integration it is more important to observe that the background is reduced to zero intensity, except for some spikes. Processing of this chromatogram is easier than the nominal mass chromatogram with a high background and high noise level. Review of the data file shows that this is the case for a large number of peaks in the data. The performance of the LTQ-FT is excellent. The mass accuracy is not influenced dramatically by the signal intensity because the linear trap controls the number of ions injected in the FT-ICR. Accurate mass data is also obtained with (Q)-ToF instruments but these systems suffer from a limited dynamic range. The measured accurate mass can easily shift by 40 or more mmu for intense signals. This gives HR chromatograms that are correct at peak start and end, but show a gap around the peak maximum (saturation of the detector).

Accurate mass is even more beneficial to peak alignment than it is for peak picking. The ion chromatograms contain fewer peaks, simplifying the task of correct classification of each of the peaks.

The advantages of high resolution and accurate mass are massive. Data processing is easier and it will improve the quality and reliability of the clean data generated. Therefore, we see an important role for this type of instrumentation in routine

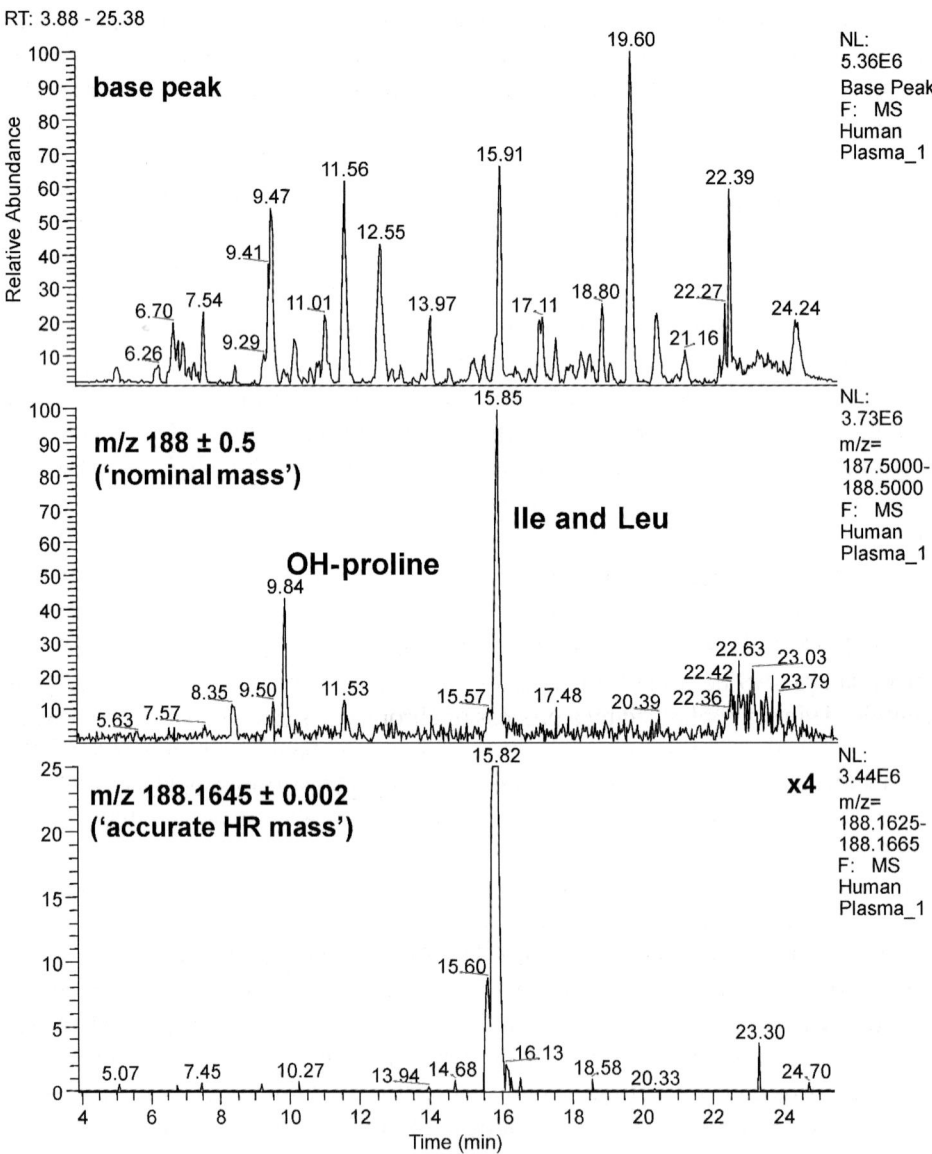

Figure 9.6 Chromatograms of human plasma metabolites after butylation of the carboxylic acid groups. The data was acquired with a ThermoFinnigan LTQ-FT instrument. The top trace is the base peak chromatogram. The middle trace is the low resolution chromatogram of *m/z* 188 with a 1 Da window. The bottom trace is the high resolution chromatogram of *m/z* 188.1645 (leucine and iso-leucine) with a window of 0.004 Da. The accurate mass chromatogram contains fewer peaks and no background and noise simplifying peak picking, integration and peak alignment.

LC-MS metabolomics and LC-MS proteomics and not only for identification and structure elucidation work. In practice, we observed many drawbacks of running high-throughput analysis on a quadrupole or ion trap instrument and the identification of unknowns on another system, e.g., Q-ToF and QStar. The use of different instruments with different ion source designs and ionisation characteristics is problematic. Peaks observed with, for instance, an LCQ sometimes are not observed with a Q-ToF because one system gives the $[M + H]^+$ ion and the other the $[M + NH_4]^+$ ion or the $[M + Na]^+$ ion. In other cases additional peaks appear, making it no longer certain which peak corresponds to the unknown. Combining routine analysis and identification on the same instrument is strongly advised, e.g., Q-ToF, LTQ-FT or TSQ Quantum with accurate mass option. Routine analysis provides quantitative information and at the same the accurate mass of each compound without having to make additional measurements. Further identification work can proceed by performing accurate mass MS/MS or MS^n on the same instrument.

9.7 Identification of metabolites

In the Metabolic Systems Biology approach one normally deals with a large number of samples to be processed. These samples undergo a process of preparation (sample clean-up), followed by a process of chromatographic and/or spectroscopic analysis. The data ("metabolite profiles") are then further analysed by multivariate statistical techniques, aiming at the detection of differences in detector signal intensities in the profiles of the different samples. These differences in intensity result from an altering concentration of one or more individual metabolites. Eventually, these "differentially expressed" metabolites may serve as a biomarker related to a phenotype of a subset of the samples.

The statistical methods are very powerful in identifying differences in large sets of very complex spectra; however, they only indicate where differences in the spectra are observed. The logical next question is then: what compounds are responsible for these differences? In other words, what is the identity of these compounds? These questions are not always easy to answer. The identity of a compound is obtained when it behaves identically under all chemical and physical conditions to an authentic sample (reference compound). In practice, adequate data is collected by different chemical and physical means (spectroscopy, chromatography) to explain the structural features of the compound and to exclude all alternative structures. Below a general strategy is discussed for the identification of individual metabolites as part of an LC-MS profile. This strategy is outlined in Table 9.1.

The first information on the identity of a metabolite is obtained from the biology of the sample. Analysing mammalian-derived samples, it is not very likely that the metabolite to be identified is a completely new structure. On the contrary, in

Table 9.1: General strategy for identification of individual constituents of an LC-MS metabolite profile

Source of information	Type of information	No. of possible candidates
Sample type	Sample specific subset of metabolites, biology of sample, chemotaxonomy	10^3–10^5
Sample pretreatment	Enrichment of specific classes of metabolites, derivatisation	10^2–10^4
Chromatography	Retention, polarity, charge, size	10–10^3
Spectroscopy	UV/vis spectrum, (accurate) mass, structural information by MS/MS spectra, molecular formula	< 10
Co-analyses with authentic sample	Confirmation of identity	< 3
Purified metabolite	2D-NMR spectra, stereochemistry	1

microbial or plant samples novel (secondary) metabolites are frequently identified. For this type of samples, it is then important to learn from chemotaxonomy about the type of secondary metabolites (alkaloids, quinones, etc.) to be expected. For eukaryotes (yeast, mammals) and many microorganisms the biochemical processes are described in great detail and the set of possible candidate structures for the metabolite to be identified is limited and confined. The identification concerns, thus, in most cases the elucidation of a known, well-described, structure. Essential reference spectra (libraries) and authentic samples are (commercially) available.

The Kyoto Encyclopedia of Genes and Genomes (KEGG [19]) LIGAND database comprises more than 10,000 molecular structures, ranging in mass up to about 5629 (insulin). It contains quite a few xenobiotics (drugs) and microbial and plant secondary products. Excluding these compounds, one is left with about 4000 primary metabolites. The Dictionary of Natural Products with more than 170,000 entries is a very rich source for identification of (secondary) metabolites. The dictionary is partly accessible via the Internet; for example, searches can be made on accurate masses [20].

Each tissue or body fluid comprises a specific subset of these metabolites. For example, cholesterol is a major constituent of plasma, while it is not detected in urine. However, it is not only the number of metabolites that determines the complexity of a sample. Extreme differences in concentration of the metabolites cause additional bottlenecks in detection and identification. For example, the dynamic range of steroid concentrations in plasma is at least a factor of 10^9. Removing major constituents as albumin and immunoglobulins from plasma does not reduce the complexity in numbers much, but offers the great advantage that minor constituents are more readily detected.

A further reduction in the number of candidates for the metabolite to be identified results from the sample pretreatment. After protein precipitation with organic solvents the lipid composition of the sample is also affected. After liquid–liquid extraction a wide range of polar compounds is excluded. Derivatisation of target molecules, however, generates new complexity in the sample due to the formation of different species of the metabolite and, inherent to the required chemistry, the formation of artefacts.

After sample pretreatment, the analytical method delivers further information on the identity. Reversed-phase chromatography facilitates the detection of nonpolar metabolites, while the "injection peak" is rich in polar metabolites.

The most informative piece of information is obtained from the detection system of the LC system. A UV/VIS spectrum provided by a photodiode array detector is very informative, but the sensitivity of the detector is not always sufficient to play a role in the identification process. In the off-line mode, NMR spectroscopy is the method of choice for structure elucidation. LC-NMR(-MS, reviewed in [21]) still has limitations and is not commonly used. On the contrary, LC-MS is now widely used because of technological (robustness), economical, and logistical (bench-top machines) developments. State-of-the-art mass spectrometry now generates accurate masses (i.e., less than 5 ppm difference between measured and theoretical mass) and structural information by MS^n spectra.

Accurate mass is without doubt the most relevant data in identifying a metabolite. Within the range of 5 ppm the number of possible elemental compositions is limited to just a few. This number can be further reduced by the following measures:

- Based on the ^{13}C-isotope peak an estimation can be made of the number of carbon atoms in the molecule;

- Based on an unusual A + 2 peak the presence of an A + 2 type element (as sulphur) can be recognised;

- An adequate estimation of the Ring and Double Bonds Equivalents (RDBE, also referred to as "degree of unsaturation" or "index of hydrogen deficiency") rules out a lot of candidates with a molecular formula based on nonrealistic chemistry;

- Application of the nitrogen rule;

- By using internal standards the accuracy of the system under the experimental conditions can be monitored. This accuracy may serve as a guideline for the mass tolerance range of the unknowns.

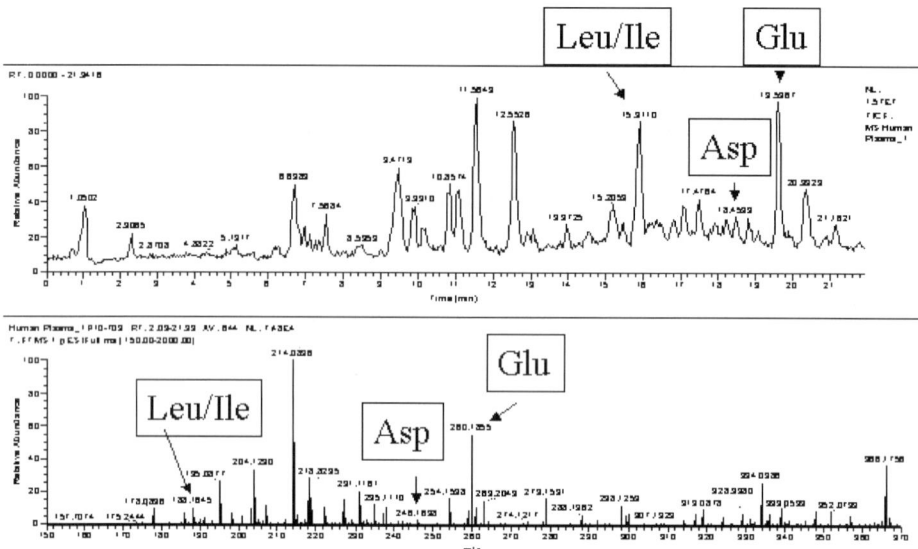

Figure 9.7 A typical chromatogram (top: TIC) of the LC-MS analysis of (butylated) amino acids. The mass spectrum (bottom) is an average (over the whole time range) full scan spectrum. Some amino acids are indicated, the signals of Asp and Glu indicated double butylated species. The sample was plasma from a healthy volunteer. The MS was a ThermoFinnigan LTQ-FT.

In our studies on metabolic profiling as part of our Systems Biology programme an LC-MS method was developed for the analysis of amino acids in plasma. The method is based on the separation of the butylated amino acids by reversed-phase chromatography and subsequent detection by ion-trap mass spectrometry. Besides, the amino acids, numerous other compounds are detected by this method. The identity of these unknowns is now searched for. A typical chromatogram is shown in Fig. 9.7 (see also Fig. 9.6). For some analyses, we were able to use the novel ThermoFinnigan LTQ-FT mass spectrometer (ThermoElectron) as detector. In this spectrometer a linear ion-trap is combined with a Fourier-Transform Ion Cyclotron Resonance MS, based on a 7 T superconducting magnet. Some features of this apparatus are LC/MS^n analyses, accurate mass determination, high resolution and high sensitivity. The additional value of accurate mass determination in the process of identification of metabolites is discussed below.

The full-scan MS data (Fig. 9.7) presents very accurate mass values; the isotope peaks may be used for estimation of the number of carbons. For butylated phenylalanine (bPhe) an exact mass of 222.1489 Da was found (Fig. 9.8). Using the parameters: maximum mass tolerance 5 ppm and RDBE -1 to 25, the closest match of molecular formula was $C_{13}H_{19}O_2N_1-H^+$, whose simulated mass spec-

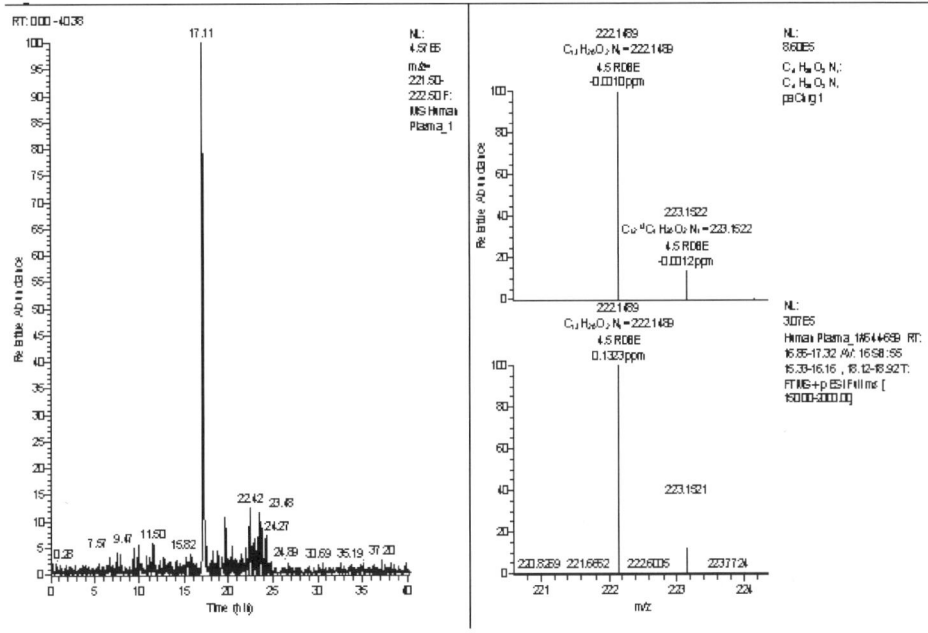

Figure 9.8 Accurate mass determination of butylated-Phe. Left: mass trace m/z 222, right top: simulated mass spectrum, right bottom: measured spectrum.

Table 9.2: Accurate mass determinations of some butylated amino acids

Butylated amino acid	Retention time	Protonated mass		Difference	
		Measured	Theoretical	(Da)	(ppm)
Lys	2.21	203.1753	203.1754	−0.0001	−0.52
Arg	5.04	231.1814	231.1816	−0.0002	−0.66
Mct	13.73	206.1209	206.1209	0.0000	−0.13
Leu/Ile	15.88	188.1645	188.1645	0.0000	−0.03
Phe	17.08	222.1488	222.1489	−0.0001	−0.25
Asp (2*butylated)	18.55	246.1700	246.1700	0.0000	0.06

trum fitted excellently to bPhe (error 0.13 ppm). Similar results were found for other amino acids (Table 9.2).

When identifying an unknown, by using the accurate mass and the other identification tools, a short list of molecular formulas can be made. A database, such as KEGG, may then provide relevant identities to these chemistries. For practical reasons it is advantageous to have a database in spreadsheet format, containing monoisotopic masses (rather than nominal or average masses) and the masses of

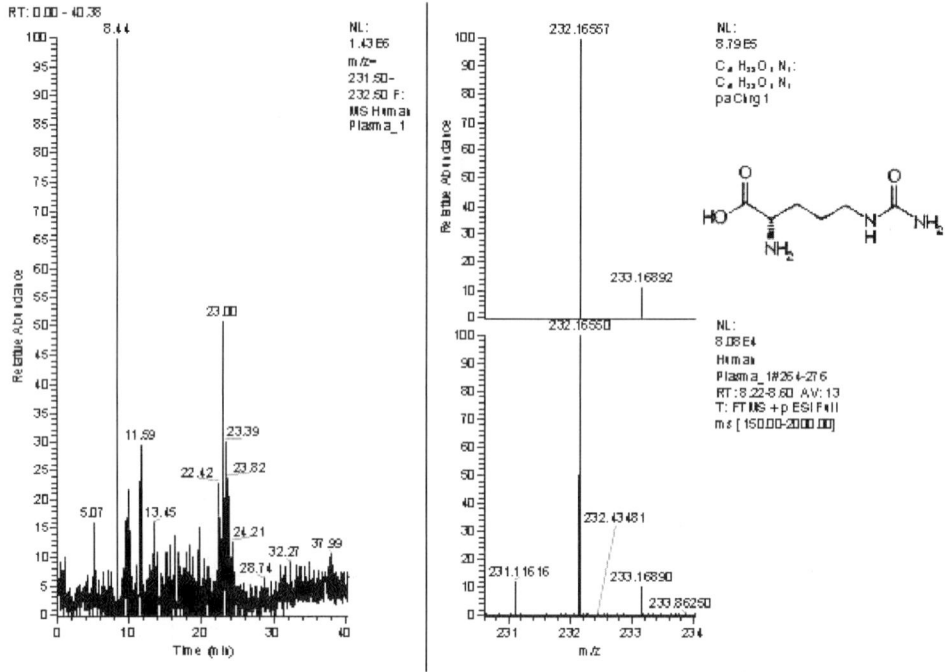

Figure 9.9 Ion trace m/z 232 and full scan MS spectrum (bottom right) of an unknown metabolite. The accurate mass fitted with the simulated spectrum (top right) of butylated citrulline.

adducts (protonated, sodiated, etc.) and derivatives (in case a derivatisation reaction is part of the sample preparation). For example, an unknown metabolite with a protonated mass of 232.16550 eluted at 8.4 min (Fig. 9.9). The accurate mass indicated a molecular formula of $C_{10}H_{22}O_3N_3$ (mass 232.16557, error < 0.3 ppm). The KEGG database did not give a reasonable match for this mass, but keeping in mind the derivatisation step in the sample pretreatment, butylated citrulline is a good candidate. Citrulline is an amino acid and its detection in this amino acid method is not unexpected. Further evidence of its identity can be obtained from an MS/MS spectrum, with a characteristic loss of -56 (butene).

An MS/MS spectrum provides structural information, which either confirms the identity or further reduces the number of possible candidates. Despite the fact that LC-MS/MS spectra are dependent on the experimental conditions, the generation of specific libraries for each method is extremely useful. In practice, the first thing done after recording a spectrum is to search the library for rapid identification.

As the example of citrulline showed, the combination of accurate mass and MS/MS is very powerful, but it is not sufficient for identification. Even if only one candidate is found, additional evidence is required to support the identification.

The first thing to do is to analyse an authentic sample (a standard) of the compound, recording retention time, accurate mass, and MS/MS spectra under similar conditions. Subsequently, the sample is enriched with the standard and reanalysed. Chromatographic behaviour (co-elution) and, if sensitivity allows, peak purity testing with photodiode array provide further evidence. This whole process is than repeated using another chromatographic system preferably with a very different separation mechanism.

In the case of citrulline, both the D- and the L-form exist. Only the L-form is found in nature, but from a pure chemical point of view the D-form is still an option. Chromatographic separation of the isomers or specific derivatisation chemistry may then give clarification. Stereochemistry plays a prominent role in the chemistry of, for example, carbohydrates and steroids. Ultimately, for the elucidation of these structures, the availability of pure compound is required for performing 2D-NMR studies. Preparative LC runs may produce the compound, preferably on the milligram scale. The availability of the sample and the concentration of the compound in the sample determine the feasibility of this operation.

To conclude, by using LC/MS/MS with accurate mass determination, the major part of the metabolites constituting a metabolite profile from a biological sample can be identified. In cases when no unambiguous identification can be made, a short list of (stereo)isomers is obtained, which implies a 10^n ($n = 2$–5) fold reduction in candidates for the metabolite to be identified.

9.8 Concluding remarks

Mass spectrometry plays a key role in Systems Biology, especially for fingerprinting of the proteome and metabolome in body fluids and tissues. A major challenge is the data preprocessing and the normalisation procedures, as information is easily lost in this stage. By applying nonlinear correction and alignment procedures small shifts in the experimental reproducibility can be corrected without the loss of the initial resolution or accuracy of the data. This allows subtle differences to be detected in fingerprints by using a suite of multivariate statistical tools [5]. In addition, in the evaluation of biomarker fingerprints related to, for instance, a certain disease state nonlinear correlation techniques allow the discovery of biological pathways involved creating biological system knowledge from the information generated at the different biological levels. However, for that step, identification of the relevant components is a prerequisite and with the availability of robust FTMS instrumentation not only can the identification process be significantly enhanced even at low concentration levels but also novel profiling methods can be designed. The novel developments at both the instrument level and the software level (data preprocessing, biostatistics, bioinformatics) will greatly enhance the discovery power of Systems Biology in the future.

Acknowledgments

We thank ThermoFinnigan (The Netherlands and Germany), in particular, Dr Wolfgang Metelmann-Strupat for the opportunity to use and for technical support on the ThermoFinnigan LTQ-FT mass spectrometer.

References

1.   T. Ideker, et al., Integrated genomic and proteomic analyses of a systematically perturbed metabolic network. *Science* **292** (2001) 929–934.

2.   H. Kitano, Systems biology: a brief overview. *Science* **295** (2002) 1662–1664.

3.   N.S. Baliga, et al., *Proc. Natl. Acad. Sci. USA A* **99** (2002) 14913–14918.

4.   L. Hood, Leroy Hood expounds the principles, practice and future of systems biology. *Drug Discov. Today* **8** (10) (May 2003) 436–438.

5.   J. van der Greef, E. Davidov, E.R. Verheij, J. Vogels, R. van der Heijden, A.S. Adourian, M. Oresic, E.W. Marple, and S. Naylor, "The role of metabolomics in drug discovery: a new vision for drug discovery and development". In: G.G. Harrigan and R. Goodacre (Eds.), *Metabolic Profiling: Its Role in Biomarker Discovery and Gene Function Analysis*, Kluwer Academic, Boston/Dordrecht/London, 2003, pp. 170–198.

6.   R. Aebersold and S.D. Patterson, *Electrophoresis* **16** (1995) 1791–1814.

7.   C.L. Gatlin, G.R. Kleemann, L.G. Hays, A.J. Linkand, and J.R. Yates III, Protein identification at the low femtomole level from silver-stained gels using a new fritless electrospray interface for liquid chromatography-microspray and nanospray mass spectrometry. *Anal. Biochem.* **263** (1998) 93–101.

8.   C.L. Gatlin, J.K. Eng, S.T. Cross, J.C. Detter, and J.R. Yates III, Automated identification of amino acid sequence variations in proteins by HPLC/microspray tandem mass spectrometry. *Anal. Chem.* **72** (2000) 757–763.

9.   J.R. Yates III, *J. Mass Spectrom.* **33** (1998) 1–19.

10.   A.J. Link, et al., Direct analysis of protein complexes using mass spectrometry. *Nat. Biotechnol.* **17** (1999) 676–682.

11.   M. Mann, R.C. Hendrickson, and A. Pandey, *Annu. Rev. Biochem.* **70** (2001) 437–473.

12.   M.P. Washburn, D. Wolters, and J.R. Yates III, Large-scale analysis of the yeast proteome by multidimensional protein identification technology. *Nat. Biotechnol.* **19** (2001) 242–247.

13.   O. Pieper, C.L. Gatlin, A.J. Makusky, P.S. Russo, C.R. Schatz, S.S. Miller, Q. Su, A.M. McGrath, M.A. Estock, P.P. Parmar, M. Zhao, S.-T. Huang, J. Zhou, F. Wang, R. Esquer-Blasco, N.L. Anderson, J. Taylor, and S. Steiner, The human serum

proteome: display of nearly 3700 chromatographically separated protein spots on two-dimensional electrophoresis gels and identification of 325 distinct proteins. *Proteomics* **3** (2003) 1345–1364.

14. J.C. Lindon, E. Holmes, and J.K. Nicholson, Pattern recognition methods and applications in biomedical magnetic resonance. *Prog. Nucl. Magn. Res. Spectr.* **39** (2001) 1–40.

15. B.A. 't Hart, J.T.W.E. Vogels, G. Spijksma, H.P.M. Brok, C. Polman, and J. van der Greef, 1-H NMR spectroscopy combined with pattern recognition analysis reveals characteristic chemical patterns in urine of MS patients and non-human primates with MS-like disease. *J. Neurol. Sci.* **15** (2003) 212(1–2): 21–30.

16. R.J. Lamers, J. de Groot, E.J. Spies-Faber, R.H. Jellema, V.B. Kraus, N. Verzijl, J.M. te Koppele, G.K. Spijksma, J.T.W.E. Vogels, J. van der Greef, and J.H.J. van Nesselrooij, Identification of disease- and nutrient-related metabolic fingerprints in osteoarthritic Guinea pigs. *J. Nutrition* (2003) 1776–1780.

17. J.W.T.E. Vogels, A.C. Tas, J. Venekamp, and J. van der Greef, Partial linear fit: a new NMR spectroscopy preprocessing tool for pattern recognition applications. *J. Chemometrics* **10** (1996a) 425–438.

18. M. Gaspari, J.W.T.E. Vogels, F. Wulfert, A.C. Tas, K. Venema, S. Bijlsma, R. Vreeken, and J. van der Greef, Novel strategies in mass spectrometric data handling. *Adv. Mass Spectrom.* **15** (2001) 283–296.

19. KEGG: http://www.genome.ad.jp/kegg/.

20. *Dictionary of Natural Products*, Chapman and Hall/CRC Press, http://www.chemnetbase.com/scripts/dnpweb.exe?welcome-main.

21. O. Corcoran and M. Spraul, LC-NMR-MS in drug discovery. *Drug Discov. Today* **8** (2003) 624–631.

Advances in Mass Spectrometry, Volume 16
A.E. Ashcroft, G. Brenton and J.J. Monaghan (Editors)

CHAPTER 10

Mass Spectrometry in Forensic Science

José R. Almirall*, Tatiana Trejos, Andria Hobbs,
Jeannette Perr, and Kenneth G. Furton

International Forensic Research Institute, Florida International University,

University Park, Miami, FL 33199, USA

Abstract

Advances in mass spectrometry have led to developments in the forensic analysis of a variety of evidentiary materials. As detection limits improve and techniques allow for better selectivity, the quality of the "evidence" improves. An overview of the advances in organic and inorganic mass spectrometry is presented covering a number of different forensic applications. Detection and identification of explosives resides and flammable and combustible liquid residues are two examples of organic mass spectrometry applications. Comparisons between ion mobility spectrometry (IMS) and the more selective ion trap mass spectrometry (ITMS) techniques for the detection and identification of explosives will be presented along with other recent advances in both IMS and ITMS including the analysis of drugs of abuse by CI-IT-MS. Inductively coupled plasma mass spectrometry (ICP-MS) and laser ablation (LA) ICP-MS have emerged as excellent tools to characterize materials that are commonly found as transfer evidence. Several researchers have demonstrated the utility of elemental analysis of small glass fragments in the association of glass materials found at a crime scene to a known source of the glass in order to associate a person to a particular crime. The multielement capability and the sensitivity of ICP-MS combined with the simplified sample introduction of laser ablation prior to ion detection provides for an excellent and relatively nondestructive technique for elemental analysis of glass fragments and other types

---

*Director, Forensic Science Graduate Program; Assistant Professor, Department of Chemistry.

of evidence. Examples of the application and utility of LA-ICP-MS for glass and paint evidence are presented. Several examples of the potential for the future impact of mass spectrometry in forensic science are also presented.

10.1 Introduction

The term "forensic" refers to the application of law or to the legal setting. Forensic chemistry is the application of chemistry, typically analytical chemistry, to legal disputes. Exciting discoveries in analytical chemistry and increasing activity in the field of forensic chemistry throughout the 1800s led to the formalization of the profession and to the first formal course of study in 1910 at the University of Lausanne, Switzerland [1]. The use of sensitive chemical tests to identify poisons such as arsenic and nicotine from human tissue was reported in the mid-1800s by chemists Mathew Orfila and Jean Servais Stas [2]. As developments in analytical chemistry and, in particular mass spectrometry, improve the measurements of analytes, "evidence" can be developed from these analyses.

Interest in forensic science has increased in the last 15 years, mainly due to the recent developments in molecular biology and biochemistry that have given rise to the powerful tool of "DNA Fingerprinting." As a result, there has been a growing need or interest to offer academic courses in forensic science around the world. A recent survey of academic programs in forensic science revealed that 89 colleges and universities in the U.S. claimed to offer a program of study in forensic science [1] and a recent report has stated that more than 90 institutions in the U.K. [3] have relaunched chemistry options as forensic science. This interest from academic institutions and students is leading to activity in research and development of forensic science tools and the U.S. is now beginning to fund the forensic research effort at a meaningful level. This paper will cover the recent advances in the application of mass spectrometry to forensic science problems.

The identification of controlled substances is routine with the aid of GC/MS. Tandem mass spectrometry (GC/MS/MS) experiments have been reported for the sensitive (part-per-trillion level) analysis of drugs of abuse from body fluids [4]. The identification of ignitable liquid residues (the term used for the compounds or mixtures of compounds often used to "accelerate" a fire) is also routinely aided by GC/MS and the application of tandem mass spectrometry has been shown to facilitate the analysis of fire debris evidence, especially when the analyte concentration is very low or when the sample contains interfering species not entirely resolved by chromatography [5]. Sample differentiation by stable-isotope-ratio mass spectrometry have been reported [6] and a recent European network named the Forensic Isotope Ratio Mass Spectrometry Network is currently undertaking a number of projects including the analysis of explosives, drugs of abuse, soils, and microbes [7].

Elemental analysis of materials has been facilitated by the sensitive methods of ICP-MS and laser ablation ICP-MS. Glass analysis using LA-ICP-MS has been reported [8] to provide excellent discrimination potential between different sources of glass. The technique has also been applied to the analysis of paint samples in order to better discriminate between similar paint samples [9].

10.2 Ignitable liquid residue and explosives analysis by GC/MS/MS

10.2.1 Ignitable liquid residue analysis

An ignitable liquid (IL) is any liquid that can fuel a fire. Ignitable liquids are considered accelerants if it can be shown that the liquid was used in order to intentionally accelerate a fire. Ignitable liquid residues (ILR) are residues of ignitable liquids that have been absorbed on substrates that typically have been burned in a fire and show similar chromatographic patterns to standard ignitable liquids run under similar instrumental conditions. Different substrates, when burned, can produce pyrolysis products whose hydrocarbons can interfere with the interpretation of chromatographic data by co-eluting with the components of interest [10]. The interpretation of ILR is commonly conducted using methods provided by the American Society of Testing and Materials (ASTM) [11–18]. The current ASTM method of choice for extraction of ILR from fire debris samples is passive headspace concentration using activated charcoal strips (ACS). In this method adsorption of the ILR is followed by elution with carbon disulfide (CS_2) or other suitable solvent [11]. There are currently two preferred ASTM standards for the identification and interpretation of ignitable liquid residues based on separation by gas chromatography and detection using either a flame ionization detector (FID) [17] or a mass spectrometric detector (MS) [18]. These methods are used to separate and identify the components of an ILR into various classes as seen in Table 10.1.

Table 10.1: Common ignitable liquid residue classes (source: ASTM, 2002)

Common ILR classes	Examples
1	Light petroleum distillates (C_4–C_{11})
2	Gasoline (C_4–C_{12})
3	Medium petroleum distillates (C_8–C_{12})
4	Kerosene (C_9–C_{17})
5	Heavy petroleum distillates (C_9–C_{23})
0	Miscellaneous (oxygenated solvents, isoparaffins, etc.)

Table 10.2: Common explosives subgroups

Common explosive classes	Examples
Aliphatic nitrates	Nitromethane
Aromatic nitrates	DNT, DNB, TNT
Nitrate esters	PETN, EGDN, NG
Nitramines	RDX, HMX, tetryl
Inorganic salts	Nitrates (NH_4NO_3), perchlorates (KOCl)

Recent advances in the analysis of ILR include the application of solid phase microextraction (SPME) most often combined with gas chromatography.

10.2.2 Explosives

An explosive is a combination of a fuel and oxidant that detonates to give a rapid release of gaseous products and energy. Organic high explosives such as TNT or TATP contain both the fuel and oxidant into the same organic molecule. Table 10.2 lists the classifications of explosives into 5 different categories. The military explosive C4 contains RDX and other compounds. Similarly, the high explosive known as SEMTEX contains a combination of PETN and RDX with oils.

10.2.3 Ion mobility spectrometry (IMS)

Ion mobility spectrometry (IMS) has been successfully applied to a wide variety of forensic specimens including drugs and explosives. IMS is a highly sensitive technique that distinguishes ionic species based on their reduced mobilities and is often used for the detection of trace organics, such as drugs of abuse or explosives, under atmospheric pressure conditions. Filter paper is placed in the desorber unit where the sample is introduced as a vapor into the spectrometer. The molecules of the vapor are carried by a stream of dried, filtered air into the reaction region where β particles emitted by ^{63}Ni ionize the sample. The product ions, positive or negative depending on instrumental mode, are gated into the heated drift region. All other ions and neutrals are swept away by the calibrant gas. Under the influence of a controlled electric field created by the drift rings and against a counter-flow of ambient air drift gas and calibrant gas, the product ions move towards the collector electrode. The time it takes the product ions to reach the collector electrode is referred to as the ion's drift time and is used to calculate the reduced ion mobility of the compound. The spectrometer is operated in the negative ion mode for explosive detection with methylene chloride as the calibrant gas. One current limitation of IMS is the limited ability to detect vapor chemicals from specimens due in part to sampling limitations. The combination of SPME with IMS has great potential to expand the method to volatile headspace components as well as less volatile parent compounds.

10.2.3.1 Methodology: ignitable liquid residue analysis

The experimental conditions of the MS/MS experiments are listed in Table 10.3 with the parent ion for all the aromatics selected as the trapped ion and m/z 85 as the trapped ion, an excitation storage level of m/z 37, and a excitation amplitudes of 0 V for the alkanes. The following were components in the standard accelerant mixture (SAM): C_8 through C_{23} n-alkanes, pristane, phytane, p-xylene, o-xylene, ethylbenzene, p-cymene, 1,2,4-trimethylbenzene, 1,2,3-trimethylbenzene, cumene, 2-methylnaphthalene, and 1-methylnaphthalene. CI experiments were conducted using methanol or acetonitrile reagent gases. The experimental parameters were optimized using an ion-trap detector (Varian 2000) by setting the transfer line temperature to 280 °C, the manifold temperature to 80 °C, and the ion trap temperature to 120 °C. A split ratio of 18:1 was used with an injector port temperature of 280 °C.

10.2.3.2 SPME-IMS of explosives

A SPME/IMS interface was designed and machined using aluminium stock, a rocker switch (Carling Switch 12 A 125 VAC 1/2 HP), a fuse (NTE 2A Thermal Breaker), Teflon wire (NTE 18 AWG Teflon, 200 °C), a prefabricated metal box, six screws, a barrier switch, a resistor (Ohmite, 500 Ω, 25 W), a grounded outlet cord, cardboard, and glass wool. A Whatman® 90 mm diameter 4 qualitative filter paper was spiked with 5 μL of varying 2,4,6-trinitrotoluene (TNT) concentrations. Each sample was run in triplicate with blanks run in between. The desorber unit remained at a constant 213 °C at the center and 150 °C at the edges. Runs were 7 s in duration. A StableFlex Divinylbenzene/Carboxen™/Polydimiethylsiloxane (DVB/CAR/PDMS) SPME fiber was used to extract the headspace created by 0.2626 g of TNT in a 10 mL vial for 15 min, 30 min, and 1 h at ambient room temperatures. The fiber was then inserted into the SPME inlet at 213 °C for the duration of the run.

10.2.4 Results and discussion

10.2.4.1 ILR analysis

The detection levels (LODs) for the MS/MS technique were determined by measuring the signal and noise of each component in the SAM as well as the background for triplicate 50 ppm runs for each of the six methods: EI-MS (electron ionization), EI-MS/MS, CI MeOH (chemical ionization with methanol) MS, CI MeOH MS/MS, CI ACN (chemical ionization with acetonitrile) MS, and ACN MS/MS are summarized in Table 10.3. The background signal was subtracted from the signal recorded for the component to give the true signal. The LOD was determined using the true signal, the noise from the original signal, and the actual

Table 10.3: Experimental conditions and level of detection calculation results

Compound	CAS registry number	Molecular weight (g/mol)	Formula	Boiling point (°C)	Density (g/mL)	Parent ion (m/z)	Excitation storage level (m/z)	Excitation amplitude (V)	EI-MS Level of detection (ng)	EI-MS/MS Level of detection (ng)	CI MeOH MS Level of detection (ng)	CI MeOH MS/MS Level of detection (ng)	CI ACN MS Level of detection (ng)
n-Octane	111-65-9	114.23	C_8H_{18}	125.66	0.7025	115	37	0	0.040293	0.642129	0.002136	0.075992	0.004990
Ethylbenzene	100-41-4	106.17	C_8H_{10}	136.20	0.8670	106	48	51.5	0.020876	0.001878	0.009187	0.008728	0.000438
p-Xylene	106-42-3	106.17	C_8H_{10}	138.35	1.4958	106	48	52.5	0.034135	0.019286	0.016854	0.040476	0.001369
o-Xylene	95-47-6	106.17	C_8H_{10}	144.40	1.5055	106	48	50.5	0.036245	0.020582	0.011762	0.004138	0.001664
n-Nonane	11-84-2	128.26	C_9H_{20}	150.80	0.7176	129	37	0	0.019669	0.567675	0.001225	0.103500	0.005638
Cumene	98-82-2	120.19	C_9H_{12}	152.40	0.8618	120	48	43	0.016251	0.004326	0.010223	0.007797	0.000907
1,2,4-Trimethylbenzene	95-36-3	120.19	C_9H_{12}	169.35	0.8758	120	48	49	0.011082	0.001543	0.008781	0.010443	0.000809
n-Decane	124-18-5	142.28	$C_{10}H_{22}$	174.10	0.7300	143	37	0	0.012905	0.288791	0.001582	0.136307	0.003092
1,2,3-Trimethylbenzene	526-73-8	120.19	C_9H_{12}	176.10	0.8944	121	48	50	0.016734	0.022650	0.015780	0.004448	0.000449
p-Cymene	99-87-6	134.22	$C_{10}H_{14}$	177.10	0.8573	134	48	43	0.009481	0.147836	0.015230	0.079543	0.000336
n-Undecane	1120-21-4	156.31	$C_{11}H_{24}$	195.90	0.7401	157	37	0	0.004130	0.316207	0.000890	0.041945	0.003649
n-Dodecane	112-40-3	170.33	$C_{12}H_{26}$	216.30	0.7487	171	37	0	0.003912	0.156388	0.000559	0.020102	0.004201
2-Methylnaphthalene	629-50-5	184.36	$C_{13}H_{28}$	235.40	0.7564	143	48	53	0.007278	0.002100	0.009166	0.068764	0.001210
n-Tridecane	91-57-6	142.20	$C_{11}H_{10}$	241.05	1.0058	185	37	0	0.005450	0.078585	0.000381	0.054270	0.005140
1-Methylnaphthalene	90-12-0	142.20	$C_{11}H_{10}$	244.64	1.0202	143	48	53.5	0.008580	0.003822	0.008883	0.066342	0.001270
n-Tetradecane	629-59-4	198.39	$C_{14}H_{30}$	253.70	0.7628	199	37	0	0.001985	0.079352	0.000150	0.008315	0.001553
n-Pentadecane	629-62-9	212.41	$C_{15}H_{32}$	270.63	0.7658	213	37	0	0.002547	0.115154	0.000085	0.007937	0.001460
n-Hexadecane	544-76-3	226.44	$C_{16}H_{34}$	287.00	0.7733	227	37	0	0.001712	0.024399	0.000112	0.007221	0.001135
n-Heptadecane	629-78-7	240.47	$C_{17}H_{36}$	301.80	0.7780	241	37	0	0.001652	0.063884	0.000061	0.011876	0.001483
Pristane	1921-70-6	268.52	$C_{19}H_{40}$	296.00	0.7827	268	37	0	0.001303	0.074300	0.000077	0.014060	0.001717
n-Octadecane	593-45-3	254.49	$C_{18}H_{38}$	316.10	0.7768	255	37	0	0.001454	0.038174	0.000057	0.012475	0.001791
Phytane	638-36-8	283.60	$C_{20}H_{42}$	310.00	0.7910	283	37	0	0.001276	0.039081	0.000075	0.015303	0.001811
n-Nonadecane	629-92-5	268.52	$C_{19}H_{40}$	329.70	0.7774	269	37	0	0.002106	0.188345	0.000055	0.018519	0.002713
n-Eicosane	112-95-8	282.55	$C_{20}H_{42}$	343.00	0.7886	283	37	0	0.003234	0.369595	0.000076	0.099126	0.003432
n-Heneicosane	629-94-7	296.57	$C_{21}H_{44}$	356.50	0.7917	297	37	0	0.006765	0.275209	0.000155	0.074377	0.005469
n-Docosane	629-97-0	310.60	$C_{22}H_{46}$	368.60	0.7944	311	37	0	0.012566	0.640636	0.000305	0.103318	0.008620
n-Tricosane	688-67-5	324.63	$C_{23}H_{48}$	380.20	0.7785	324	37	0	0.030954	1.185252	0.002126	0.183416	0.015699

mass injected by Eq. (10.1). The LOD was better for aromatics using MS/MS than MS. CI improves the LOD over EI due to a decrease in background. The MeOH CI method improves the LOD for aromatic compounds while the ACN CI method improves the LOD for the alkanes over EI.

$$\frac{\text{True signal}}{\text{Actual mass injected (ng)}} = \frac{3 \times \text{noise}}{\text{Level of detection (ng)}} \tag{10.1}$$

10.2.4.2 IMS of explosives

A prototype SPME/IMS interface was constructed out of aluminum and has been shown to be effective, efficiently transferring analytes from SPME fibers into the IMS source. The data collected from the above experiments was used to construct a calibration curve shown in Fig. 10.1. The amount extracted and introduced by the SPME fiber exceeds the linear range of the IMS, even at very short extraction times. The extraction temperature can be optimized over ambient temperature for increased sensitivity for IMS methods but good results have been obtained thus far for TNT.

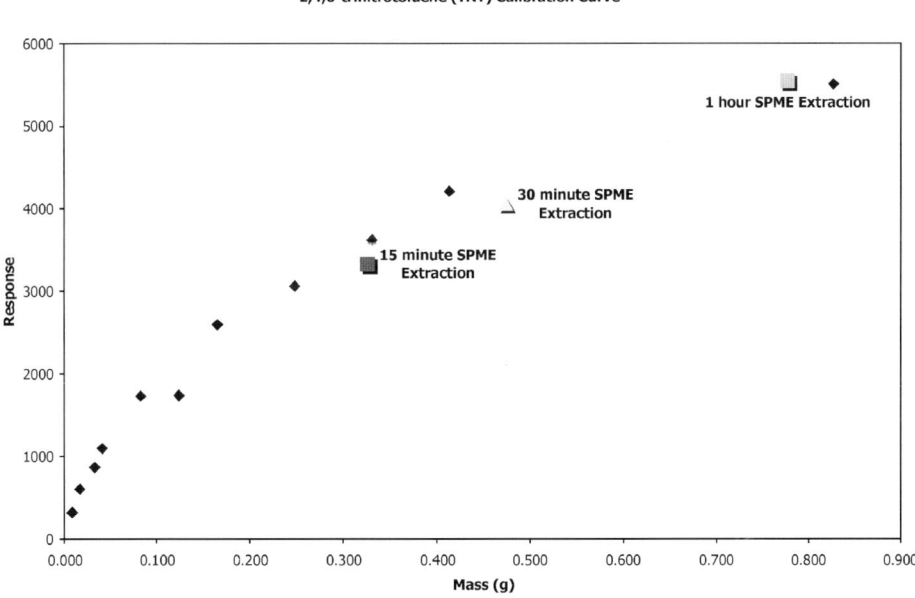

Figure 10.1 IMS calibration curve for TNT.

10.3 Glass and paint analysis by laser ablation inductively-coupled plasma mass spectrometry (LA-ICP-MS)

10.3.1 Glass analyses

Glass and automotive paints are common types of trace evidence found at crime scenes, in cases such as hit-and-run accidents [19]. Glass examiners measure the physical and optical properties of glass such as color, thickness, and refractive index (RI). However, the informing power of refractive index measurements has decreased as the glass manufacturing industry has improved its quality control. Elemental analysis is now commonly performed in forensic laboratories by ICP-OES, ICP-MS, X-ray fluorescence (XRF), or scanning electron microscopy (SEM-EDS) and it has been demonstrated that ICP provides better detection limits and discrimination power than the other elemental techniques [19–32].

ICP-MS can be coupled with a laser ablation (LA) system in order to offer a new tool for trace elemental analysis in the forensic field. LA-ICP-MS combines the advantages of the minimum removal of sample by LA with the excellent sensitivity (low detection limits) and precision associated with ICP-MS [19–23]. LA-ICP-MS has been evaluated as a method for the forensic analysis of glass and paint and in comparison with conventional acid digestion methods, LA can reduce costly and time-consuming sample preparation and minimize contamination.

In order to evaluate the relative discrimination power of the technique, the results of the LA analysis from 91 samples originating from two different sets of glasses were compared, in a statistical analysis, with the results obtained from the conventional digestion external calibration methods (EC) and solution isotope dilution (ID) analyses of the same samples.

A paint LA-ICP-MS method was developed to complement the methods used to characterize the organic components in paints and coatings currently in use by forensic laboratories. The technique can be considered essentially "nondestructive" since the sample preparation does not require manipulation. Additionally, very small amounts of material are removed (~ 300 ng). The possibility of sequential analysis of all the paint layers without the need for extensive sample preparation is of interest to forensic examiners.

A summary of the results from the application of this innovative sample introduction technique to glass and paint analysis, including the advantages and limitations, is presented.

10.3.2 Instrumentation

The ICP-MS used in this study was an Agilent, model HP-4500 Plus Shield Torch (Agilent Technologies, Palo Alto, CA, USA), equipped with a CETAC ASX-500 autosampler (CETAC, Omaha, NE, USA). Laser ablation analyses were performed with a CETAC (Omaha, NE, USA) Laser Ablation System, model LSX-200 Plus, Q switched Nd:YAG, 266 nm, 4.8 mJ output energy laser. A Leica microscope, Model L2 was used to mount the glass fragments in labeled paper squares for LA analysis. A GRIM II (Foster and Freeman, Worcestershire, UK) system equipped with a FP82 hot stage (Mettler, Grifensee, Switzerland) was used for refractive index measurements. A Scanning Electron Microscope (SEM/EDX) JSM-5900-LV (JEOL, Japan) from the FCAEM (Florida Center for Analytical Electron Microscopy at FIU) was used for the imaging of the craters on glass in order to calculate the ablated mass from volume measurements.

10.3.3 Glass analysis by LA-ICP-MS

The surfaces of the glass fragments were slightly scratched with a sand paper of 3600 mesh and then washed three times with deionized water followed by washing with 0.8 molL^{-1} HNO_3 for 30 min under the ultrasonic bath. The samples were rinsed with deionized water and then let dry overnight. The dried fragments were mounted under the microscope into a small piece of "tacky blue" mounting medium of ~ 1 cm^2 square. For the CFS subset, the samples were observed under UV-lamp (365 and 254 nm) and the nonfluorescent side was selected for the ablation to avoid the introduction of Sn from the surface of glass manufactured by the float process.

A single point calibration curve using NIST 612 and an internal standardization using isotope ^{29}Si was used for quantification. The headlamps and automobile window were analyzed using the single line and depth profile ablation mode,

Table 10.4: Optimized parameters used for the analysis of glass and paint samples by LA-ICP-MS

Parameters	Sample set		
	CFS–auto windows	Headlamps	Paints
RF power	1302 W	1302 W	1302 W
RF matching	1.95 V	1.95 V	1.95 V
Argon as carrier gas after cell	1.05 L/min	0.95 L/min	0.95 L/min
Ablation cell volume	50.2 mL	50.2 mL	50.2 mL
Helium as carrier gas through cell	0.95 L/min	0.95 L/min	0.95 L/min
Ablation mode	Depth profile	Single line	Single spot
Spot size	50 μm	50 μm	100 μm
Energy output	100%	100%	50%
Frequency	10 Hz	10 Hz	5 Hz
Preablation	No	Yes	No

respectively. The optimized parameters used for the analysis of glass samples by LA are summarized in Table 10.4.

10.3.4 Refractive index

The standard test method described in ASTM 1967-98 was used for the determination of the refractive index measurements at 589 nm. Each reported refractive index value was the result of the mean of the measurements from five different fragments.

10.3.5 Glass sampling

The automotive windows subset consisted of 46 auto window samples provided by the Centre of Forensic Sciences (CFS), Toronto, Canada, from actual casework samples that were submitted to the laboratory. The set consisted of 18 windshields and 28 windows (side or rear). The headlamp subset consisted of a total of 45 headlamp samples collected from a variety of automobiles in a junkyard representing 20 years of manufacturing dates and included 34 lenses and 11 reflectors. Quantitative analysis on all 91 samples was conducted by both EC and LA methods. Isotope dilution analysis was performed only on those samples that were not distinguishable by the reported EC method.

10.3.6 Statistical analysis

All statistical analyses were performed by either the use of SYSTAT for Windows 8.0 (SPSS Science, Chicago, IL) or Excel 2000 (Microsoft Corp, v9.0.2719). The pairwise comparisons were carried out using analysis of variance (ANOVA) using the General Linear Model (GLM) tool from SYSTAT followed by the Tukey's honestly significant different test (HSD) [32,33]. For sets of few pairs, such as the ones analyzed by ID, a simple t-test was used instead of ANOVA. Data reduction of laser ablation data was performed using the GLITTER software (GEMOC, Macquarie University, Australia).

10.3.7 Paint analyses

10.3.7.1 Instrumentation

The CETAC LSX 200 Plus laser ablation system (CETAC Technologies, Omaha, NE, USA) was coupled to an Agilent (HP) 4500 Plus Shield Torch ICP-MS benchtop system (Agilent, Palo Alto, CA, USA).

10.3.7.2 Paint analysis by LA-ICP-MS

The element menu for the paint was comprised of magnesium (Mg), titanium (Ti), chromium (Cr), manganese (Mn), iron (Fe), copper (Cu), zinc (Zn), rubidium (Rb), strontium (Sr), tin (Sn), cesium (Cs), barium (Ba), lanthanum (La), cerium (Ce), lead (Pb), as well as bismuth (Bi). These elements were chosen because they have been found to be present with the greatest frequency and to vary significantly between layers and samples. A Pb calibration curve using NIST SRMs 2570, 2574, and 2575 (developed for the analysis of lead in paint by portable XRF analyzers) was constructed for lead quantification. The search for a suitable internal standard is under way, so the quantitative values do not currently account for instrumental drift. The layers of each sample were ablated sequentially in a single run by depth profiling from the top of the sample to the bottom, and each sample was analyzed in triplicate.

The instrument response (in intensity counts) for each element in each layer was calculated with the use of CETAC's Geopro software by integrating the transient signal for each layer and subtracting the gas background. The counts for Pb were interpolated from the calibration curve constructed from the XRF standards and used for Pb quantification. Element profile plots were constructed in Microsoft Excel to be used as a "fingerprint" for each layer. The plots display the additional elemental data obtained and aid the further discriminate between the samples.

10.3.7.3 Paint sampling

Eighteen casework-like automotive paint samples were collected from various sources. Each sample consisted of three or four layers on a steel panel. The layer sequence was clearcoat–colorcoat–electrocoat for the three-layered systems or clearcoat–colorcoat–primer–electrocoat for the four-layered systems. The only sample preparation involved cutting each sample into small enough portions to be placed inside the ablation chamber (~ 2 cm^2).

10.3.8 Results and discussion

10.3.8.1 Glass analysis

One of the aims of this work was to demonstrate that the analysis of glass by LA-ICP-MS could provide comparable power of discrimination, precision, and accuracy to the well-known techniques of external calibration and isotope dilution ICP-MS. The repeatability of LA was tested in seven fragments from a single windshield glass sheet. The seven fragments were selected randomly and each fragment was measured in triplicate. Figure 10.2 shows that good repeatability was obtained from the analysis of different pieces of glass originating from the same source.

The laser ablation technique was shown to provide very low detection limits for elements of forensic interest permitting the quantitative analysis on the order of subpicogram detection and quantification (see Table 10.2). The relative standard deviations on Table 10.5 represent the deviations for the measurements of the elements at the concentration level present in the reference material NIST 612

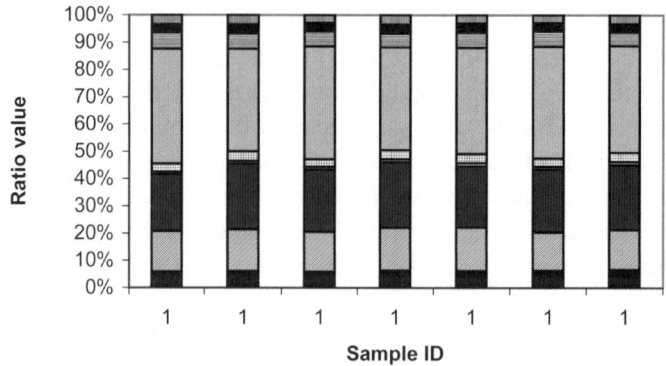

Figure 10.2 Repeateability of 7 fragments of an automobile windshield. For each sample, the colored pattern represent one elemental ratio, for each of the seven fragments from the top to the bottom: U/Th, Pb/Th, Ce/La, Hf/Zr, Sn/Ba, Sr/Zr, Mn/Rb, Ti/Mn, and Al/K.

Table 10.5: Laser ablation-ICP-MS limits of detection for SRM 612

Element	LOD (μg g^{-1})	LOD (fg)	% RSD ($\sim$ 40 μg g^{-1} in glass)
Ti	4.1	1150	5.3
Mn	0.8	226	3.4
Ga	0.4	103	7.6
Sr	0.2	51	3.9
Zr	0.4	124	6.9
Sn	1.1	294	7.6
Cs	0.1	28	7.2
Ba	1.2	322	6.2
Ce	0.1	35	5.8
Hf	0.6	180	3.0
Pb	0.2	57	7.6

($\sim$ 40 μg g^{-1}). All the selected elements presented relative standard deviations below 10%.

Photomicrography as well as scanning electron microscopy (SEM) imaging were used to calculate the volume of the ablated material and estimate the ablated mass in the standard NIST 612, using the formula

$$\text{Density of glass} = \frac{\text{ablated mass}}{\text{removed volume}}.$$

The density of the glass SRM 612 was approximated to 2.4 g cm^{-3}.

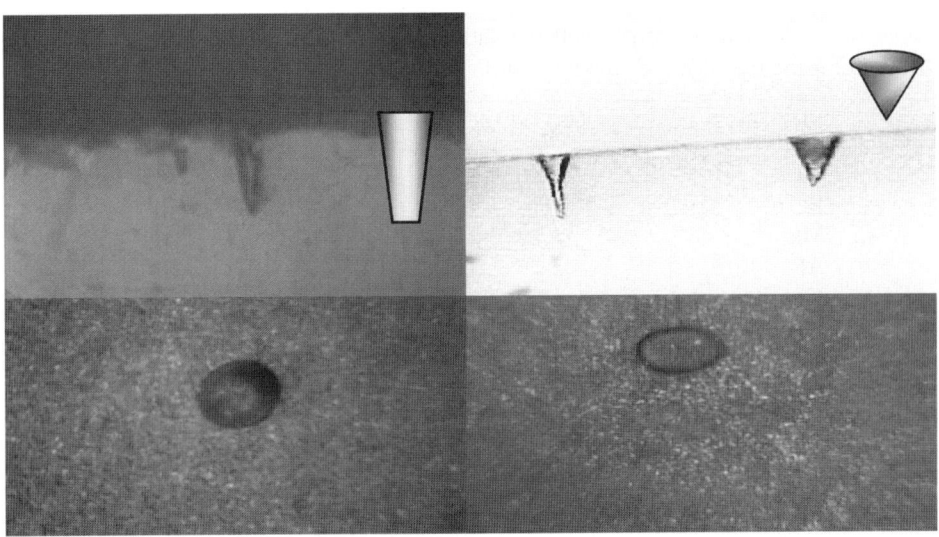

Figure 10.3 Comparison of the shape of the craters produced by the LSX 500 and the LSX 200. Left side: above: lateral view of 50 and 100 µm spot size, respectively; below: 50 µm spot size. Right side: above: LSX 200 lateral view 100 and 50 µm spot size; below: 50 µm spot size.

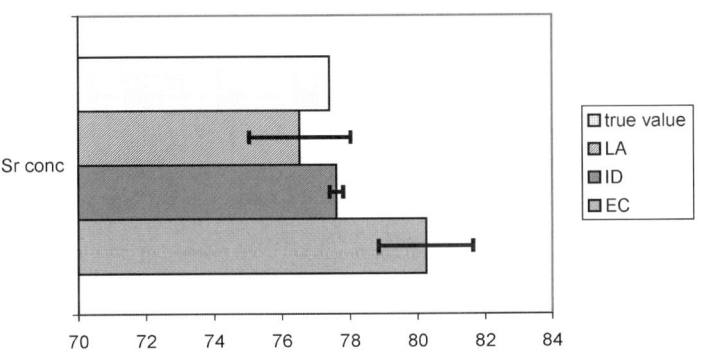

Figure 10.4 Comparison of the precision (error bars) and accuracy of the methods of LA, ID, and EC-ICP MS for the determination of Sr in SRM 612. LSX 500 was used for analysis by LA.

Figure 10.3 shows the photomicrography image of a crater resulting from the analysis of the NIST 612 using two different laser systems, whose main difference is that the LSX 500 has a flat beam profile instead of a Gaussian beam profile. The flat beam profile design provided more symmetric shape of the craters, more sensitivity, and better relative standard deviations between runs (see Table 10.6).

Table 10.6: Comparison of percentage relative standard deviation (% RSD) obtained for SRM 612 using the LSX 200 and LSX 500

Element	LSX 200	LSX 500
Li	8.5	1.6
B	10.5	2.7
Mg	6.7	1.4
Al	3.3	1.0
K	26	2.8
Ca	3.3	0.5
Ti	10.5	4.4
Mn	9.0	1.3
Fe	16	4.7
Co	6.5	0.7
Ga	6.2	2.2
Rb	9.9	1.1
Sr	8.8	1.5
Zr	9.9	1.0
Sn	12.2	1.8
Sb	9.7	1.4
Cs	9.6	0.8
Ba	11.3	1.4
La	11.2	1.0
Ce	10.5	0.5
Hf	9.6	2.1
Pb	9.7	2.6
Bi	10.6	2.5
Th	11.5	1.0

Figure 10.4 shows that laser ablation presented good accuracy and precision when compared with EC and ID-ICP-MS. Similar results were obtained for the other elements analyzed in this study.

10.3.8.2 Paint analysis

The paint analysis resulted in several sets of data that could be used to associate and discriminate between samples. The first is a time resolved plot of the responses due to analyte concentrations in the samples. The second is the Pb quantification data, and the third is an element profile plot used to display all elements under investigation.

The plots allow the analyst to view the instrument response of the elements in the element menu and to show the transitions of the layers during the ablation. Figure 10.5 illustrates the layer transitions and demonstrates that the elemental

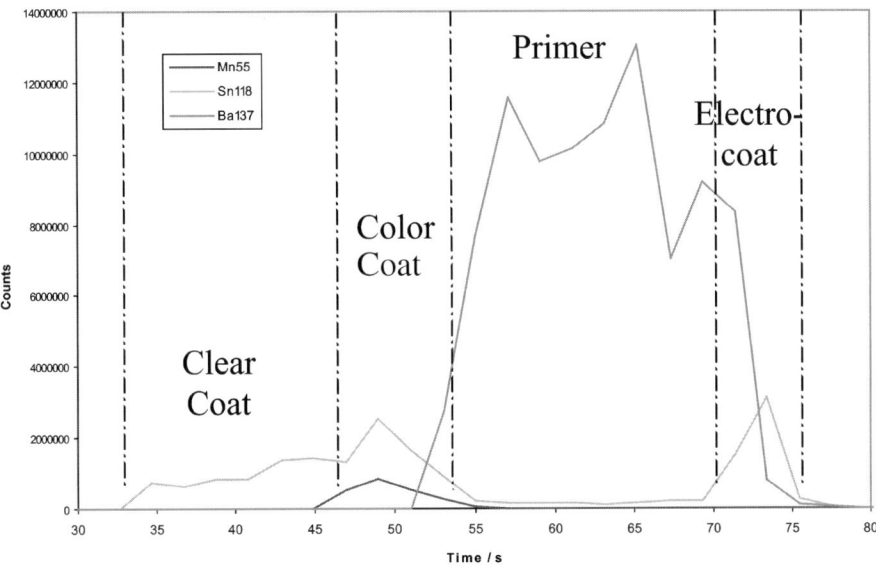

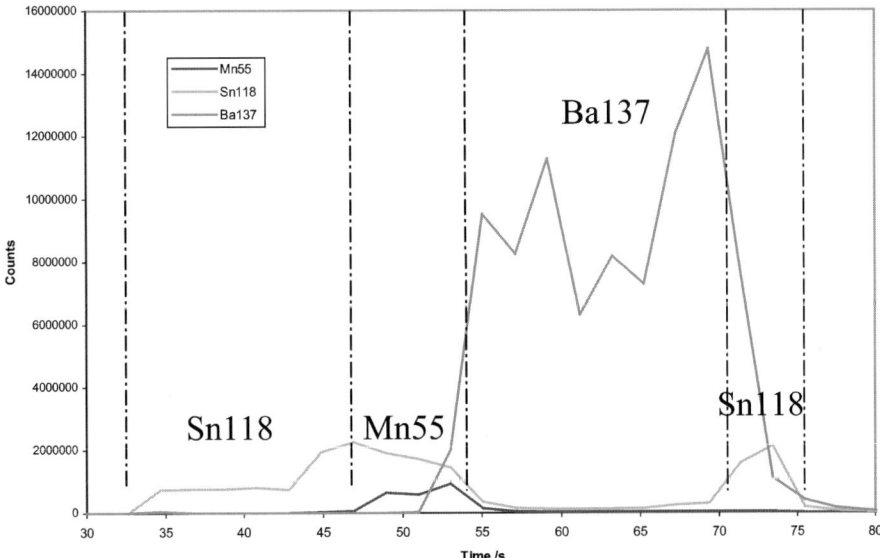

Figure 10.5 Time resolved plots of a single sample. The plot demonstrates ability to visualize the layer transitions even with only three elements and also demonstrates reproducibility.

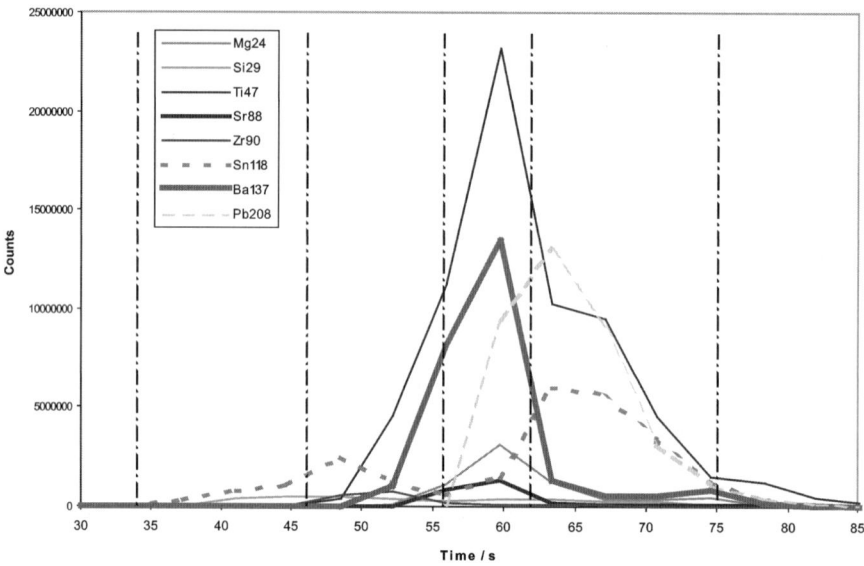

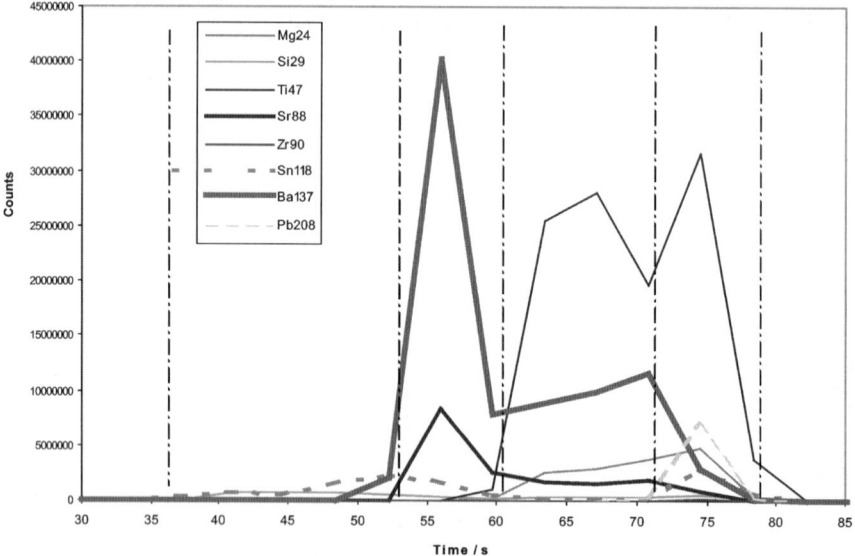

Figure 10.6 Time resolved plots of two different automobile paint samples. The plots show the pattern of the layers, the elements present and the relative intensities, which show the difference between two paints. The vertical lines mark the transition between the layers, with the first line marking the onset of ablation and the last line marking the end of the ablation.

characteristics and intensities for two measurements from a single sample are consistent from one plot to the next. Figure 10.6 shows the time-resolved plots of two different paint samples. The elements and intensities are different from the first plot to the second. Visual comparisons of the time-resolved plots of the automotive paint samples generated as a result of this study result in the association of samples originating from the same source and for distinguishing samples originating from different sources.

In addition to the time-resolved plots, lead quantification data was obtained. NIST standard reference materials 2570, 2574, and 2575 are available for Pb quantification. From the calibration curve, the limit of detection (LOD) and the limit of quantification (LOQ) were determined to be 0.012 mg/cm^2 (3 $*$ background) and 0.028 mg/cm^2 (7 $*$ background), respectively, which were approximated to be 0.05 and 0.12 ppm. No additional standards for paints and coatings were found. A pairwise comparison was completed, in which each sample is compared with each other for a possible 153 $[n(n-1)/2]$ total comparisons. Based on a 95% confidence interval, 37 pairs or 24% of the possible comparison pairs were indistinguishable. Based on a 99% confidence interval, 39 pairs or 26% of the possible comparison pairs were indistinguishable. The inability to distinguish approximately 25% of the samples illustrated the need to examine additional elements; however, the ability to distinguish approximately 75% of the pairs using only a single element was very encouraging for the applicability of this method. However, instrumental drift has not been accounted for, leading to a potential error in the measurements.

The final type of information gathered in the analysis is the multielemental semiquantitative data, which can be used to construct element profile plots. The responses of the additional elements measured in the element menu are displayed in bar graphs, which will show each element's relative instrumental intensity to all the others. Element profile plots can be created for each layer of each sample and compared with all others for association or discrimination. Figure 10.7 shows the reproducibility of a single sample (labeled A) and the discrimination of two samples (A and B).

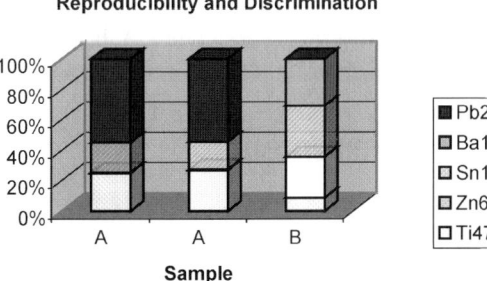

Figure 10.7 Two analyses of sample A are shown to demonstrate the reproducibility of the element ratio plots, and B can be compared with A to demonstrate the use of the plots for discrimination. The plots show the results of the electro-coat layer for these two samples.

Table 10.7: Summary of results of pairwise comparisons by refractive index, EC, ID, and LA-ICP-MS for automobile window and headlamp sets

Subset	CFS	Headlamps
Number of samples	46	45
Number of comparison pairs	1035	990
Indistinguishable pairs by RI	471	102*
Indistinguishable pairs by EC-ICP-MS + RI ($p < 0.05$)	1	0
Indistinguishable pairs by ID-ICP-MS ($p < 0.05$)	1	0
Indistinguishable pairs by LA-ICP-MS + RI ($p < 0.05$)	3	0

*102 pairs out of 861 possible pairs were not distinguished when compared by the RI measured.

10.3.9 Conclusions

The analysis of ignitable liquids from fire debris and of explosives residues has been facilitated by the use of sensitive measurements by GC/MS, GC/MS/MS, and IMS. The detection of the odor signatures (volatile components) from the explosives can be used to detect devices. SPME has been coupled to an IMS in order to preconcentrate the odor signatures on the SPME fiber for subsequent desorption into the IMS.

The analysis of glass and paint materials by LA-ICP-MS has been shown to produce excellent evidence of association of these materials when two fragments (of either glass or paint) have originated from the same source. The glass studies present excellent discrimination power for the technique when comparing glass samples from different sources. Table 10.7 summarizes the discrimination power of each of the techniques in comparison with the other. This suggests that obtaining a matching profile for a comparison pair by this sensitive technique results in very good evidence of association between the pair. The LA technique has been shown to provide the same power of discrimination as for solution ICP-MS methods but the LA method provides significant advantages, most importantly, much reduced sample preparation prior to analysis and that the technique is essentially nondestructive. The technique also shows great potential for the analysis of paint chips with multiple layers. The LA-ICP-MS technique has excellent potential for application in materials characterizations in forensic cases.

Acknowledgments

The authors thank the Centre of Forensic Sciences in Toronto, Canada for supplying casework samples for analysis. Initial research on the application of elemental analysis by ICP-MS of glass evidence in our group was funded by the National Institute of Justice. The Technical Support Working Group (TSWG) under contract DAAD05-00-C-7128 to FIU funded the more recent work on standard method development, isotope dilution method development and statistical analyses of database samples. CETAC, Inc. provided a partial instrument grant. We also thank the National Institute of Standards and Technology (NIST) for providing the paint SRMs, and Atlas Material Testing Solutions in South Florida and Ford for providing automotive paint samples. We also thank the Florida Center for Analytical Electron Microscopy (FCAEM) at Florida International University for use of the SEM-EDS in the study. GE Ion Track, Inc. and Dr Chin Wu are gratefully acknowledged for IMS instrument loan and useful discussions. Funding is contributed from the National Forensic Science Technology Center (NFSTC) toward this research; partial support was also obtained from Supelco Inc. and Field Forensics Inc.

References

1. J.R. Almirall and K.G. Furton, Trends in forensic science education: expansion and increased accountability, *J. Anal. Bioanal. Chem.* **376** (2003) 1156–1159.

2. S. Kind and M. Overman, *Science Against Crime*, Aldus Books, London, UK, 1972.

3. J. Henry, "Firms seek quality register after doubts on science degrees", *The Sunday Telegraph*, April 6, 2003, p. 19.

4. R.A. Yost, D.D. Feterolf, R.J. Hass, D.J. Harvan, A.F. Weston, P.A. Skotnicki, and N.M. Simon, *Anal. Chem.* **56** (1984) 2223.

5. J.R. Almirall and J. Perr, "The use of compound specific MS/MS for the identification of ignitable liquid residues in fire debris analysis". In: J. Yinon (Ed.), *Advances in Forensic Applications of Mass Spectrometry*, CRC Press, 2003, pp. 181–201.

6. R.H. Liu, "Sample differentiation by stable-isotope mass spectrtometry". In: M.H. Ho (Ed.), *Analytical Methods in Forensic Chemistry*, Ellis Horwood, 1990, pp. 40–51.

7. The forensic isotope ratio mass spectrometry (FIRMS) network web site: http://www.forensic-isotopes.rdg.ac.uk/index.htm, accessed on August 1, 2003.

8. T. Trejos, S. Montero, and J.R. Almirall, Analysis and comparison of glass fragments by laser ablation inductively coupled plasma mass spectrometry (LA ICP MS), *J. Anal. Bioanal. Chem.* **376** (2003) 1255–1264.

9. A. Hobbs and J.R. Almirall, A novel method for the forensic characterization of paints and coatings using laser ablation inductively coupled plasma mass spectrometry (LA-ICP-MS). *J. Anal. Bioanal. Chem.* **376** (2003) 1265–1271.

10. J.R. Almirall and K.G. Furton, Identification and characterization of pyrolysis products in fire debris analysis. *J. Anal. Appl. Pyrolysis*, first published online on July 9, 2003.

11. "ASTM E1412-00 standard practice for separation of ignitable liquid residues from fire debris samples by passive headspace concentration with activated charcoal". In: *ASTM Annual Book of Standards, vol. 14.02*, ASTM International, West Conshohocken, PA, 2002.

12. "ASTM E1413-00 standard practice for separation and concentration of ignitable liquid residues from fire debris samples by dynamic headspace concentration". In: *ASTM Annual Book of Standards, vol. 14.02*, ASTM International, West Conshohocken, PA, 2002.

13. "ASTM E1388-00 standard practice for sampling of headspace vapors from fire debris samples". In: *ASTM Annual Book of Standards, vol. 14.02*, ASTM International, West Conshohocken, PA, 2002.

14. "ASTM E1385-00 standard practice for separation and concentration of ignitable liquid residues from fire debris samples by steam distillation". In: *ASTM Annual Book of Standards, vol. 14.02*, ASTM International, West Conshohocken, PA, 2002.

15. "ASTM E1386-00 standard practice for separation and concentration of ignitable liquid residues from fire debris samples by solvent extraction". In: *ASTM Annual Book of Standards, vol. 14.02*, ASTM International, West Conshohocken, PA, 2002.

16. "ASTM E2154-01 standard practice for separation and concentration of ignitable liquid residues from fire debris samples by passive headspace concentration with solid phase microextraction (SPME)". In: *ASTM Annual Book of Standards, vol. 14.02*, ASTM International, West Conshohocken, PA, 2002.

17. "ASTM E1387-01 standard test method for ignitable liquid residues in extracts from fire debris samples by gas chromatography". In: *ASTM Annual Book of Standards, vol. 14.02*, ASTM International, West Conshohocken, PA, 2002.

18. "ASTM E1618-01 standard test method for ignitable liquid residues in extracts from fire debris samples by gas chromatography-mass spectrometry". In: *ASTM Annual Book of Standards, vol. 14.02*, ASTM International, West Conshohocken, PA, 2002.

19. J.R. Almirall, "Elemental analysis of glass fragments". In: B. Caddy (Ed.), *Forensic Examination of Glass and Paint*, Taylor & Francis, London, 1999, pp. 65–80.

20. D.C. Duckworth, C.K. Bayne, S.J. Morton, and J.R. Almirall, *J. Anal. Atom. Spectrom.* **15** (7) (2000) 821–828.

21. J.R. Almirall, "Glass as evidence of association". In: M. Houck (Ed.), *Mute Witness; When Trace Evidence Makes the Case*, Academic Press, San Diego, 2001, pp. 139–155.

22. J.R. Almirall, J. Buckleton, J. Curran, and T. Hicks, "Examination of glass". In: J. Curran, J. Buckleton, and T. Hicks-Champond (Eds.), *Forensic Interpretation of Glass Evidence*, CRC Press, Boca Raton, 2000, pp. 1–24.

23. J. Buscaglia, *Anal. Chim. Acta* **288** (1994) 17–24.

24. J.R. Almirall, M.D. Cole, G. Gettinby, and K.G. Furton, "Science and justice". *J. Forensic Sci. Soc.* **38** (2) (1998) 93–100.

25. J. Curran, C. Triggs, J.R. Almirall, J. Buckleton, and K. Walsh, *Sci. Just.* **37** (4) (1997) 241–245.

26. J. Curran, C. Triggs, J.R. Almirall, J. Buckleton, and K. Walsh, *Sci. Just.* **37** (4) (1997) 245.

27. J.R. Almirall, D.C. Duckworth, C.K. Bayne, S.J. Morton, D.H. Smith, R.D. Koons, et al., "Elemental analysis of forensic glasses by ICP-MS". In: K. Higgings (Ed.), *Investigation and Forensic Science Technologies, Proceedings of SPIE*, Boston, MA 3576, 1999, pp. 18–23.

28. S. Montero, A. Hobbs, T. French, and J.R. Almirall, *J. Forensic Sci.* **48** (5) (2003) 1101–1107.

29. D.C. Duckworth, S.J. Morton, C.K. Bayne, S. Montero, R.D. Koons, and J.R. Almirall, *J. Anal. Atom. Spectrom.* **17** (7) (2002) 662–668.

30. R.D. Koons, C.A. Peters, and P.S. Rebbert, *J. Anal. Atom. Spectrom.* **6** (1991) 451–456.

31. A. Hobbs and J.R. Almirall, *J. Anal. Bioanal. Chem.* **376** (2003) 1265–1271.

32. P. Kuisma-Kursula, *X-ray Spectrosc.* **29** (2000) 111–118.

33. A. Zurhaar and L. Mullings, *J. Anal. Atom. Spectrom.* **5** (1990) 611.

34. T. Parouchais, I.M. Warner, L.T. Palmer, and H. Kobus, *J. Forensic Sci.* (41) (1996) 351.

35. D.C. Duckworth, C.K. Bayne, S. Morton, and J.R. Almirall, *J. Anal. Atom. Spectrom.* **15** (2000) 821–828.

36. U.S. Department of Justice, Federal Bureau of Investigation, Scientific Working Group on Materials Analysis (SWGMAT), *Forensic Sci. Comm.* **1** (2) (1999 July).

37. A.J. Mank and P.R. Masson, *J. Anal. Atom. Spectrom.* **14** (1999) 1143–1153.

38. A. Raith, R. Godfrey, and C. Hutton, *Fresenius' J. Anal. Chem.* **354** (1996) 163–168.

39. E. Schroeder, M. Hamester, and M. Kaiser, *Appl. Surf. Sci.* **127–129** (1998) 292–298.

40. J. Stix, G. Gauthier, and J. Ludden, *Canadian Mineralog.* **33** (1995) 435–444.

41. N.J. Pearce, W.T. Perkins, J.A. Westgate, M.P. Gorton, S.E. Jackson, and C.R. Neal, *Geostand. Newslett.* **21** (1) (1996) 115–144.

Advances in Mass Spectrometry, Volume 16
A.E. Ashcroft, G. Brenton and J.J. Monaghan (Editors)

CHAPTER 11

The Ion Mobility Mass Spectrometry Method and Its Application to Duplex Formation of Oligonucleotides and Aggregation of Proteins

Thomas Wyttenbach, Erin Shammel Baker, Summer L. Bernstein,
Alessandra Ferzoco, Jennifer Gidden, Dengfeng Liu,
and Michael T. Bowers

Department of Chemistry and Biochemistry, University of California at Santa Barbara,
Santa Barbara, CA 93106, USA

Abstract

Ion mobility spectrometry coupled to mass spectrometry provides a powerful tool to explore the three-dimensional shape of polyatomic ions. The ion structure is obtained by measuring collision cross sections in a high-pressure drift tube filled with helium and comparing it with model structures obtained by various theoretical methods such as molecular modeling. The temperature of the drift tube is generally adjustable (typically from 80 to 800 K) providing a unique opportunity to address topics such as the thermal motion of floppy molecules, the unfolding process of folded structures, and the kinetics of structural interconversion and dissociation processes. The theoretical background and the concepts of these ion mobility based experiments are briefly reviewed in this article. Furthermore, results of some examples are given including oligonucleotide duplexes and proteins involved in protein conformational disorders such as the amyloid-β peptide (Alzheimer's disease) and α-synuclein (Parkinson's disease).

11.1 Introduction

The area of ion mobility research to obtain structural information started to emerge in the early 1990s and was pioneered by the groups of Bowers [1,2] and Jarrold [3,4]. Most of the early works has been reviewed by Clemmer and Jarrold [5], the more recent work by Wyttenbach and Bowers [6]. One major point of interest in the field is clearly the structure of biological molecules in the presence and absence of solvation molecules. Molecules as small as glycine and as large as α-synuclein, a protein composed of 140 amino acids, were studied to address issues such as zwitterion formation in peptides, proteins, and oligonucleotides; protein folding; peptide helix formation; the dynamics of interconversion between conformations; and the effects of hydration [6]. In this chapter we will briefly outline the concepts of the ion mobility method to obtain structural information of ions and then demonstrate the capability of the method with some examples.

11.2 Concepts

Ions drifting through a buffer gas under the influence of a weak uniform electric field E quickly reach an equilibrium between forward acceleration due to the electric field and the retarding effect due to collisions with the buffer gas, resulting in a constant drift velocity v_D. The drift field is weak when the steady flow of ions along the electric field is much slower than random motion leading to diffusion. The low field mobility K is the proportionality constant between v_D and E [7]:

$$v_D = KE. \tag{11.1}$$

Hence, measuring the drift time t_D for a given drift length and given E yields an experimental value for K. For a given pressure (or particle density N) and temperature T of the buffer gas the ion mobility K for a given collision cross section σ is given in Eq. (11.2)

$$K = \frac{3e}{16N} \left(\frac{2\pi}{\mu k_B T} \right)^{1/2} \frac{1}{\sigma}, \tag{11.2}$$

where e is the charge of the ion, μ the reduced mass of ion and buffer gas, and k_B the Boltzmann constant [7]. Thus, ions with compact structures have a small cross section and a large ion mobility, whereas the opposite is true for large extended structures. Hence, in ion mobility spectrometry (IMS) ions are separated by size (collision cross section) in contrast to mass spectrometry, where they are dispersed by mass (mass-to-charge ratio).

Generating model structures of the ionic molecular species under investigation is essential for interpreting ion mobility data. Computational methods are generally used to obtain candidate model geometries, although crystal structures obtained by X-ray crystallography and solution structures obtained by nuclear magnetic resonance techniques are sometimes useful as well. Molecular mechanics/molecular dynamics (MM/MD) is often the method of choice for calculations because the size of the systems does not allow use of higher level theories. AMBER [8] is used in all the examples given in this work. Various methods [9–12] are then used to determine an orientation averaged cross section of a given model structure for comparison with experiment.

11.3 Instrumentation

The basic setup allowing for most flexibility to carry out ion mobility experiments is schematically depicted in Fig. 11.1. Ions are generated in an ion source, mass selected in the first mass filter MS1 and injected into the drift tube containing several torr of helium. If the ion source produces a continuous ion beam, the beam has to be gated in front of the drift cell. The gate then triggers the clock for measuring the drift time of the ion pulse. Sophisticated equipment like a quadrupole ion trap [13–15] or an ion funnel [16] can be used as a gate and ion storage device to convert a continuous ion beam into a pulsed one without significant ion loss.

Ions exiting the drift tube are mass analyzed in mass spectrometer MS2, an important feature if reactions are occurring in the drift cell. Ions are generally detected after MS2 by ion counting techniques. The mass spectrometers MS1 and MS2 are typically quadrupole mass filters, and either one or the other can be run in RF-only mode for better signal but without mass selection, if desired.

Ion sources used in biological applications are typically electrospray ionization (ESI) and matrix-assisted laser desorption/ionization (MALDI). The machines used in the examples given below are a MALDI-TOF (time of flight) instrument equipped with a drift cell and quadrupole mass filter at the back end [17,18] and an ESI machine with ion funnel, cell, and quadrupole mass filter [16].

11.4 Oligonucleotides

The most common types of secondary structure observed in polynucleotides are double-stranded helices including Watson–Crick base pairing. Two of these types

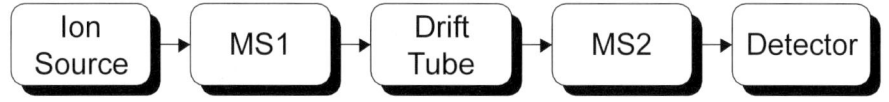

Figure 11.1 Schematic outline of the basic ion mobility–mass spectrometry setup using mass selection before (MS1) and after (MS2) the drift tube.

of helices are the A helix, most commonly found in ribonucleic acid (RNA), but also in deoxyribonucleic acid (DNA) samples crystallized under low humidity conditions, and the B helix, typically observed in DNA. Both the A and the B helix are right-handed, but the A-form is fat and slow rising, whereas the B-form is skinny and rises faster. A third type of double helix is the skinny left-handed Z helix, which is found in samples containing a substantial fraction of 5-methylcytosine.

An interesting question is which factors such as solvent, metal ions, and size of the system contribute to the stability of each of these types of helices. Gas phase studies, in general, are ideal to address these types of questions [19–23], but the ion mobility method has the added potential of being able to differentiate between the different types of helices based on their various shapes.

Here we are demonstrating the approach on the example of a series of cytosine (C) and guanine (G) containing DNA samples. We have carried out experiments for all of the molecules $d(CG)_n$, where $n = 1–4$, but because of space limitations we are focusing on the results of dCG and dCGCGCGCG. X-ray studies indicate that DNA segments with an alternating pyrimidine–purine sequence (as in dCG) have a strong preference for the Z helix.

Figure 11.2 shows ion mobility data of mass selected ions obtained on an instrument equipped with a MALDI source. The top panel shows a spectrum of deprotonated dCG [24]. Two features are evident, corresponding to cross sections of 142 and 151 Å^2, respectively. Molecular mechanics calculations indicate that the feature at shorter drift times corresponds to a family of structures where the rings of the bases are stacked (theory average 141 Å^2) and the feature at longer times corresponds to structures where the two bases are H-bonded side-by-side (theoretically 152 Å^2).

Figure 11.2b shows data of the deprotonated dCG duplex. Again, two features labeled 1 and 2 are evident. Molecular modeling indicates that feature 1 corresponds to structures where C of one strand is H-bonded to G of the other strand whereas the other two bases are stacked. Structures of feature 2 exhibit C–G H-bonds for both base pairs. However, the arrangement of the bases is not the typical planar arrangement found in Watson–Crick base pairing for either feature 1 or 2.

Figure 11.2c shows one feature corresponding to the sodiated dCG duplex where according to theory all four bases interact with sodium, but not with each other. Figure 11.2d contains data for copper Cu^+ bound to the deprotonated dCG duplex. The minor feature, denoted "M^+" at 230 Å^2, agrees with structures found for the sodiated complex (theory 231 Å^2). The major feature, denoted "WC" at 252 Å^2, is clearly a much more open structure and the only match with theory is found for structures with Watson–Crick base pairing (theory 249 Å^2).

The MALDI mass spectrum (not shown) obtained for the octanucleotide dCGCGCGCG shows two peaks of interest corresponding to the deprotonated sin-

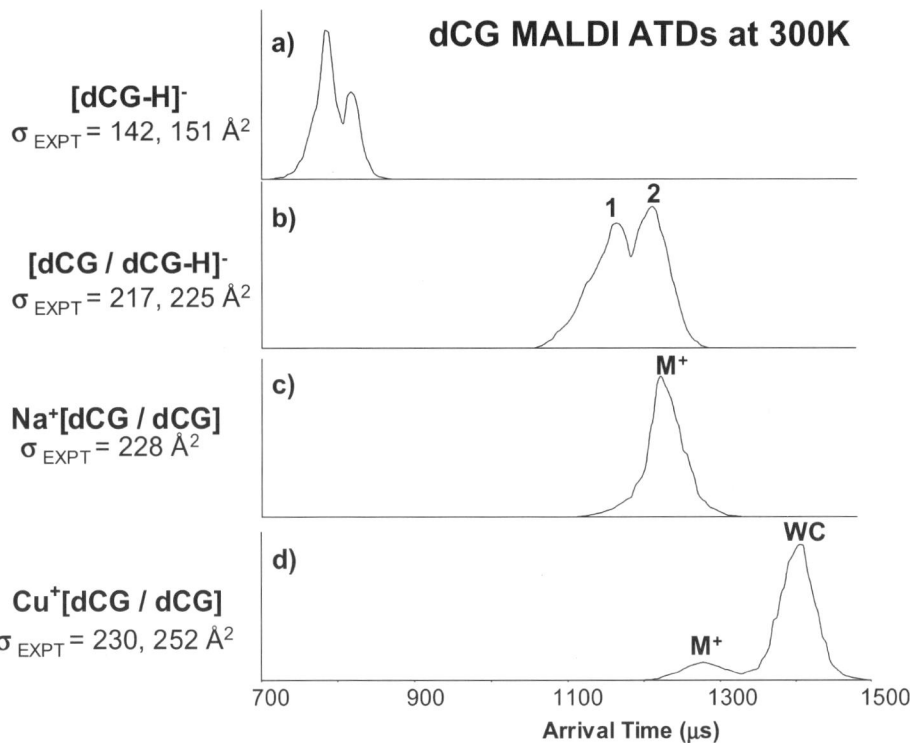

Figure 11.2 MALDI arrival time distributions (ATDs) of dCG for the (a) deprotonated single strand, (b) deprotonated duplex, (c) sodiated duplex, and (d) cuprous duplex obtained at a drift cell temperature of 300 K. Multiple peaks in the ATDs indicate more than one conformation with different collision cross sections. 1 and 2 represent the number of hydrogen bonded base pairs, M^+ stands for the metal duplex form, and WC corresponds to Watson–Crick base pairing.

gle strand and the deprotonated duplex. The only feature in the ion mobility spectrum measured for the single strand species has a cross section (332 Å^2) that agrees with a theoretical structure of 338 Å^2 where G at the 3′ end of the oligomer undergoes a Watson–Crick type of interaction with C at the 5′ end. A second Watson–Crick pairing is present between two interacting bases along the middle of the chain. The only feature (535 Å^2) in the ion mobility spectrum of the duplex has to be a fairly compact globular structure (theoretically 541 Å^2) and any type of double stranded helix can be excluded (theoretically 652–673 Å^2).

The major species present in the ESI mass spectrum of dCGCGCGCG (Fig. 11.3) are the doubly and triply deprotonated single strand oligonucleotides, $(M - 2H)^{2-}$ and $(M - 3H)^{3-}$. Also present with less intensity are the charge state −4 of the duplex and various sodiated species within each charge state

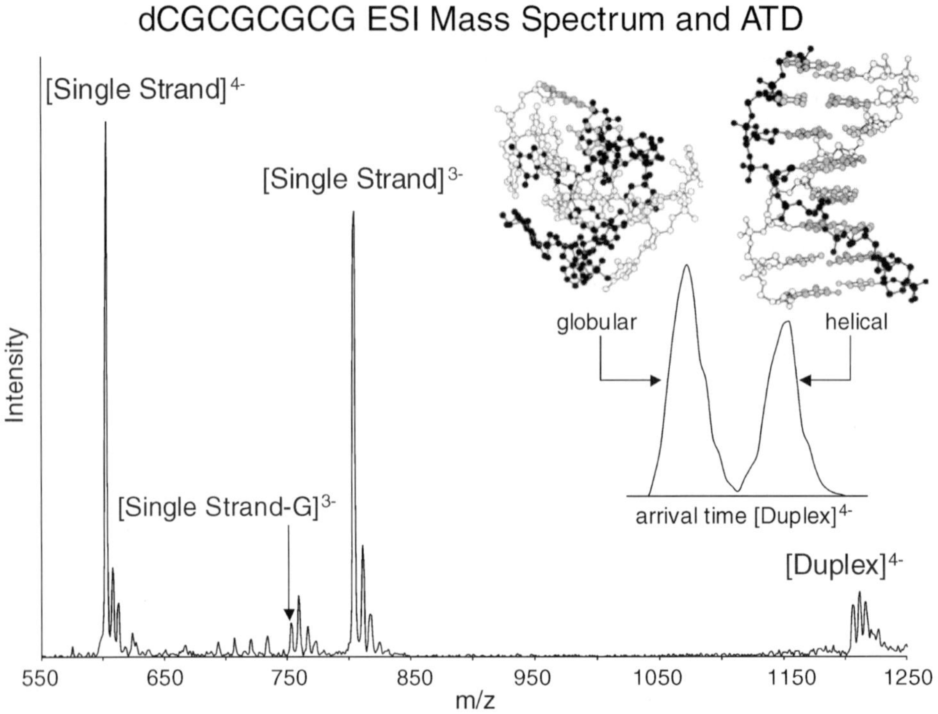

dCGCGCGCG ESI Mass Spectrum and ATD

Figure 11.3 The ESI mass spectrum and ATD are illustrated for the octanucleotide along with the lowest energy theoretical structures calculated for the globular and helical duplex forms (the Z-form helix is shown). The backbone of strand 1 is colored white, while strand 2 is black. Base pairs involved in Watson–Crick base pairing are gray. Hydrogens have been omitted for clarity.

$(nM + iNa - jH)^{z-}$, $j = z + i$. The mass-to-charge ratio of the −4 duplex is the same as that of the −2 single strand, but the spacing between adjacent sodiated species is 5.5 units yielding a charge of −4.

The ion mobility data of the −4 duplex shows two clearly resolved features (536 and 664 Å^2). The feature at shorter drift time agrees with globular structures obtained by molecular mechanics (541 Å^2), and the feature at longer times with helical structures. The calculated cross section for a dCGCGCGCG duplex A-form helix is 673 Å^2, that for the B-form 652 Å^2, and that for the Z-form 668 Å^2. Best agreement with experiment is obtained for the Z helix, but all types of helices are within the combined error bars of experiment and theory.

In summary, all of the single strand ions d(CG)$_n$, $n = 1$–4, fold into loops with Watson–Crick pairing between two bases, one at each oligomer end. All of the oligonucleotide duplexes examined by MALDI are globular. However, Cu$^+$

stabilizes Watson–Crick pairing in the dCG duplex. And finally, oligonucleotide duplexes examined in ESI experiments are globular, but in the case of the octanucleotide duplex both the globular and the helical conformer are observed.

11.5 Proteins

In this section we are going to show results of two proteins that are both related to protein conformational disorders, a topic getting a lot of attention recently [25–27]. In this group of diseases, peptides, or proteins convert from their normally soluble forms to insoluble fibrils or plaques, which accumulate in a variety of organs including the brain. This type of disease includes Alzheimer's and Parkinson's diseases, Creutzfeldt–Jakob disease, and type II diabetes. These diseases can be sporadic, inherited, or infectious, and often surface only late in life. Each disease is associated with a particular protein. Diseased patients are thought to have an excess of a misfolded form of the protein leading to aggregation and the aggregates of the protein are thought to be the direct or indirect origin of the pathological conditions associated with the disease [25–27]. The two proteins included in this study are those associated with Alzheimer's and Parkinson's diseases, the amyloid-β protein and α-synuclein, respectively.

Alzheimer's disease (AD) is the leading cause of dementia in elderly people, affecting nearly 4 million people worldwide. Clinically, AD is characterized by memory loss, cognitive decline, and behavioral changes. Abundant synapse loss, neuroinflammation, and neuronal death are found in AD brains [28]. Genetic and biochemical evidence indicate the amyloid-β protein (Aβ) as the main cause of AD. Aβ is produced in two major alloforms, 40 and 42 amino acids long, designated Aβ40 and Aβ42, respectively. Despite the small structural difference between the two alloforms, they display distinct biological and biophysical behavior. Aβ42 fibrillizes faster than Aβ40 [29] and is more neurotoxic [30]. In addition, Aβ40 and Aβ42 assemble into distinct oligomer populations [31]. This distinct oligomerization may be related to the particularly strong link of Aβ42 with AD. We seek to understand the oligomerization process and the structural transitions Aβ undergoes during its assembly. In this study we report preliminary results using ion mobility mass spectrometry and associated computer modeling to probe the structure of wild type and mutant Aβ peptides.

The ion mobility technique is a useful technique to study the conformation and oligomerization of Aβ because it can separate conformers with the same mass-to-charge ratio. Experimental parameters such as injection energy and drift cell temperature can be varied to study the relationships and transitions between monomeric and oligomeric species and in favorable cases this yields activation energies for various processes.

Negative ion ESI mass spectra of Aβ42 samples show two major peaks corresponding to the three- and four-fold deprotonated peptide. The ion arrival time distributions (ATDs) of both charge states exhibit multiple features with three clearly resolved features in charge state -3. Model structures created by molecular modeling indicate that the feature at longest drift time, which is most intense at high injection energies, agrees with fairly compact globular monomer structures. Typical model structures obtained for the Aβ42 dimer $(2M - 6H)^{6-}$ have a cross section that is between the other two ATD features at shorter times suggesting that those two components are two distinctly different dimer conformers. The most extended dimer structures calculated agree with the middle ATD feature, whereas the most compact theoretical structures agree with the ATD feature at shortest drift times. Hence, the ion mobility results in combination with molecular mechanics data suggest that charge state -3 is actually composed not only of -3 monomers but also of two types of -6 dimers, all species having an identical mass-to-charge ratio.

Substitution of Phe[19] by Pro has been shown to affect the oligomerization process for Aβ42 [32] and in our studies causes a significant difference in mass spectra and ATDs compared with its wild type (wt) counterpart. Negative ion mass spectra for the Pro[19] mutant are shown in Fig. 11.4 as a function of cell temperature. The spectra contain the triply and quadruply deprotonated monomers, as well as a -5 dimer, that is very weak when spraying fresh solutions, but becomes one of the most intense peaks after 24 h incubation time. This latter feature is not readily observable in the wt peptide. The ATD for the -5 charged dimer is composed of two peaks D_1 and D_2 (Fig. 11.4). The more compact conformer D_1 appearing at shorter time is the dominant peak at low temperatures and the more open conformer D_2 dominates at higher temperatures. It appears that the decrease in the -5 dimer signal observed in the mass spectra in Fig. 11.4 with increasing temperature is primarily due to dissociation of D_1 in the temperature range from 380 to 440 K and due to dissociation of D_2 in the temperature range above 440 K.

α-Synuclein, the protein associated with Parkinson's disease, is composed of 140 amino acid residues. This relatively large protein appears to have the rather unique property of not assuming any defined secondary structure under physiological conditions. Hence, neither X-ray crystallography nor NMR methods were successful in supplying any structural information. Alternative methods such as those employed in Gray's lab [33–35] or those used here might be the only chance of getting structural information about α-synuclein.

ESI mass spectra obtained when spraying an α-synuclein solution at pH 7 show a charge state distribution extending from charge state $+7$ to $+15$ with a maximum around $+11$. At pH 2 the distribution is narrower ($+9$ to $+13$) but interestingly still peaking at $+11$. In the negative ion mode no spectra could be obtained by spraying basic or neutral solutions, but interestingly acidic solutions yielded mass

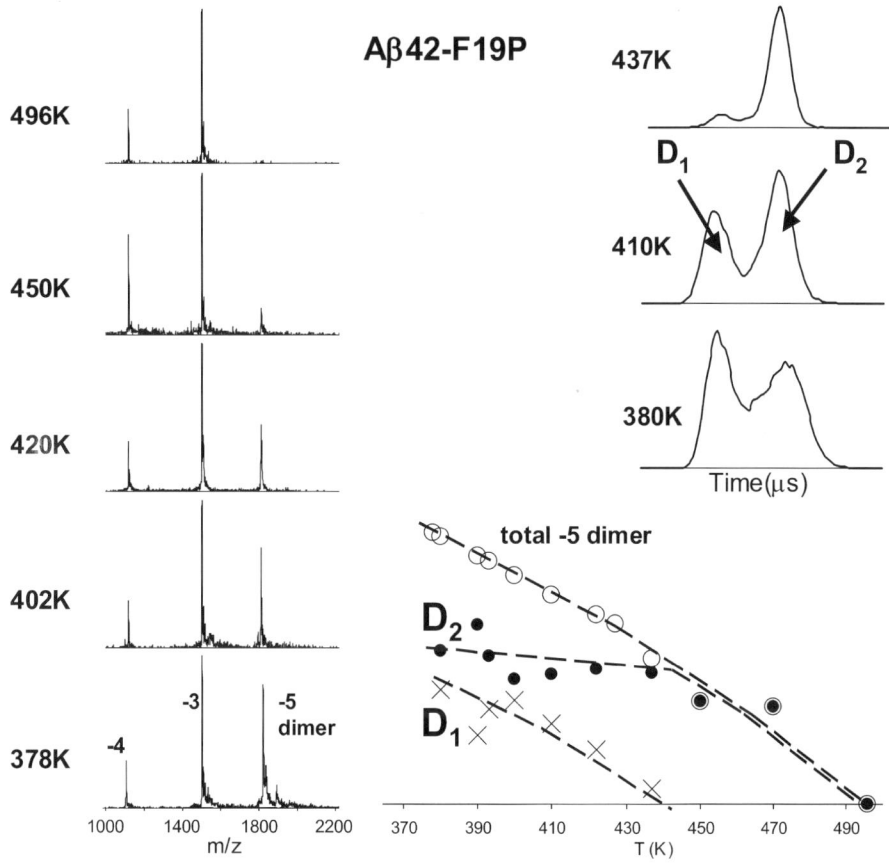

Figure 11.4 Mass spectra (left panel) and ATDs (right panel, top) of the amyloid-β mutant F19P as a function of temperature. The decrease of the two components D_1 and D_2 of the -5 dimer signal with increasing temperature is illustrated in the lower half of the right panel.

spectra with good intensity and a charge state distribution ranging from -6 to -13, peaking at -10. Also present (with relatively small intensity) in both positive and negative ion mode are mass peaks corresponding to odd charge states of protein dimers.

ATDs recorded for any of the mass peaks look generally rather complex, composed of multiple monomer and/or dimer components. For instance, the $+19$ dimer is fairly compact at low cell injection energies (30 V), but more extended when annealed by 80 V collisions. Figure 11.5 shows cross sections deduced from ion mobility data for the different features seen in the ATDs of different charge states for both positive and negative ions. α-Synuclein ions with little charge are very com-

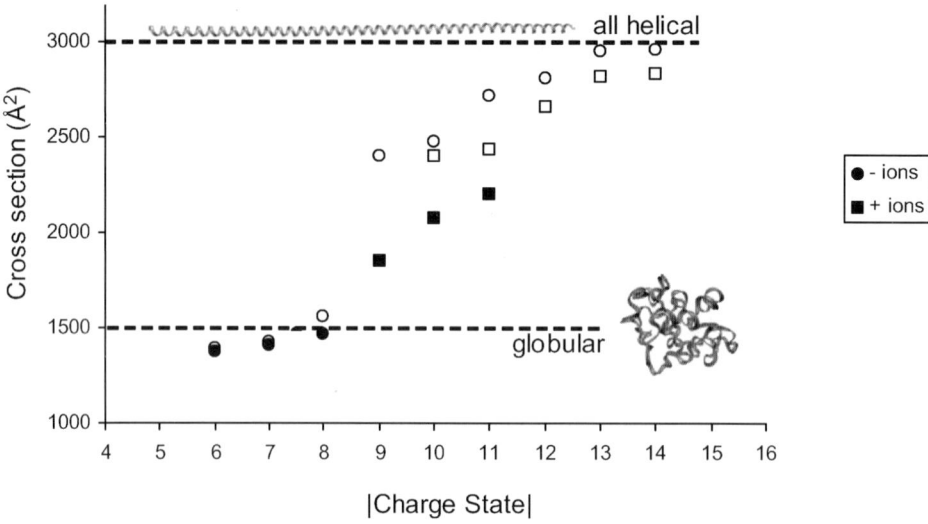

Figure 11.5 Experimental cross sections of various features in the ATDs obtained from an α-synuclein solution as a function of charge state. Squares indicate positive ions, circles negative ions, closed symbols are a family of compact structures, and open symbols represent a family of extended structures. Theoretical cross sections for a globular and a fully α-helical structure are given as dashed lines for comparison.

pact in agreement with a calculated globular structure. As ions are more and more charged, Coulomb repulsion drives them to open up and finally they reach a cross section at charge state 14 that is comparable with the cross section of an extended, fully α-helical conformation. The range of cross sections covered is remarkable: the cross section of charge state 14 is more than twice that of charge state 6.

In conclusion, we have observed the first step to aggregation, the dimers, for both the Alzheimer's peptide amyloid-β and the Parkinson's protein α-synuclein. Mass spectrometretric methods allow determination of solution dimerization rates, since the extent of dimers observed depends in certain cases on the incubation time. Temperature dependent studies will yield energetic information of the dimerization process. First results of our molecular modeling efforts on amyloid-β and α-synuclein are available and capture some of the main features observed experimentally. Further improvements are necessary and underway. Ion mobility data reveal multiple families of conformations for both proteins and for the dimers of the two proteins. There is good evidence (e.g., pH dependence) that proteins can be trapped in a solution phase conformation as they are lifted into the gas phase. Gas phase annealing experiments can be used to convert the preferred solution conformation into the preferred gas phase conformation, which is readily detected by the ion mobility method. And finally, we would like to pose the interesting question whether ion mobility data correlate with solution aggregation properties as a function of mutations. Measuring mutants of amyloid-β and α-synuclein will be an active field of research in our group.

Acknowledgments

This research was supported by the National Science Foundation under Grant CHE-0140215. We gratefully acknowledge our collaborators Gal Bitan and David Teplow (amyloid-β) and Jennifer Lee and Harry Gray (α-synuclein).

References

1.   G. Von Helden, M.T. Hsu, P.R. Kemper, and M.T. Bowers, *J. Chem. Phys.* **95** (1991) 3835.

2.   G. Von Helden, P.R. Kemper, N.G. Gotts, and M.T. Bowers, *Science* **259** (1993) 1300.

3.   M.F. Jarrold and V.A. Constant, *Phys. Rev. Lett.* **67** (1991) 2994.

4.   M.F. Jarrold and J.E. Bower, *J. Chem. Phys.* **96** (1992) 9180.

5.   D.E. Clemmer and M.F. Jarrold, *J. Mass. Spectrom.* **32** (1997) 577.

6.   T. Wyttenbach and M.T. Bowers, *Top. Curr. Chem.* **225** (2003) 207.

7.   E.A. Mason and E.W. McDaniel, *Transport Properties of Ions in Gases.* Wiley, New York, 1988.

8.   D.A. Case, D.A. Pearlman, J.W. Caldwell, T.E. Cheatham III, W.S. Ross, C.L. Simmerling, T.A. Darden, K.M. Merz, R.V. Stanton, A.L. Cheng, J.J. Vincent, M. Crowley, V. Tsui, R.J. Radmer, Y. Duan, J. Pitera, I. Massova, G.L. Seibel, U.C. Singh, P.K. Weiner, and P.A. Kollman, *AMBER 7*, University of California, San Francisco, 1999.

9.   T. Wyttenbach, G. Von Helden, J.J. Batka, D. Carlat, and M.T. Bowers, *J. Amer. Soc. Mass Spectrom.* **8** (1997) 275.

10.   T. Wyttenbach, M. Witt, and M.T. Bowers, *J. Amer. Chem. Soc.* **122** (2000) 3458.

11.   M.F. Mesleh, J.M. Hunter, A.A. Shvartsburg, G.C. Schatz, and M.F. Jarrold, *J. Phys. Chem.* **100** (1996) 16082.

12.   A.A. Shvartsburg and M.F. Jarrold, *Chem. Phys. Lett.* **261** (1996) 86.

13.   C.S. Hoaglund-Hyzer and D.E. Clemmer, *Anal. Chem.* **73** (2001) 177.

14.   C.S. Hoaglund, S.J. Valentine, and D.E. Clemmer, *Anal. Chem.* **69** (1997) 4156.

15.   C.S. Creaser, J.R. Griffiths, and B.M. Stockton, *Eur. J. Mass Spectrom.* **6** (2000) 213.

16.   T. Wyttenbach, P.R. Kemper, and M.T. Bowers, *Int. J. Mass Spectrom.* **212** (2001) 13.

17. J. Gidden, P.R. Kemper, E. Shammel, D.P. Fee, S. Anderson, and M.T. Bowers, *Int. J. Mass Spectrom.* **222** (2003) 63.

18. E. Shammel, J. Gidden, D.P. Fee, P.R. Kemper, S. Anderson, and M.T. Bowers, *Int. J. Mass Spectrom.* **227** (2003) 205.

19. P.D. Schnier, J.S. Klassen, E.F. Strittmatter, and E.R. Williams, *J. Amer. Chem. Soc.* **120** (1998) 9605.

20. V. Gavelica, F. Rosu, C. Houssiaer, and E. De Pauw, *Rapid Commun. Mass Spectrom.* **14** (2000) 464.

21. V. Gavelica and E. De Pauw, *J. Mass Spectrom.* **36** (2001) 397.

22. S. Sakamoto and K. Yamaguchi, *Tetrahedron Lett.* **44** (2003) 3341.

23. R. Sudha and R. Zenobi, *Helv. Chimica Acta* **85** (2002) 3136.

24. J. Gidden and M.T. Bowers, *Eur. Phys. J. D* **20** (2002) 409.

25. C.M. Dobson, *Trends Biochem. Sci.* **24** (1999) 329.

26. J.-C. Rochet and P.T.J. Lansbury, *Curr. Opin. Struct. Biol.* **10** (2000) 60.

27. R.W. Carrell and B. Gooptu, *Curr. Opin. Struct. Biol.* **8** (1998) 799.

28. N.M. Mesulam, *Neuron* **24** (1999) 521.

29. C.J. Barrow and M.G. Zagorski, *Science* **253** (1991) 179.

30. J.G. Hu, *Brain Res.* **785** (1998) 195;
    D. Paris, T. Town, T. Mori, T.A. Parker, J. Humphrey, and M. Mullan, *Neurobiol. Aging* **21** (2000) 183;
    L.Q. Tong, P.L. Thornton, R. Balazs, and C.W. Cotman, *J. Biol. Chem.* **276** (2001) 17301;
    Y.S. Gong, *Soc. Neurosci. Abstr.* (2001);
    H.W. Wang, *Brain Res.* **924** (2002) 133;
    M.P. Lambert, *J. Neurochem.* **79** (2001) 595.

31. G. Bitan, M.D. Kirkitadze, A. Lomakin, S.S. Vollers, G.B. Benedek, and D.B. Teplow, *Proc. Natl. Acad. Sci. USA* **100** (2003) 330.

32. G. Bitan and D.B. Teplow, Private communication, 2003.

33. H.B. Gray and J.R. Winkler, *Ann. Rev. Biochem.* **65** (1996) 537.

34. J.C. Lee, K.C. Engman, F.A. Tezcan, H.B. Gray, and J.R. Winkler, *Proc. Natl. Acad. Sci. USA* **99** (2002) 14778.

35. J.C. Lee, I.J. Chang, H.B. Gray, and J.R. Winkler, *J. Mol. Biology* **320** (2002) 159.

Advances in Mass Spectrometry, Volume 16
A.E. Ashcroft, G. Brenton and J.J. Monaghan (Editors)

CHAPTER 12

A Tiered Approach to Metabolite Characterization Studies in Support of Drug Discovery

K.A. Cox, D. Grotz, D. Rindgen, and D.J. Weston

Department of Drug Metabolism, Schering Plough Research Institute, 2015 Galloping Hill Road,
Kenilworth, NJ 07033-0539, USA

Abstract

A tiered system is proposed for addressing metabolite characterization studies in support of drug discovery programs. Given the highly specific nature of metabolite characterization for new chemical entities (NCEs), the tiered system is designed to answer the most critical questions first, returning essential information to the discovery team in a rapid fashion, while saving more time-consuming questions for later in the discovery process. The first look at the metabolic profile of an NCE involves dosing the compound orally into a rat and collecting bile and urine. Samples are run under standard LC conditions and subjected to constant neutral loss (CNL) scans to determine, by the detection of conjugated metabolites, whether there is evidence for potentially toxic metabolites. CNL scans also provide the most comprehensive look at metabolites without requiring any compound specific information. MS/MS scans are then conducted to confirm the identities of any possible conjugated metabolites and also to detect any common phase I metabolic transitions such as hydroxylation or demethylation. If the NCE contains a radiolabel or an isotopically characteristic halogen, additional information can be provided. In most cases, this rapid evaluation of the NCE provides critical data to the discovery team about the metabolically labile sites on the molecule and about any potentially toxic metabolites. Once an NCE has progressed further in the discovery process, a more in-depth evaluation of metabolites is conducted. LC conditions are optimized to separate metabolites. Precursor ion scans provide a more comprehensive evaluation of the full metabolic fate of an NCE. MS^n and accurate mass experiments can help determine the empirical formula and propose a metabolite structure rather than simply localizing a metabolic alteration to a section of the molecule.

12.1 Introduction

Metabolism studies that support drug discovery programs are facing greater challenges than ever before [1]. The cost and time involved in developing a successful drug continue to rise, and the failure of a potential drug candidate late in the development process results in a tremendous loss of resources [2]. Drug candidates spend comparatively little time in the discovery phase relative to the period of time required for development and clinical studies. Therefore, the challenge posed to discovery scientists is to find out as much as possible about NCEs within the discovery time frame in an effort to minimize costly attrition later in the development process [3]. High-throughput techniques have been successfully implemented to screen large libraries of compounds to look for activity against a particular therapeutic receptor [4]. In vitro assays have been developed to predict in vivo metabolic parameters such as absorption [5], enzyme inhibition [6], and induction [7] in an effort to screen large numbers of compounds in a rapid and resource-efficient manner. High-throughput in vivo pharmacokinetic techniques have also been developed to obtain critical pharmacokinetic parameters using cassette dosing [8] or sample pooling prior to analysis [9] coupled with multiple analyte detection. However, it is particularly difficult to standardize metabolite characterization studies to meet the challenges of a high-throughput environment. Every compound exhibits a unique metabolic profile dependent on its structure, the system and species selected to metabolize the compound and what matrix is selected for evaluation of metabolites. In vitro systems such as microsomes, hepatocytes, and liver slices provide a higher-throughput matrix; however, it is difficult to generate a complete picture of the metabolism that will occur in vivo. Since every system is unique, it is difficult to design an analytical methodology that is general enough to adequately characterize all samples, yet is specific enough to capture all the potential metabolic pathways.

We have previously described a multidimensional methodology, combining the unique capabilities of triple quadrupole, ion trap and time-of-flight (TOF) mass spectrometers, as well as software programs, to fully evaluate the metabolic profile of NCEs [10]. While this methodology is comprehensive, it can still be time consuming and may not be necessary for every NCE submitted for metabolite characterization in support of drug discovery. Often, metabolite characterization is needed to provide a quick look at NCEs generated early in discovery programs.

These NCEs are not likely to be the final drug candidate, but rather are early structural analogs, designed to understand the overall behavior of a particular structural series. Early feedback about metabolically labile sites or potentially toxic metabolites is crucial to the discovery team in order to direct future synthetic pathways. Through the use of characteristic MS/MS scanning techniques, combined with the advantages offered by metabolite ID software programs, we have developed a tiered system designed to answer the most critical questions first, and provide information about the general metabolic profile for an NCE rapidly.

This 'first look' at the metabolic profile, designated as Tier I, is not comprehensive but, in most cases, provides valuable feedback to the discovery team in a timely manner. Since the goal is to generate orally administered drugs, the metabolites formed after oral dosing are the most relevant. Bile and urine are collected from a rat dosed orally with the NCE. Once the samples are generated, they are injected directly onto an HPLC-ESI triple quadrupole system. Since this is the first evaluation of metabolites for the NCE, the routes of metabolism are not yet known, so no sample cleanup is employed to avoid the risk of losing metabolites in the cleanup process. Standard LC conditions are utilized. Electrospray ionization is utilized to minimize unintentional fragmentation of the protonated molecules upon ionization.

Samples are subjected to constant neutral loss scans that are characteristic for the presence of glucuronide (176 Da), glutathione (129 Da), and sulphate (80 Da) conjugated metabolites. Conjugates usually possess less intrinsic biological or chemical activity than the corresponding unconjugated xenobiotic, and this is typically a detoxification mechanism for elimination of drug related products from the body [11]. Analytically, glucuronide, glutathione, and sulphate conjugates can be detected by characteristic constant neutral losses where the mass that is lost is dependent on the conjugate, not the drug (Fig. 12.1). Therefore, conjugated metabolites can be detected without any prior knowledge of the parent drug or previous metabolic pathways. In addition, since these scans are only specific for the conjugate, this approach also can provide indirect evidence of Phase I (P450) metabolism. More than 95% of the drugs in the market are metabolized by P450s, uridine diphosphate glucuronosyltranferases (UDPGTs) and sulfotransferases, and these metabolites are typically excreted in the bile. Thus evaluation of conjugated metabolites in bile and urine often can provide a fairly comprehensive view of the metabolic fate of an NCE.

Glucuronidation is quantitatively the most important conjugation reaction of xenobiotics mediated by UDPGT [12]. It is a low-affinity and high-capacity reaction. Although, glucuronidation is typically a detoxification pathway, if the glucuronide is conjugated through an acyl moiety to the corresponding carboxylic acid aglycone, the resulting acyl-glucuronide can undergo acyl migration to form reactive intermediates, capable of covalently binding to proteins. This intermediate can interfere with the normal protein function or introduce an immunogenic

Glururonide Conjugate (CNL 176)

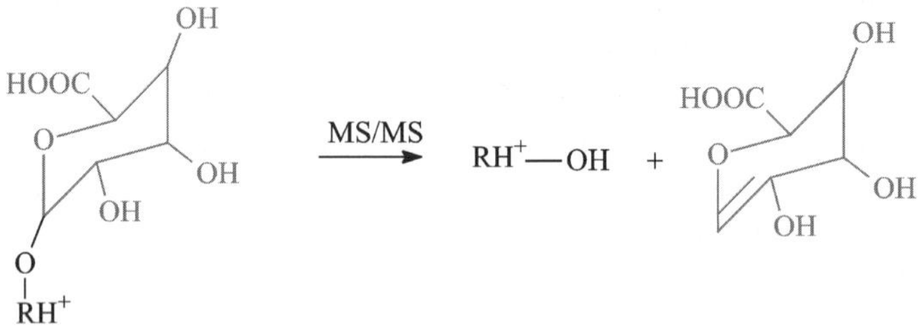

Sulphate Conjugate (CNL 80)

Glutathione Conjugate (CNL 129)

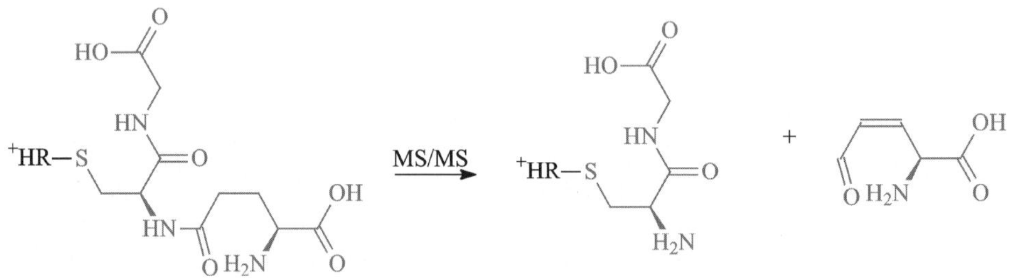

Figure 12.1 Characteristic constant neutral loss transformations for detecting glucuronide, sulphate, and glutathione conjugates by mass spectrometry.

effect. Again, the characteristic CNL scan for loss of 176 Da will detect and help characterize these potentially toxic metabolites.

Glutathione conjugates can be detected by a CNL 129 scan. Glutathione is present in the body and acts as a detoxification mechanism for the elimination of electrophilic entities. Glutathione S-transferase (GST) protects cells from oxidative stress and chemical-induced toxicity by catalyzing the glutathione conjugation reaction with electrophilic, and potentially toxic, xenobiotics [13]. Thus, detection of a glutathione metabolite, although this species is not itself toxic, is indicative of a potentially reactive precursor and can have serious consequences for the fate of the NCE.

After detection of conjugated metabolites in CNL scans, the structures of these putative metabolites are confirmed by performing the respective product ion scans in order to obtain structurally characteristic fragments. In addition, targeted product ion scans are conducted to confirm the presence of any common metabolite transitions such as hydroxylation and demethylation, as well as any relevant metabolic transformations common to the particular program area.

Metabolite ID software has advanced significantly in the past decade and excellent programs are available for the three major MS platforms. The software is capable of subtracting a control sample from the sample of interest. This can be extremely useful when dealing with complex matrices such as bile and urine where the matrix ions can mask metabolites and parent drug completely. In addition, the software programs can be used to evaluate a total ion chromatogram to search for characteristic isotopic patterns such as those generated by compounds containing Cl or Br atoms. Again, this is a powerful tool to extract drug-related ions from a complex matrix. The software can also search the total ion chromatogram for common or expected metabolites such as oxidation, demethylation and carboxylic acid formation and automatically set up and conduct MS/MS scans to characterize these metabolites. All of these software tools are intended to complement, not fully replace, manual data interrogation and, although they require carefully chosen initial parameters in order to be effective, they possess the potential to be extremely useful and time efficient.

Once an NCE has progressed past the initial stages in the discovery process and significant resources are being put forth to progress it as a potential drug candidate, a more comprehensive metabolite characterization is required. This puts the compound into Tier II. In this stage, precursor ion scans are performed to detect as many drug-related metabolites as possible. Precursor ions are chosen not only based on characteristic fragments of the protonated parent NCE, but also based on alterations to these fragments reflecting potential metabolic alterations. For example, the addition of 16 Da to a characteristic precursor ion could reflect hydroxylation to that portion of the parent molecule. Constant neutral loss scans can also be utilized to detect metabolites possessing characteristic structural features that are similar to the parent drug. Again, software programs can evaluate the resulting reconstructed ion chromatogram to highlight potential metabolites and to set up and conduct the appropriate MS/MS scans for confirmation.

As more discovery resources are put toward progressing the NCE, it will be dosed in at least one additional nonrodent species for drug metabolism and pharmacokinetic evaluation. The metabolic profile of an NCE is evaluated in additional species in Tier II.

Typically, the NCE will be radiolabeled at this stage (typically with ^{3}H), allowing for a quantitative evaluation of metabolites based on a radiochromatogram. There is a desire to determine relative amounts of metabolites in addition to their identities, so LC conditions are optimized at this stage to separate co-eluting metabolites. The radiotrace will also reveal whether any metabolites containing the radiolabel have not been characterized based on the precursor, targeted product and CNL scans. These metabolites typically involve cleavage or some other major alteration of the parent drug. Software programs can again be utilized here to help identify these metabolites.

Once an NCE is chosen as the lead drug candidate, more specific metabolite characterization is necessary. This stage is designated as Tier III. At this stage, the NCE is subjected to the most comprehensive metabolite characterization possible. MS^n experiments along with accurate mass MS and MS/MS experiments are performed to localize the site of metabolic alteration as much as possible. Often, a metabolite of particular interest is isolated from the biological matrix and subjected to NMR for definitive structural elucidation.

In addition, the metabolic profile of the NCE is determined in plasma at this stage. Identification of circulating metabolites is often critical to explain the pharmacokinetic and/or the pharmacodynamic profile. An NCE may show efficacy that is inconsistent with what is predicted based upon the known concentration of the parent drug. These inconsistencies could be due to the presence of an active metabolite. Knowledge of these metabolites will also dictate how the analysis of samples will be conducted in development and clinical studies. If significant metabolites are present, they must be monitored throughout the development of the drug.

12.2 Results and discussion

12.2.1 Tier I

An NCE was generated early in a discovery program. As part of the initial evaluation of a promising structural series, this compound was submitted for Tier I metabolic evaluation to get an idea of the metabolically labile sites and to determine whether or not any potentially toxic metabolites were generated. This particular NCE contained a chlorine atom and a tritium label. In accordance with the Tier I protocol, the NCE was dosed orally into a rat and bile and urine were collected 0–24 h post dosing. The bile sample was injected directly into an HPLC MS/MS system fitted with a radioflow detector and the resulting radiochromatogram is shown in Fig. 12.2. All the LC/MS/MS systems are coupled with on-line radio flow detectors that provide radiochromatograms to complement the ion chromatograms when radiolabeled compounds are available. Often, NCEs generated early in the

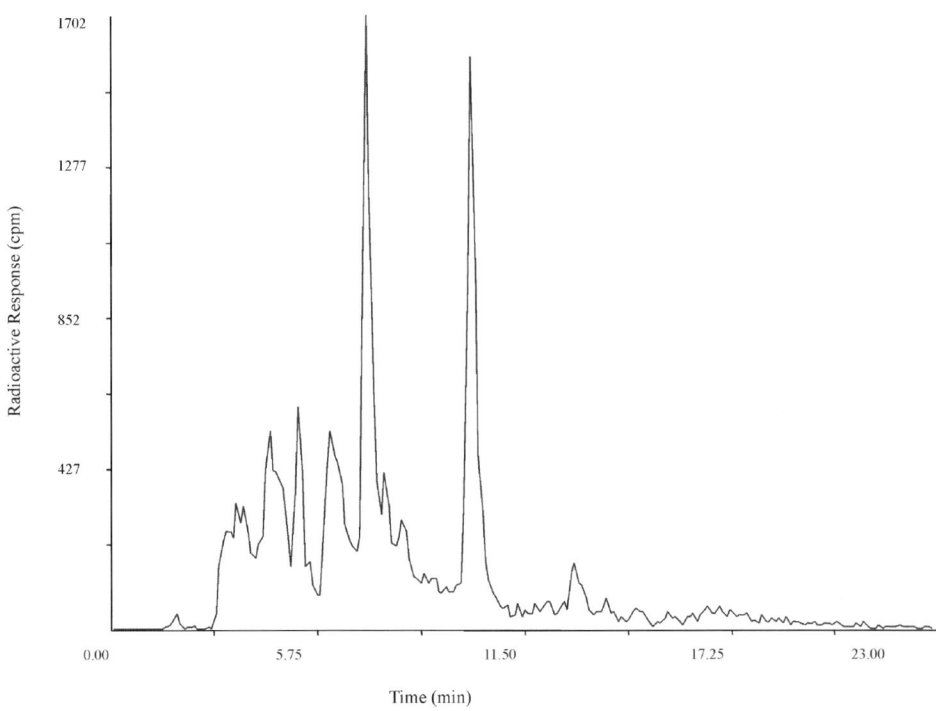

Figure 12.2 Radiochromatogram of an NCE in rat bile (0–24 h post dosing); cpm = counts per minute.

discovery process do not contain a radiolabel; however, this particular case was shown for illustrative purposes to track how accurately the methodology captures the major metabolic pathways.

Bile is an extremely complex matrix and, without the aid of a sample cleanup procedure, the total ion chromatogram typically provides little to no useful information about the routes of metabolism. With no knowledge of metabolic pathways, any sample cleanup procedure runs the risk of losing important metabolites. However, control bile (no drug present) was also injected under the same LC conditions and was subtracted from the original sample in an effort to minimize the contribution from the matrix.

This compound also contains a Cl atom, so the MS software program was employed to obtain a cluster analysis trace. The software searches the total ion chromatogram to find any evidence for the characteristic cluster pattern generated by the $^{35}Cl/^{37}Cl$ isotopic pattern.

The sample was also injected on a triple quadrupole LC/MS system and the characteristic CNL scans were performed to detect glucuronide (CNL 176), glutathione (CNL 129), and sulphate (CNL 80) conjugates. The reconstructed ion

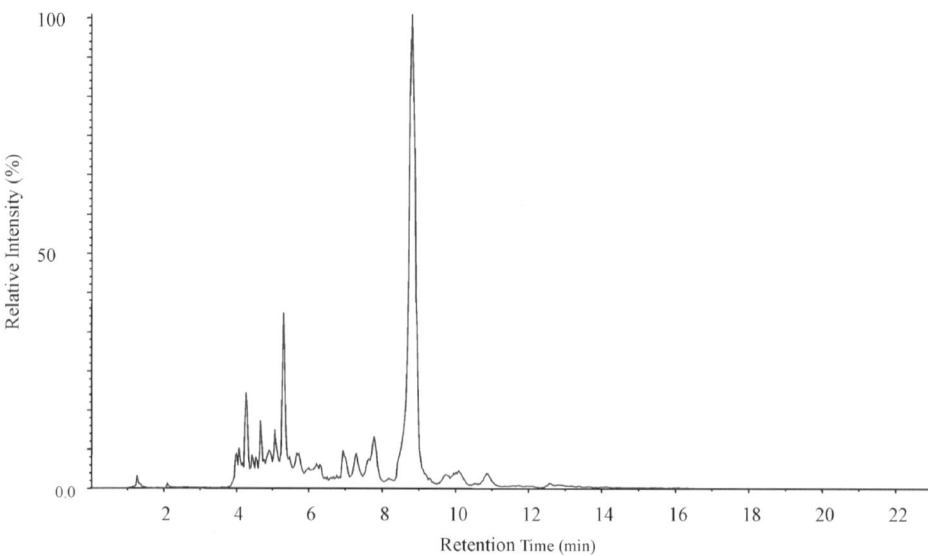

Figure 12.3 Reconstructed ion chromatogram from a CNL 176 experiment of an NCE in rat bile (0–24 h post dosing).

chromatogram for the CNL 176 experiment is shown in Fig. 12.3. Note that the peak observed at approximately 9 min is also a major peak observed in the radiochromatogram (Fig. 12.2), indicating that this major metabolic transition involved glucuronide conjugation.

Cluster analysis performed on the CNL 176 trace highlighted 24 possible ions that had the characteristic isotopic pattern for a Cl atom (Table 12.1). Note that the effectiveness of the cluster analysis interrogation is highly dependent on the initial parameters supplied by the operator. A higher noise rejection threshold would result in fewer hits.

No glutathione or sulphate metabolites were detected. Product ion scans were performed to confirm the identity of the putative glucuronide conjugates and additional MS/MS scans were conducted to search for common metabolic transitions (mono and dihydroxylation, demethylation, carboxylic acid formation, etc.). Two monohydroxylated metabolites and one dihydroxylated metabolite were detected in this manner. These preliminary, general scanning techniques, combined with software functions of background subtraction and cluster analysis, allowed the characterization of the majority of the metabolites formed by this NCE in the rat. No toxic metabolites were detected and the major metabolic pathways were identified—as confirmed by radiochemical detection. The major metabolic pathway for this NCE was hydroxylation, followed by glucuronidation (Fig. 12.4).

Typically, the major metabolic pathways can be identified in this manner. This NCE was one of the first compounds to be synthesized for this particular discovery

Table 12.1: Chlorine ion cluster analysis performed on the CNL 176 trace for NCE in rat bile (0–24 h post dosing)

Retention time (min)	Transformation name	Proposed transformation	Index
3.99	Loss of 67.0	$-243+176$	1
4.26	Loss of 147.0	$-323+176$	2
4.31	Loss of 117.0	$-293+176$	3
4.45	Loss of 131.0	$-307+176$	4
4.68	Loss of 131.0	$-307+176$	5
4.86	Loss of 131.0	$-307+176$	6
5.08	Gain of 30.2	$-146+176$	7
5.32	Gain of 29.2	$-145+176$	8
5.32	Gain of 45.2	$-131+176$	9
5.35	Gain of 11.2	$-165+176$	10
5.71	Loss of 147.4	$-323+176$	11
6.23	Gain of 13.2	$-163+176$	12
6.97	Gain of 208.2	$+32-176$	13
7.63	Gain of 208.2	$+32+176$	14
7.79	Gain of 124.2	$-52+176$	15
8.52	Loss of 145.6	$-322+176$	16
8.56	Gain of 228.2	$+52+176$	17
8.63	Gain of 222.4	$+46+176$	18
8.81	Gain of 194.2	$+18+176$	19
8.81	Gluc + oxidation	$+16+176$	20
8.83	Glucuronidation	$+176$	21
8.83	Gain of 174.2	$-2+176$	22
9.74	Loss of 65.4	$-241+176$	23
10.86	Gain of 204.2	$+28+176$	24

program. Early characterization of the metabolic pathways for this structural series facilitated the synthesis of future compounds.

12.2.2 Tier II

As an NCE becomes a possible lead compound, more information is required about its metabolic profile. LC conditions must be optimized in order to resolve co-eluting metabolites, and a more comprehensive metabolic characterization is conducted. Samples are subjected to precursor ion scanning where any ions showing fragments characteristic of the protonated parent NCE will be detected. Again, software programs can be utilized to monitor these data 'on the fly,' highlighting any potential metabolites and conducting product ion scans to confirm their identity (Fig. 12.5). In Fig. 12.5, the top trace reflects a data-dependent precursor ion

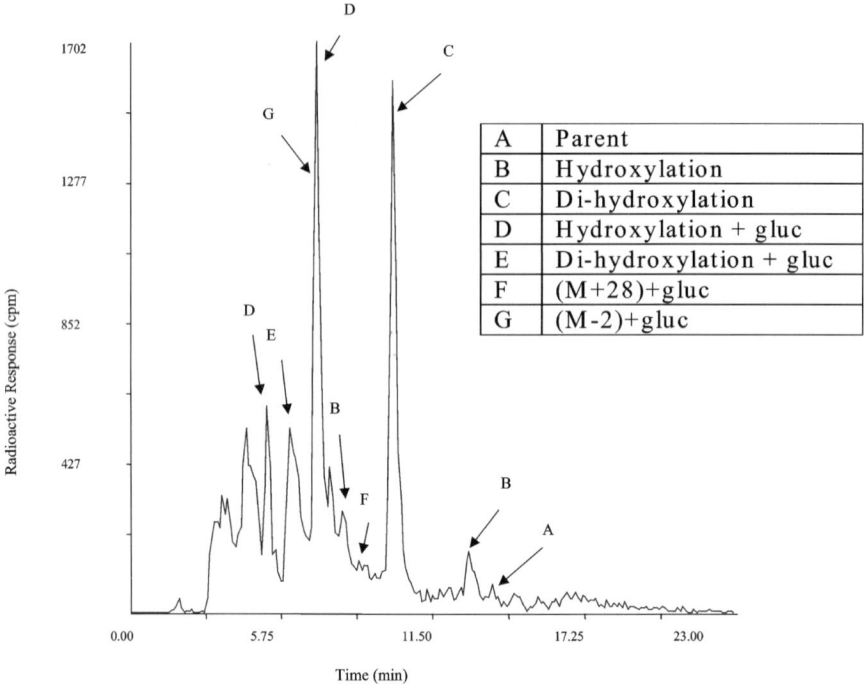

A	Parent
B	Hydroxylation
C	Di-hydroxylation
D	Hydroxylation + gluc
E	Di-hydroxylation + gluc
F	(M+28)+gluc
G	(M-2)+gluc

Figure 12.4 Radiochromatogram of an NCE in rat bile (0–24 h post dosing) with characterized metabolites labeled; cpm = counts/minute.

scan in which precursors of m/z 203 and 232 are monitored. Once the software detects a 'hit' from either of these two product ions, the triple quadrupole mass spectrometer switches from precursor ion scanning to an MS/MS experiment in order to capture the full product ion spectrum. Cluster analysis can also be utilized to evaluate precursor ion scans. Precursor scans provide a selective look at a very broad data set. When a selective MS scan such as a precursor ion scan is coupled with subsequent cluster analysis, the results can often be very informative. However, in order to conduct a cluster ion analysis of a precursor ion data set when monitoring a product ion containing the halogen, the resolution must be opened on the third quadrupole (Q3) in order to capture the desired cluster pattern (typically a 2 u window).

At this stage, the NCE is typically radiolabeled and comparative metabolic profiles can be obtained in rodent and nonrodent animal species. The NCE is incubated in human hepatocytes in order to get a preliminary look at human metabolism and determine how it compares with the metabolism observed in animals. At this point in the development of the NCE, every effort is made to identify metabolites that are present at $\geqslant$ 10% abundance in the radiotrace. In addition, the structures of

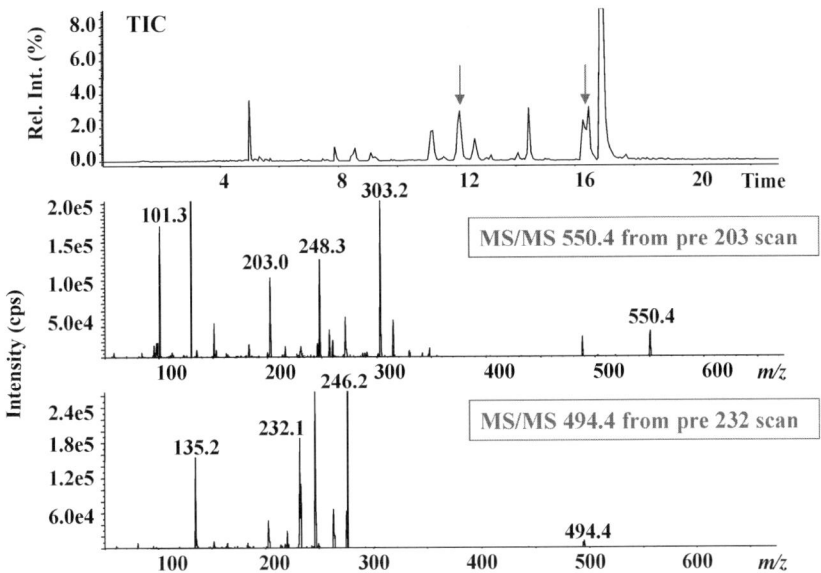

Figure 12.5 Data dependent experiment using two precursor ion scans (*m/z* 203 (11.5 min) and 232 (16 min)) as survey scans.

potentially toxic metabolites are fully evaluated regardless of their relative abundance. This detailed structural investigation is more time consuming and typically requires additional MS techniques such as accurate mass and/or MS^n and takes the compound into Tier III.

12.2.3 Tier III

Tier III is typically reserved for late stage discovery compounds and the studies that require the most amount of time, but also often are the most critical for progressing an NCE through the final stages of the discovery process. In this stage, MS^n experiments are employed whenever it is necessary to narrow the site of metabolic alteration to a specific portion of the molecule. Figure 12.6 shows the results of an MS^3 experiment on a glucuronide metabolite. In the first stage of fragmentation, the glucuronide conjugate is cleaved. The second stage of MS/MS results in characteristic fragment ions providing information about the site of hydroxylation for this particular metabolite.

Accurate mass MS and MS/MS experiments can distinguish nominally isobaric masses aiding in reducing a list of possible empirical formulae for a proposed metabolic structure. Accurate determination of fragment ion masses further narrows the list of possible formulae and can lead to definitive determination of an unknown structure. It is important to note that in most cases, accurate mass data

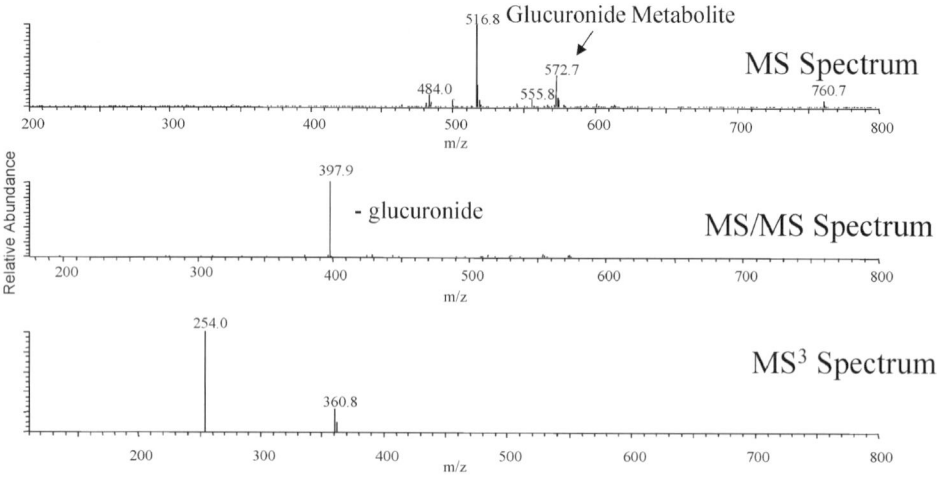

Figure 12.6 MS3 experiment depicting loss of the glucuronide conjugate in the first stage and characteristic product ions in the second stage of MS/MS.

Table 12.2: Proposed modifications that would result in a net +4 modification to the parent molecule along with net mass additions and mass measurement accuracy values

Possible alteration	Net mass addition	Mass measurement accuracy (ppm)
M − C2H6 + O2H2	3.95853	+157
M − C + O	3.994915	+4.5
M + 4H	4.0313	−151

will not *identify* a metabolic structure. Its utility lies in its ability to distinguish between proposed empirical formulae for nominally isobaric ions. For example, a novel metabolite was detected in which the metabolic transition resulted in a net addition of 4 mass units to the molecule. The site of modification was narrowed down to a small portion of the molecule by accurate mass MS/MS experiments. Several modifications were proposed that would result in a net nominal modification of +4 mass units to the parent molecule. However, the exact masses of these proposed transformations are quite different (Table 12.2). The accurate mass LC/MS/MS experiment for this metabolite involved using external and internal lock mass references for both the intact protonated molecule and a characteristic fragment ion. The protonated parent molecule for the (M + 4) metabolite was measured against an external lock mass in full scan MS and MS/MS routines. To localize the metabolic transformation on the parent molecule, a characteristic fragment ion with a known mass was used as an internal lock mass for the MS/MS experiment. In this experiment, confident accurate mass confirmation of candidate formulae should normally fall within a five ppm measurement window. Thus, the data generated was consistent with a net modification of (M − C + O). This proposed structural transformation was confirmed by NMR analysis. While this type of distinction is not possible for traditional ion traps or quadrupole mass spectrometers, time of flight instruments operate routinely with a resolution of 5–10 K FWHM, so the Δm values in this experiment are easily distinguishable.

The final metabolite characterization required for an NCE prior to its progression into a development program is the characterization of metabolites circulating in plasma. If circulating metabolites are also active against the therapeutic receptor, this could lead to a lack of correlation between observed efficacy and measured levels of the parent drug. In addition, if metabolites are circulating at levels that are $\geqslant 25\%$ of the parent concentration, they must be monitored throughout the preclinical and clinical lifetime of the drug. Initial characterization of circulating metabolites is an analytical challenge because of their low concentrations. Circulating metabolites typically represent less than 1% of the total dose and animals need to be dosed at very high levels of radioactivity in order to obtain a radiotrace of circulating metabolites. In the absence of high levels of radioactivity, we can take advantage of highly sensitive LC-MS/MS instruments, combined with software capabilities for detection and characterization of these low level metabolites.

Commercial time-of-flight mass spectrometers are equipped with multi channel plate array detectors for enhanced sensitivity. In addition, many commercial time-of-flight instruments are equipped with the capability to perform precursor and neutral loss scanning experiments. These selective scan modes, combined with the accurate mass capabilities of a time of flight instrument and the sensitivity of an array detector, make these systems ideal for the characterization of circulating metabolites.

12.3 Conclusions

Currently in drug discovery, we are charged with developing drug metabolism and pharmacokinetic profiles on new chemical entities that are as complete as possible. However, we are only given a limited amount of time and resources to accomplish this. This is particularly difficult for studies that are not readily amenable to high throughput, such as metabolite characterization. In this case, efficiency can be maximized by identifying and addressing the critical questions first. We have developed a tiered system that is designed to provide some level of metabolite characterization for all NCEs. The amount of time spent characterizing a particular NCE is dependent on the needs of that particular program and how far advanced the compound is in the discovery process. The tiered system is structured to utilize LC/MS/MS techniques along with metabolite ID software to answer the most critical questions first. Constant neutral loss scans are employed on bile and urine samples to determine whether conjugated metabolites are present. This, combined with product ion scans for expected or common metabolites, can typically provide a fairly comprehensive metabolic profile in a short period of time. Precursor ion scans and tailored CNL scans provide additional information. Studies that require more time such as MS^n or accurate mass experiments and characterization of low level metabolites circulating in plasma are more open-ended in terms of the time and resources required, so they are not performed until the NCE is fairly advanced through the discovery process. By utilizing the hardware and software technology efficiently and effectively, we are able to support the metabolite characterization needs of many different discovery programs simultaneously.

References

1.   R.E. White, *Ann. Rev. Pharmacol. Toxicol.* **40** (2000) 133.

2.   J. Caldwell, *Pharm. Sci.* **2** (1996) 117–119.

3.   A.D. Rodrigues, *Med. Chem. Res.* **8** (1998) 422.

4.   P.B. Fernandez, *Curr. Opin. Chem. Biol.* **2** (1998) 597.

5.   S. Yamashita, Y. Tanaka, Y. Endoh, Y. Taki, T. Sakane, et al., *Pharm. Res.* **14** (1997) 486.

6.   C.L. Crespi, V.P. Miller, and B.W. Penman, *Anal. Biochem.* **248** (1998) 188.

7.   J.M. Silva, P.E. Morin, S.H. Day, B.P. Kennedy, P. Payette, et al., *Drug Metab. Dispos.* **26** (1998) 490.

8.   J. Berman, K. Halm, K. Adkison, and J. Shaffer, *J. Med. Chem.* **40** (1997) 827.

9.   K.A. Cox, K. Dunn-Meynell, W.A. Korfmacher, L. Broske, A.A. Nomeir, C.C. Lin, M. Cayen, and W.H. Barr, *Drug Discov. Today* **4** (1999) 232.

10.   N.J. Clarke, D. Rindgen, W.A. Korfmacher, and K.A. Cox, *Anal. Chem.* **73** (2001) 430A.

11.   M.W. Sinz and T. Podoll, "The mass spectrometer in drug metabolism". In: D.T. Rossi and M.W. Sinz (Eds.), *Mass Spectrometry in Drug Discovery*, Dekker, New York, 2002.

12.   D.J. Clarke and B. Burchell, "The uridine diphosphate glucuronosyltransferase multigene family: function and regulation". In: F.C. Kauffman (Ed.), *Conjugation– Deconjugation Reactions in Drug Metabolism and Toxicity*, Springer-Verlag, Berlin, 1994.

13.   A.K. Daly, "Pharmacogenetics". In: T.F. Woolf (Ed.), *Handbook of Drug Metabolism*, Dekker, New York, 1999.

Advances in Mass Spectrometry, Volume 16
A.E. Ashcroft, G. Brenton and J.J. Monaghan (Editors)

CHAPTER 13

Recent Developments in Polymer Characterization Using Mass Spectrometry

Charles N. McEwen

DuPont Corporate Center for Analytical Sciences, DuPont Experimental Station,
Wilmington, DE 19808, USA

13.1 Introduction

The ultimate goal of polymer characterization whether biological or synthetic is to relate composition to property and function. The difficulty of this task can easily be recognized for protein polymers because we know that property/function is related to structure, kinetics of motion, interaction with other molecules, the molecular environment, etc. The three-dimensional structure and the way the protein folds into its 3-D structure may be related not only to its composition but to its interaction with other proteins, its environment, and possibly other yet to be determined criteria. In addition, the protein may have multiple functions, some or all of which are unknown at the present time. Ultimately, we may want to design proteins that have properties that perform a certain desirable function without undesirable side reactions.

For synthetic polymers, a goal of the polymer chemist is also to design polymers with desirable properties that will perform certain well-defined functions. The complexity of this task is also high. During the technical revolution of the past century polymer chemist were most concerned with making new polymers and finding applications that were commercially viable. The challenge in polymer characterization has been to find tools that can measure composition or properties that relate in some fashion to the function of the polymer. Thus, it is important to know the monomer composition, the concentration of end groups, the molecular weight distribution, viscosity, tensile strength, etc. Before the advent of matrix assisted laser desorption/ionization [1], mass spectrometry played a small but sometimes significant role in polymer characterization, mostly through mass spectral analysis of the volatile products from thermal decomposition of the polymer.

As the technical revolution advances into our new century, the demands on polymer characterization will grow as the need for materials with tighter limits on properties (i.e., designer polymers) increase. There are numerous examples, but one that stares you in the face is the printed page. As the resolution requirements for printing increase, the demand on the chemistry used in both the inks and the paper increases. This is most apparent in printing high-resolution color photographs, which requires both skilled engineering and intricate chemistry. Polymers with well-defined properties are critical to this application. Equally critical today is the need for speed to market. Reaching the market first or answering a competitor's advance is increasingly important for the corporate bottom line. Finally, but often most important, the polymers need to be cost-wise competitive.

Answering the need for inexpensive polymers with well-defined properties developed in a timely manner has resulted in efforts to synthesize polymers using combinatorial methods and in attempts to better understand the relationships between composition and properties. In either case, with the advent of MALDI and electrospray (ES) ionization techniques, mass spectrometry has become an exceptionally important analytical tool. The ability to obtain a great deal of compositional information, and the potential to obtain molecular weight distribution information rapidly is of great importance for combinatorial/parallel polymer synthesis methods. Clearly, for mass spectrometry to reach its potential in this arena, it will be necessary to make advances in the technology. A wider variety of polymer types will need to be analyzed, especially polymers with little or no ionizable functionality as well as more intractable polymers. Improving the signal to noise for polymers with complex structures will be necessary to better define the composition of polymers. Relative quantitation will need to be improved for, for example, oligomers of different mass, oligomers with end groups of different composition, or copolymer oligomers with different monomer content. The ability to rapidly obtain reliable molecular weight information along with compositional information will greatly increase the value of mass spectrometry for characterizing libraries of newly synthesized polymers.

Another approach to synthesizing designer polymers is to have such a deep understanding of composition/property relationships that one knows a priori the target structure. For the polymer chemist having this understanding, building the desired polymer becomes a matter of finding the best synthetic scheme. As mentioned above, mass spectrometry along with other analytical techniques can provide information that will (hopefully) lead to improved understanding of composition/property relationships. As this knowledge increases, mass spectrometry will be an important tool in guiding the synthetic effort of building the polymer with the correct molecular weight distribution and having the required copolymer and/or end group composition.

Mass spectrometry will continue to be an important tool in patent litigation and competitive analyses. As polymer properties become more critical to the application, small changes in composition may distinguish a good from a competitively disadvantaged polymer. Determining the composition differences requires analysis methods that are capable of "seeing" all components in a highly reproducible manner.

From the increasing number of papers involving the use of mass spectrometry for the analysis of polymers, it is clear that the tools currently in our possession are highly useful. Much of the research discussed in these papers involves novel ways to use the current tools to obtain a more complete insight into the composition of polymeric systems. However, if mass spectrometry is to make inroads in polymer characterization, new breakthroughs are needed that allow us to explore polymers in more depth and with more accuracy or to even explore polymer systems not currently amenable to mass spectrometric methods. The good news is that there are new frontiers that need exploring and much that needs to be learned. Below are brief discussions of a few selected areas of research into polymer characterization using mass spectrometry that have potential to advance the field.

13.2 MALDI without solvents

Traditionally, MALDI sample preparation required that both the matrix and sample be soluble in a suitable solvent. Fortunately, MALDI can tolerate a wide range of solvents. Nevertheless, the success of polymer characterization often depends on the polymer being highly soluble in the selected solvent. For poorly soluble samples, the most time consuming feature of the analysis is getting the sample in solution. This often entails finding an appropriate solvent by a trial and error process. Frequently, overnight shaking and heating is necessary to get the polymer into solution. Additionally, poorly soluble samples are often more difficult to analyze by MALDI than samples that readily dissolve. For samples in which no available solvent is satisfactory, analysis by mass spectrometry is only possible using pyrolysis methods or for relatively low molecular weight polymers using laser desorption, field desorption, desorption chemical ionization, or secondary ion mass spectrometry techniques.

A solvent free MALDI sample preparation method was introduced by Skeleton et al. [2] for poorly soluble polyamides having molecular weights up to 4000 Da. Przybilla et al. [3] demonstrated a similar solvent-free method for insoluble giant polycyclic aromatic hydrocarbons. These authors demonstrated that new matrix materials could be used in this sample preparation. Trimpin et al. [4] evaluated and improved the solvent-free sample preparation method, concluding that the method produced essentially identical results to the solvent-based method for polymer distributions between 2 and 100 kDa. The solvent-free sample preparation method was also shown by Trimpin et al. [5] to be applicable to MALDI MS analysis of the insoluble fraction of poly(9,9-diphenyl-2,7-fluorene) obtained by Soxhlet extraction of that polymer for 5 days using toluene. The success of the solvent-free sample preparation method strongly suggests that obtaining a homogeneous mixture of sample and matrix at the molecular level is not necessary for successful MALDI analysis [4].

Further development is needed to simplify and to extend the solvent-free method to a wider variety of insoluble and poorly soluble polymers. Convenience of use and in some cases the amount of sample required are factors that limit the acceptance of this method. A simple microscale neat MALDI sample preparation method would have the potential to supplant the solvent-based method for some polymer types.

13.3 Analysis of nonpolar polymers

Hydrocarbon polymers are an important class of industrial materials. Analysis of highly unsaturated hydrocarbon polymers such as polystyrene, polybutadiene, and polyisoprene can be accomplished using MALDI MS by adding a silver or copper salt to the matrix [6,7]. However, analyses of hydrocarbon polymers having few or no sites of unsaturation are problematic by MALDI MS. Low molecular weight hydrocarbon oligomers such as those found in parafins and microcrystalline waxes were shown by Pruns et al. [8] to ionize by Ag^+ cationization in laser desorption MS after solvent-free treatment with AgTFA. This work was similar in some respects to a report by Kahr and Wilkins [9] in which silver cationization of polyethylene was observed using laser desorption FTMS. Recently, Yalcin et al. [10] used a coarse cobalt powder as a bottom layer substrate and silver nitrate as the cationization agent to produce intact polyethylene ions up to 5000 molecular weight using laser desorption. Field desorption mass spectrometry was shown by several groups to be capable of ionizing low molecular weight (< 3600) polyethylene oligomers [11,12]. Recently, Schaub et al. [13] demonstrated the utility of combining field desorption ionization and high resolution FT-ICR for the analysis of mid-boiling crude oil distillates. Pyrolysis photoionization mass spectrometry and desorption chemical ionization have recently been applied to the characterization of nonpolar polymers [14]. Pyrolysis has long been a useful tool for polymer characterization using mass spectrometry [15]. Combined with soft ionization methods such as photoionization, the method provides considerable composition information for copolymers in a very short time. This makes pyrolysis a potentially powerful method for the analysis of combinatorially produced polymer samples. This method does not require the polymer to be soluble.

Bauer et al. [16] as well as Lin-Gibson et al. [17] have shown that covalently attaching an organic moiety to the end of polyethylene oligomers to produce cationized molecules enables MALDI mass spectra to be obtained for oligomers having molecular weights as high as 15,000. The molecular weight moments for the polymer distributions obtained using these derivatization methods were lower than expected. However, covalent cationization of polystyrene gave results in good agreement with classical methods, even for a polymer with $M_n = 40000$, demonstrating that the cationization method itself was not limiting [18]. Ji et al. [19] used sulfonation of polyisobutylene end groups to produce ionizable oligomers. These authors reported good agreement for molecular moment determination for lower molecular weight polyisobutylene samples when comparing the MALDI MS method and conventional methods of molecular weight distribution analysis. Ionization of long-chain n-alkanes in the gas phase using transition metal ions is also being studied for the purpose of eventually applying this technology to the analysis of saturated hydrocarbon polymers [20,21]. Early results seem promising, but any MALDI method that is developed to analyze nonpolar polymers such as polyethylene will have to deal not only with ionization issues but also with sample preparation issues. Many of these materials are in the difficult to dissolve category, often requiring hot, less volatile solvents. Finding conditions to co-crystallize matrix, polymer and the cationization salt can be difficult. The solvent-free methodology applied to these polymers would be a major advance.

13.4 End group characterization and quantitation

One of the most important functions of mass spectrometry in polymer analysis is end group characterization and quantitation. Polymer properties often correlate with end group composition. The end group composition becomes a larger fraction of the whole as the mass of the oligomer decreases. Thus, the end group composition over the mass range in which composition can be determined by mass spectrometry is most important in determining polymer properties. For an unmodified homopolymer, the combined mass of end groups is easily determined by MALDI MS. Combined with some knowledge of how the polymer was synthesized or with information from other analytical techniques, the end group composition can usually be inferred. Nandam et al. [22], for example, used a combination of MALDI MS, FT-IR, and thermochemical calculations to unequivocally assign the end groups of poly(methylstyrene peroxide). Additional information on end group composition can also be obtained from mass spectrometry by using accurate mass measurement and fragmentation [23]. For co- and ter-polymers, end group composition analysis becomes more problematic. Nevertheless, end group composition analysis has been an active field and there are numerous publications dealing with the subject.

End group quantitation has been less well studied. Although not designed to look at end groups, a study by Larsen et al. [24] demonstrated that equimolar tert-butyl capped poly(methylmethacrylate) oligomers gave essentially the same ion current in the MALDI analysis even though the ratio of repeat unit to end group mass changed by a factor of two. Analysis in our laboratory of approximately equimolar solutions of polyamide 6,6 cyclic oligomers as well as linear oligomers having different end groups showed only minor differences in total ion abundance for the various ends. Thus, within the error of preparing low mass ($<$ 1300) equimolar compositions of cyclic oligomers and linear oligomers with acid–acid, amine–amine, and acid–amine ends, all species gave nearly equal ion current when applied to the MALDI target in equimolar mixtures. On the other hand, large differences in desorption/ionization efficiency were observed for stiff polyphenylene type polymers having different end groups. Puglisi et al. [25] studied the relationship of end groups to ionization efficiency. These authors found that the MALDI MS ionization observed for equimolar blends of nylon 6 and polybutyleneterephthalate (PBT) could be made to favor one polymer or the other simply by altering the end groups on one or both polymers. These authors also found that at low mass, hydroxy terminated PBT was less efficiently desorbed/ionized in the MALDI mass spectra than carboxy terminated PBT, but at higher mass, where the end groups are a less significant part of the molecule, the ionization efficiencies for both species were approximately equal.

Clearly, quantitation of end groups is problematic when it is not possible to determine a priori when and to what degree end group ionization discrimination will occur in MALDI MS. Additional studies are needed to ferret out the complex relationship between end group composition, polymer structure, matrix material, and the observed ion abundance. The above cited examples of derivatizing the ends of nonpolar polymers to effect ionization is an extreme example of the relationship of end group and observed ion abundance in MALDI. It would be surprising (to this author) if the improved ionization of these molecules had any strong relationship to position of the cationization moiety on the oligomer. It is more likely a handle for ionization. On the other hand, end group effects appear to be subtler than simply a function of ease of ionization. Chen et al. [26] noted that for an ethylene oxide/propylene oxide copolymer, chains with different ends changed in abundance with changes in matrix composition, solvent composition and even matrix to analyte ratios. These authors did note that under constant experimental conditions, relative changes in end group concentrations could be measured.

It may be possible to study some end group effects by combining separation methods with mass spectrometry. At low mass, chromatographic resolution of oligomers having different ends is possible for some polymers. At higher mass, techniques such as chromatography near the critical point of adsorption may be used to fractionate some polymers that differ only by their end groups. Weidner et al. [27] used two-dimensional chromatography to study the heterogeneity of end groups in modified nylon 6,6. Specifically, chromatography near the critical point of adsorption was used to separate oligomers by end group functionality. Chromatographic methods when coupled to mass spectrometry present an especially powerful tool for polymer characterization [28].

13.5 Signal to noise issues

In any situation where quantitative results are desired, signal to noise becomes an issue. In the just discussed case of end group quantitation, ion signals for peaks that are only two or three times background noise will not produce high precision results. However, because polymers produce repeating patterns, it is possible to sum the signals for all species within a mass range window that have the same end group mass. Doing this for repeating patterns over a selected mass range improves the precision of relative ion current measurements. Even so, improving signal to noise for polymer analysis would be very beneficial for many analyses. One example is provided in Fig. 13.1. The left mass spectrum represents one sample in a series of homopolymer samples in which the polymer chemist was interested in finding experimental conditions to eliminate certain end groups in the product. In the spectrum shown, the end groups in question had already been reduced to a fairly low level, but by using the above additive procedure, it was possible to follow the status of subsequent reactions with high confidence. The right spectrum is a triblock copolymer made with similar starting materials to the homopolymer. In fact, homopolymers made from either of the monomers used in the copolymer synthesis produced comparable spectra under similar conditions. Even though the two samples shown here were run under similar conditions, the signal to noise is much poorer for the more complex polymer. This trend continued as the polymer structure became more complex or as the polydispersity of the polymer increased. The ability to follow the progress in removing unwanted end groups in the more complex polymer is severely limited by the poor signal to noise ratio. Signal summing improves the signal to noise, but the improvement observed between 20 and 40 summed spectra will be approximately the same as will be observed between 200 and 400 or 2000 and 4000 summed spectra. Clearly, there is a limit to spectral improvement using this approach.

There has been progress in improving sensitivity in the analysis of biological polymers, but synthetic polymer analysis is usually not sample limited [29]. The issue for synthetic polymers is more one of reducing detector and chemical noise, or of enhancing signal without increasing noise. A new cryodetector discussed below promises very low electronic noise but at a very high price. Sample preparation methods such as variation in solvent composition and the use of detergents have been shown to provide dramatic improvement of signal for some peptides [30,31]. This kind of enhancement in signal would be welcome in MALDI MS of polymers.

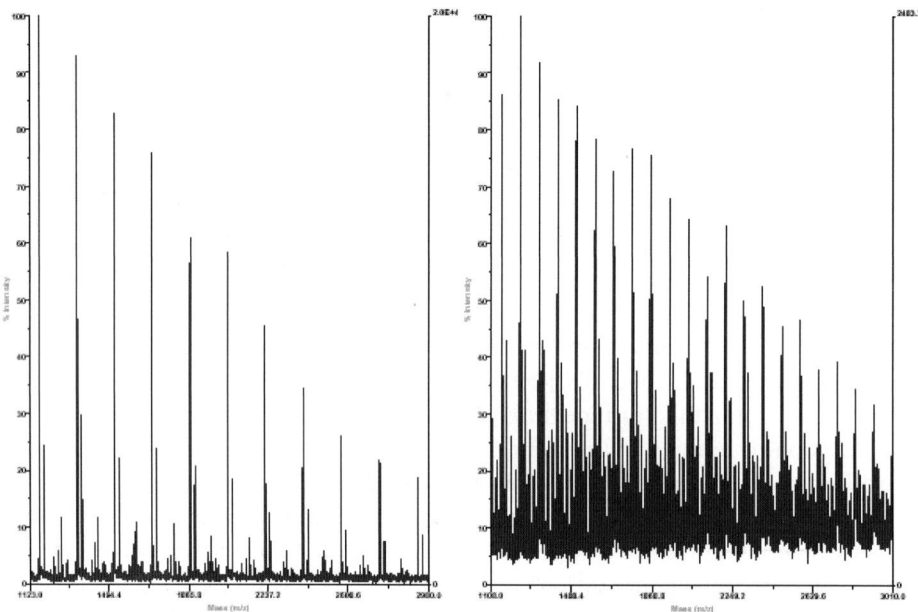

Figure 13.1 Left; MALDI mass spectrum of a homopolymer submitted for end group comparison studies. Spectrum shows good signal to noise. Right; MALDI mass spectrum of a triblock copolymer submitted for end group comparison studies showing poor signal to noise for low abundance peaks.

13.6 Polymer molecular distribution analysis

Another area where ion abundance issues may eventually be important is in using MALDI mass spectrometry for determining molecular moments of polymer distributions. It is now well ingrained that MALDI can be reasonably accurate in the analysis of narrow polydisperse polymers [6], but does not produce accurate molecular moments for polymers with polydispersity greater than about 1.2 [32]. The reason that MALDI fails to work with wide polydisperse polymers is still fraught with misconceptions. The most frequently quoted misconception is that MALDI does not desorb/ionize high mass molecules as efficiently as with low mass molecules. The loss of ionization efficiency at high mass has even been attributed to an end group effect [25]. The authors of this study concluded that the possibility of mass discrimination due to imperfect detectors could be ruled out in the case being studied. End groups in some cases may account for a mass dependent loss or gain in sensitivity, but this in no way clears all detectors of imperfection. The spectra shown in Figs. 13.2 and 13.3 are of equimolar mixtures of insulin, myoglobin, carbonic anhydrase, and bovine serum albumin. The top spectrum (Fig. 13.2) was taken with a standard MCP detector. The bottom spectrum (Fig. 13.3) was taken

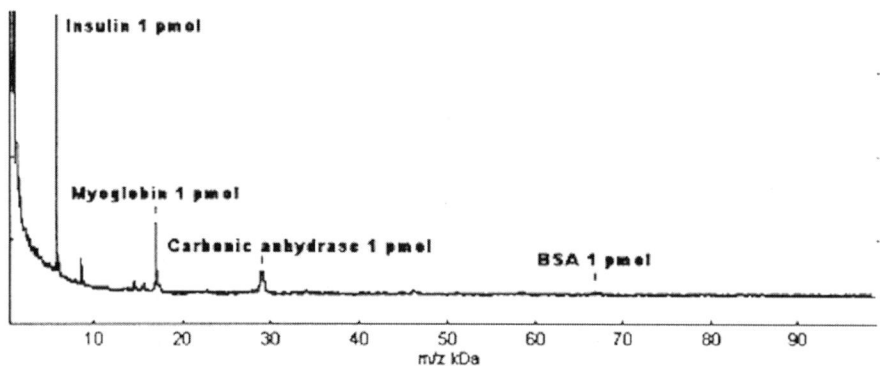

Figure 13.2 MALDI-TOF mass spectrum of an equimolar mixture of four proteins with ion detection using an MCP detector (Fig. compliments of Comet AG).

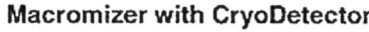

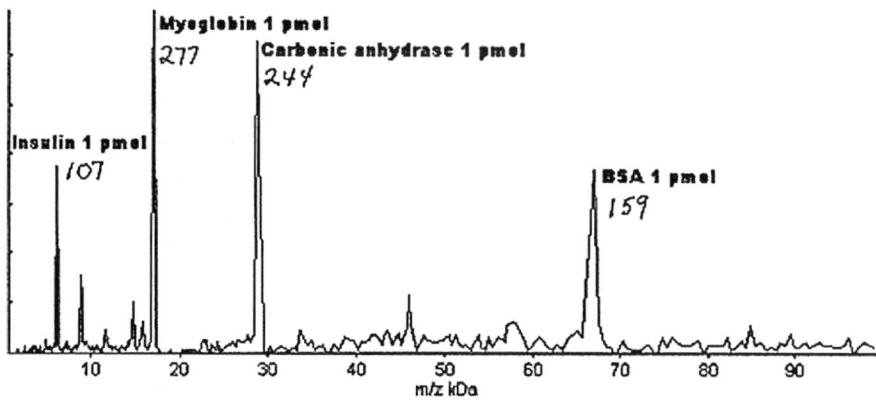

Figure 13.3 MALDI-TOF mass spectrum of an equimolar mixture of the same four proteins shown in Fig. 13.2 obtained using the the Macromizer® cryodetector (Fig. comliments of Comet Ag).

with cryodetector that is known to give a response that is independent of mass. The ion counts for the lowest mass species (ca. 5700) was 107 while that for the highest mass component (ca. 66,000) was 159. Assuming the detector works as advertised, then there is discrimination, probably in the ionization event, but it is not mass related.

Byrd et al. [33] demonstrated this point for poly(methylmethacrylate) polymers. An equimass mixture of PMMA 2400 and 52,000 gave more or less the expected ratio for total ion current with post-acceleration using an MCP detector but greatly overemphasized the high mass component with post-acceleration and a discrete dynode detector. In other words, in one detector, the high mass species were over-represented. However, as others have found with equimolar mixtures, when narrow polydisperse standards were mixed in equimass amounts to mimic a wide polydisperse polymer, both detectors underrepresented the higher mass components. There is still a debate concerning the cause of this high mass discrimination when polymer is present over a wide mass range as in a wide polydisperse polymer. The authors of the above cited study suggested detector saturation as one possible cause.

Unfortunately, solving the high mass discrimination problem will still not produce accurate molecular moments for wide polydisperse polymers unless the signal to noise, especially at high mass is greatly improved. The reason that this is the case is the importance of high-mass oligomers in the molecular moment calculations, especially for M_w and higher moments. There are several reasons why these values can be accurately obtained by size exclusion chromatography (SEC) and not easily by mass spectrometry. First, SEC detects equal volumes and these equate to a log mass scale. That is to say a much wider mass range is present in a detected volume at high mass than at low mass. The time-of-flight mass spectrometer looks at equal time windows and these equate to a square root mass scale. This approach gives some discrimination toward enhancing the high-mass signal (a wider mass range per time window at high mass), but most manufacturers use software to convert mass and abundance to a linear mass scale. SEC has the additional advantage that the detector used is typically a weight-fraction detector where molecules produce signal in proportion to their mass. Thus, in principle, 1 molecule at mass 100,000 will give about the same signal as 100 molecules of mass 1000. Mass spectrometry detectors, on the other hand, are designed to be number fraction detectors. If they were perfect number fraction detectors then 1 molecule at 100,000 would give the same signal as 1 molecule at 1000 or for that matter at any mass.

The problem for polymer analysis in molecular moment calculations, especially M_w and higher order moments, is the disproportionate contribution of the high-mass oligomers. As the polydispersity increases, the tail of the distribution spreads over an increasing mass window. Assuming ionization, transmission and detection of oligomers of all mass values are equally efficient, mass spectrometry will not give accurate molecular moments unless the signal to noise, especially at high mass, is much higher than can currently be achieved with standard detector technology. A new, but untested in this application, cryodetector has been built by Comet AG that is advertised to be a true number fraction detector with extremely low noise. It will be interesting to see if this detector improves on the polydispersity data that is obtained by MALDI MS.

It may be possible to improve the high-mass signal to noise problem by building a detector for mass spectrometry that behaves like a weight fraction detector. In other words, the detector will produce more signal for a high-mass ion than for a low-mass ion. Such a detector may need to be calibrated for the intensity axis. Because high-mass ions are detected better than low-mass ions, it may also reduce any saturation problem that exists with current detectors that are due to an overabundance of low-mass ions. That such a detector may be possible can be determined by observing results from a dual-particle detector (scientific analysis instruments) described by Thompson et al. [34] and comparing the results with data taken from a discrete dynode detector described by Byrd and McEwen [33]. Looking at the same distribution of a mixture of narrow polydisperse polymers on the same instrument under nearly identical conditions, using the dual particle detector, it is possible to select either electrons or hydride ions sputtered from the surface of the conversion dynode by the primary ion impact event. The observed ion distribution is different for the electron and hydride signals. The electron signal is stronger for matrix and low-mass polymer ions than the hydrid signal, but at higher mass the electron signal drops off more rapidly than the hydride signal. On the other hand, the discrete dynode detector observes all secondary species sputtered from the conversion dynode surface and the mass spectrum of the polymer ion distribution is skewed to higher abundances at higher mass. This can be explained if higher mass secondary species are more efficiently sputtered from the dynode surface by higher mass primary ion impact. This data suggests that detecting only the higher mass secondary ions from the conversion dynode may be more suitable for wide polydisperse polymers than current detectors. Of course, if time of analysis is not a major factor, as Montaudo et al. [35] and others [36], have demonstrated, MALDI can be combined with SEC to provide mass calibration for the SEC chromatogram.

13.7 Conclusion

The good news for those of us using mass spectrometry for polymer characterization is that we have tools to deliver important and sometimes critical information about polymer composition. The other good news is that our methods are not perfect, so there is much research that still needs to be done.

References

1.   M. Karas, D. Bachmann, and F. Hillenkamp, *Anal. Chem.* **57** (1985) 2935.
2.   R. Skelton, F. Dubois, and R. Zenobi, *Anal. Chem.* **72** (2000) 1707–1710.

3. L. Przybilla, J.-D. Brand, K. Yoshimura, H.J. Rader, and K. Mullen, *Anal. Chem.* **72** (2000) 4591–4597.

4. S. Trimpin, A. Rouhanipour, R. Az, H.J. Rader, and K. Mullen, *Rapid Commun. Mass Spectrom.* **15** (2001) 1364–1373.

5. S. Trimpin, A.C. Grimsdale, H.J. Rader, and K. Mullen, *Anal. Chem.* **74** (2002) 3777–3782.

6. C.M. Gutmann, S.J. Wetzel, W.R. Blair, B.M. Fanconi, J.E. Girard, R.J. Goldschmidt, W.E. Wallace, and D.L. Vanderhart, *Anal. Chem.* **73** (2001) 1252–1262.

7. T. Yalcin, D.C. Schriemer, and L. Li, *J. Amer. Soc. Mass Spectrom.* **8** (1997) 1220–1229.

8. J.K. Pruns, J.-P. Vietzke, M. Strassner, C. Rapp, U. Hintze, and W.A. Konig, *Rapid Commun. Mass Spectrom.* **16** (2002) 208–211.

9. M.S. Kahr and C.L. Wilkins, *J. Amer. Soc. Mass Spectrom.* **4** (1993) 453.

10. T. Yalcin, W.E. Wallace, C.M. Guttman, and L. Li, *Anal. Chem.* **74** (2002) 4750–4756.

11. R.P. Lattimer and H.R. Schulten, *Int. J. Mass Spectrom. Ion Phys.* **52** (1983) 105.

12. J.H. Gross and S.M. Weidner, *Eur. J. Mass Spectrom.* **6** (2000) 11.

13. T.M. Schaub, C.L. Hendrickson, K. Qian, J.P. Quinn, and A.G. Marshall, *Anal. Chem.* **75** (2003) 2172–2176.

14. D.L. Zoller, M.V. Johnston, K. Qian, and D.J. Lohse, *Macromolecules* **33** (2000) 5388–5394.

15. F.J. Cox, R.N. Feudale, M.V. Johnston, C.N. McEwen, and E. Hauptman, *J. Anal. Appl. Pyrolysis* **64** (2002) 305–312.

16. B.J. Bauer, W.E. Wallace, B.M. Fanconi, and C.M. Guttman, *Polymer* **42** (2001) 9949–9953.

17. S. Lin-Gibson, L. Brunner, D.L. Vanderhart, B.J. Bauer, B.M. Fanconi, C.M. Guttman, and W.E. Wallace, *Macromolecules* **35** (2002) 7149–7156.

18. S. Lin-Gibson, S.A. Bencherif, K.L. Beers, and H.C.M. Byrd, *Macromolecules* **36** (2003) 4669–4671.

19. H.N. Ji, N. Sato, Y. Nakamura, Y.N. Wan, A. Howell, Q.A. Thomas, R.F. Storey, W.K. Nonidez, and J.W. Mays, *Macromolecules* **35** (2002) 1196–1199.

20. R. Chen, T. Yalcin, W.E. Wallace, C.M. Guttman, and L. Li, *J. Amer. Soc. Mass Spectrom.* **12** (2001) 1186.

21. H.C.M. Byrd, C.M. Guttman, S. Robinson, and D.P. Ridge, In: *51st ASMS Conference*, June 8–12, 2003, TOEam 11:15.

22. A.K. Nandam, K. Ganesh, K. Kishore, and M. Surimarayanan, *Polymer* **41** (2000) 9063–9072.

23. R.J. Goldschmidt, S.J. Wetzel, W.R. Blair, and C.M. Guttman, *J. Amer. Soc. Mass Spectrom.* **11** (2000) 1095–1106.

24. B.S. Larsen, W.J. Simonsick Jr., and C.N. McEwen, *J. Amer. Soc. Mass Spectrom.* **7** (1996) 287–292.

25. C, Puglisa, F. Samperi, R. Alicata, and G. Montaudo, *Macromolecules* **35** (2002) 3000–3007.

26. R. Chen, N. Zhang, A.M. Tseng, and L. Li, *Rapid Commun. Mass Spectrom.* **14** (2000) 2175–2181.

27. S.M. Weidner, U. Just, W. Wittke, J.F. Friedrich, F. Rittig, and F. Gruber, In: *ASMS Conference*, June 8–12, 2003, TPU391.

28. G.E. Kassalainen and S.K.R. Williams, *Anal. Chem.* **75** (2003) 1887–1894.

29. S.X. Xiong, Q.X. Ding, Z.W. Ahao, W.Z. Chen, G.H. Wang, and S.J. Liu, *A New Proteomics* **3** (2003) 265–272.

30. K.O. Bornsen, M.A.S. Gass, G.J.M. Bruin, J.H.M. Vonadrichen, M.C. Biro, G.M. Kresbach, and M. Ehrat, *Rapid Commun. Mass Spectrom.* **11** (1997) 603–609.

31. P.M. Peacock, "Enhanced sensitivity of MALDI via surfactant addition". In: *ASMS Conference*, June 11–16, 2003, MPN258.

32. G. Montaudo, M.S. Montaudo, C. Puglisi, and F. Samperi, *J. Polym. Sci., Part A: Polym. Chem.* **34** (1996) 439.

33. H.C.M. Byrd and C.N. McEwen, *Anal. Chem.* **72** (2000) 4568–4576.

34. S.P. Thompson, V.C. Parr, and C.N. McEwen, In: *Proceedings of the 46th Conference on Mass Spectrometry and Allied Topics*, May 31–June 4, 1998, p. 1072.

35. M.S. Montaudo, C. Puglisi, F. Samperi, and G. Montaudo, *Rapid Commun. Mass Spectrom.* **12** (1998) 519–528.

36. S.D. Hanton and X.M. Liu, *Anal. Chem.* **72** (2000) 4550–4554.

Advances in Mass Spectrometry, Volume 16
A.E. Ashcroft, G. Brenton and J.J. Monaghan (Editors)
© 2004 Published by Elsevier B.V.

CHAPTER 14

ICP-MS: Metals and Much More

Katarzyna Wrobel [1], Katie DeNicola [2], Kazimierz Wrobel [1], and
Joseph Caruso [2,*]

[1] Instituto de Investigaciones Científicas, Universidad de Guanajuato, L. de Retana No 5,

36000 Guanajuato, Mexico

[2] Department of Chemistry, University of Cincinnati, Cincinnati, OH 45221-0172, USA

14.1 ICP-MS: a primary detection tool for element determination and speciation

Inductively coupled plasmas have long been used as ion sources for optical emission spectroscopy, but only recently have been employed for mass spectrometry, with the first commercial inductively coupled plasma mass spectrometers (ICP-MS) being introduced in the mid-1980s. Yet, in this short time, ICP-MS has quickly become recognized as a powerful tool and primary detection technique for element determination and speciation. Not only has ICP-MS been proven to be accurate and precise, but it also possesses both multielemental and isotopic capabilities and has frequently been used in situations where universal detectors, such as UV detectors, have failed. Also, ICP-MS boasts detection limits in the part-per-billion (ppb) to part-per-trillion (ppt) range for most elements and wide linear dynamic ranges, typically spanning 8 orders of magnitude [1]. As a result, this analytical tool is well suited for a variety of different sample types and has consequently been used in a number of areas, laying claim to environmental, clinical, geochemical, metallurgical, and even nuclear applications.

Yet despite its growing popularity and application, it seems that ICP-MS is an even more versatile technique than most may be aware. Current applications of ICP-MS are furnishing more information than only metal content or species. ICP-MS is being used to elucidate metabolic pathways and to study methylation

---

*Corresponding author. E-mail address: joseph.caruso@uc.edu (J. Caruso).

capacity in biological systems, and has also become a valuable tool in proteomics studies. Furthermore, the introduction of high resolution and collision/reaction cell instruments has minimized the interference problems associated with a number of elements, further increasing the types of samples that can be evaluated as well as the analytes that can be detected. This paper aims to explore the versatility of ICP-MS by focusing on these and other innovative applications and discussing the instrumental and methodological developments that continue to make ICP-MS the primary detection tool for element determination and speciation.

14.1.1 Basic principles of ICP-MS

Although more detailed discussions can be found elsewhere [2–6], it may be instructive to understand the basic principles behind ICP-MS in order to fully realize its potential. The fundamental component of ICP-MS is, of course, the plasma, which consists of a partially ionized argon gas at temperatures of about 6000–10,000 K, and functions as the ionization source for the mass spectrometer. As shown in Fig. 14.1, analysis begins when the sample, typically in liquid form, is pumped through the spray chamber and nebulizer of the sample introduction system and emerges as an aerosol. This aerosol is then introduced to the plasma and undergoes drying, vaporization, atomization, and ionization as it travels through

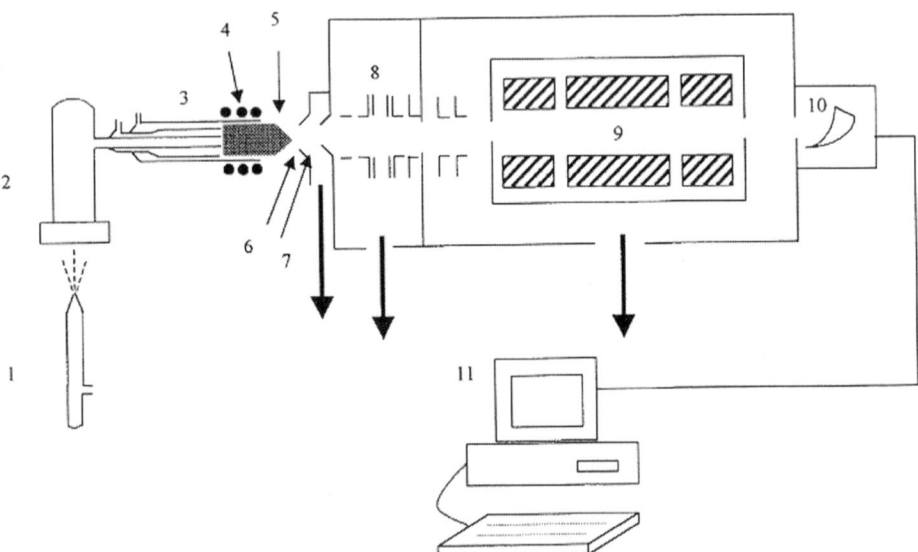

Figure 14.1 Diagram of an ICP-MS with quadrupole mass analyzer: (1) nebulizer, (2) spray chamber, (3) torch, (4) load coil, (5) plasma, (6) sampler cone, (7) skimmer cone, (8) lenses, (9) quadrupole, (10) electron multiplier, (11) computer interfaced to ICP-MS. Arrows denote areas kept under vacuum. (Courtesy of Elsevier.)

the various temperature zones. There is sufficient energy in the plasma to remove electrons from the sample's constituent elements, thus generating positively charged ions, which can be detected and measured by the mass spectrometer.

The mass spectrometer itself consists of a mass analyzer as well as a detector. The mass analyzer works to separate analyte ions from the incoming plasma based on mass to charge ratios (m/z). As will be discussed, various types of mass analyzers can be used, but the quadrupole analyzer has by far been the most popular. It consists of four cylindrical rods of equal length and diameter, usually made of stainless steel or molybdenum. Direct current (dc) is applied to two of the rods and radio frequency (rf) to the opposite pair. Specific dc-rf voltages produce either a positive or negative bias, which directs ions of a particular m/z through the rods toward the detector. All other ions are ejected. Different dc-rf voltages can be scanned, allowing other ions of interest to subsequently be delivered to the detector. The detector, whether it be a channel electron multiplier, Faraday cup, or discrete dynode electron multiplier, converts these ions into electrical signals, which can be correlated with calibration measurements in order to determine the concentration of targeted elements within the sample.

14.1.2 Elemental speciation with ICP-MS

A main advantage of ICP-MS is that it cannot only be used to determine the total elemental content of a sample but, by coupling it with various separation techniques, it can be used to establish the form in which those elements are present. Elemental speciation, or the analyses that lead to determining the distribution of an element's particular chemical species in a sample, can provide information concerning oxidation state, organometallic nature or complex form, which can be of particular use when assessing the risks and benefits that may be associated with the elemental composition of a given sample. Likewise, such speciation information is crucial in the elucidation of biochemical pathways or the detection of pharmaceutical impurities.

The coupling of liquid chromatography (LC) to ICP-MS for elemental speciation has become almost routine, thanks to its sensitivity, ruggedness, and relatively simple interface. However, factors such as the separation mechanism, the composition and flow of the mobile phase, and the type of nebulizer and sample introduction device must be carefully considered and have subsequently been topics of discussion in many papers and reviews [7–11].

Although the large assortment of available liquid chromatographic columns makes it possible to analyze a vast range of compounds, other separation techniques can also be used in conjunction with ICP-MS to obtain speciation information. Capillary electrophoresis (CE), for example, is another widely used technique, since, like LC, the eluent is in liquid form, and, therefore, readily interfaced to the ICP-MS. The gas phase eluent present in gas chromatography (GC) can also be directly coupled to an ICP-MS system by way of a heated transfer line, thus allowing for the speciation of volatile compounds. Even solid gels containing electrophoretically separated peptides and proteins can be analyzed with ICP-MS through laser ablation. Ultimately, it is the compatibility of ICP-MS with these numerous separation techniques that is responsible for its unique versatility and exceptional potential, which will be highlighted throughout this paper.

14.2 Detection of non-metal species by ICP-MS

The use of ICP-MS has usually pertained to the detection and quantification of metal- or metalloid-containing species. This is mostly due to the fact that non-metal species possess higher ionization energies and, therefore, are not efficiently able to be ionized within the plasma. For instance, the first ionization energies of halogens range from 17.4 to 10.4 eV, whereas typical first ionization energies for transition metals such as Cr, Cu, and Ag are 6.77, 7.73, and 7.58 eV, respectively.

Spectral polyatomic interferences are also responsible for some of the difficulties associated with non-metal analysis. Just like the analytes of interest, other elements within the sample matrix undergo ionization within the plasma, and can form complexes with each other or the argon of the plasma. These complexes interfere with analyte detection when they exhibit the same mass to charge ratio as the element of interest.

Sulfur and phosphorus are examples of two non-metals whose detection has been limited by such interferences. Oxygen within a sample can form $^{16}O^{16}O^+$ in the plasma, interfering with ^{32}S, the major isotope of sulfur. Likewise, the measurement of ^{31}P can be inhibited by $^{15}N^{16}O^+$ and $^{14}N^{16}OH^+$.

Fortunately, modifications both to ICP-MS instruments and to methodologies have been developed to overcome these and other difficulties associated with the analysis of non-metal species. These modifications, which are described below, have allowed typically problematic non-metals such as iodine, bromine, sulfur, and phosphorus to be readily detected and monitored by ICP-MS, thereby increasing the potential application of this analytical technique as well as its overall versatility.

14.2.1 Useful tools in the analysis of non-metal species

14.2.1.1 Isotope dilution

As mentioned, the main problem associated with halogen detection by ICP-MS is poor ionization efficiency due to the high ionization energies of these elements. Isotope dilution represents one way in which this obstacle can be circumvented. It is based on the principle that most elements in nature possess fixed isotopic ratios.

Spiking a sample with an analyte standard consisting of a different isotopic abundance can alter this natural ratio. The change in the measured isotopic ratio can then be used to elucidate the original elemental composition of the sample. The only real requirements of this technique are that the analyte of interest has more than one stable isotope and that the spike standards are isotopically pure. Furthermore, isotope dilution has become widely recognized as a definitive technique that can be used to overcome many of the problems associated with instrumental drift and incomplete extraction of analyte from the sample [12]. In the same manner, isotope dilution can also be used to compensate for the poor ionization efficiency exhibited by a number of non-metal elements.

Over the past forty years, isotope dilution mass spectrometry (IDMS) has proven to be a very effective method for the measurement of total element concentration [13–16], and has routinely been used as part of EPA Method 6800 [17]. Following the equilibration of spiked isotopes with natural isotopes, the concentration of the element in the sample, C_x, can be calculated by the following equation:

$$C_x = (C_s W_s / W_x)\left({}^1A_s - R_{1/2}{}^2A_s / R_{1/2}{}^2A_x - {}^1A_x\right),$$

where C_s = the concentration of analyte in the spike; W_x and W_s = the mass (grams) of the sample and spike, respectively; ${}^1A_x$ and ${}^2A_x$ = the natural abundances of the analyte's isotopes; ${}^1A_s$ and ${}^2A_s$ = the isotopic abundances of the spike; and $R_{1/2}$ = the altered isotope ratio as measured by the mass spectrometer. Depending on the accuracy of the mass spectrometer and known isotopic abundances, the calculated value of C_x can be anywhere from 99.5 to 99.999% accurate [18].

In addition to total elemental measurements, isotope dilution can also be applied to elemental speciation studies. Heumann was the first to describe two different methodologies by which this can be accomplished [19,20] although a number of other papers have subsequently been published [12,18,21–28]. The "species-specific" mode entails spiking the sample with a specific elemental species that is isotopically labeled with an enriched isotope. This method is useful in the validation of speciation procedures, particularly in situations where certified reference materials are not available. The "species-unspecific" mode, on the other hand, involves the addition of a simple enriched isotope of the desired element after physically separating the various species present in the sample, thereby allowing reliable quantification of unknown species.

Both these methods offer clear benefits over other methods of species quantification, including shorter analysis times and less organic solvent consumption due to lower extraction efficiency requirements. However, they do not take into consideration any species interconversion that may occur during sample preparation. For that reason much attention has been focused on speciated isotope dilution mass spectrometry (SIDMS) [17,18,29,30], a technique specifically designed to account for this potential interconversion. SIDMS requires that each of the species spiked into the sample be labeled with a different isotope, thereby allowing any interconversions to be identified and mathematically calculated. Although originally developed for the speciation of organometallic complexes, SIDMS as well as "species-specific" and "species-unspecific" methods have become particularly useful in the analysis of non-metal containing compounds whose detection by ICP-MS has not normally been viable.

14.2.1.2 High resolution instruments

While many non-metals have more than one stable isotope, making them good candidates for isotope dilution, many of these isotopes suffer from spectral interferences, eliciting the need for high-resolution ICP-MS instruments. Quadrupole mass analyzers, which are found in nearly 90% of all ICP-MS systems, offer resolutions in the range of 0.7–1.0 amu and are not adequate to differentiate many non-metal elements from argon-, solvent-, and matrix-based spectral interferences [31]. Therefore, the introduction of commercially available high resolution ICP-MS (HR-ICP-MS) instruments in the late 1980s represented a significant advance in the analysis of non-metal elements.

HR-ICP-MS instruments make use of either magnetic sector or double-focusing sector mass analyzers. Magnetic sector mass analyzers consist of a magnetic field perpendicular to the ion beam from the plasma. When exposed to this field, ions are pushed into a circular path, the trajectory of which is dependent on mass to charge ratio, permitting the separation of different elements and isotopes [32,33].

Double-focusing sector mass analyzers work in the same manner except that they also subject ions to an electrostatic field with the resultant trajectory being dependent on the energy of the ion [32,33]. In instruments-based on standard Nier–Johnson geometry the electrostatic analyzer (ESA) is placed before the electromagnet, but as shown in Fig. 14.2, those based on reverse Nier–Johnson geometry position the electromagnet first [31]. The magnetic sector and electrostatic sector are carefully designed so that their energy dispersions are equal in magnitude, but arranged so that they cancel each other. This effect is referred to as double focusing and allows high resolution to be achieved even if the ion beam is not monoenergetic. If the direction of the electrostatic field is then switched during a mass scan, ions after this change will no longer reach the exit slit. Consequently, masses (or mass ranges) of interest can be selectively detected by switching the electric sector at appropriate times during the mass scan cycle.

The resolving power, R, of these instruments can be as high as 10,000. In comparison, resolving powers in the range of 300–400 are typically offered by traditional quadrupole mass analyzers. These values are calculated according to the equation

$$R = m/\Delta m$$

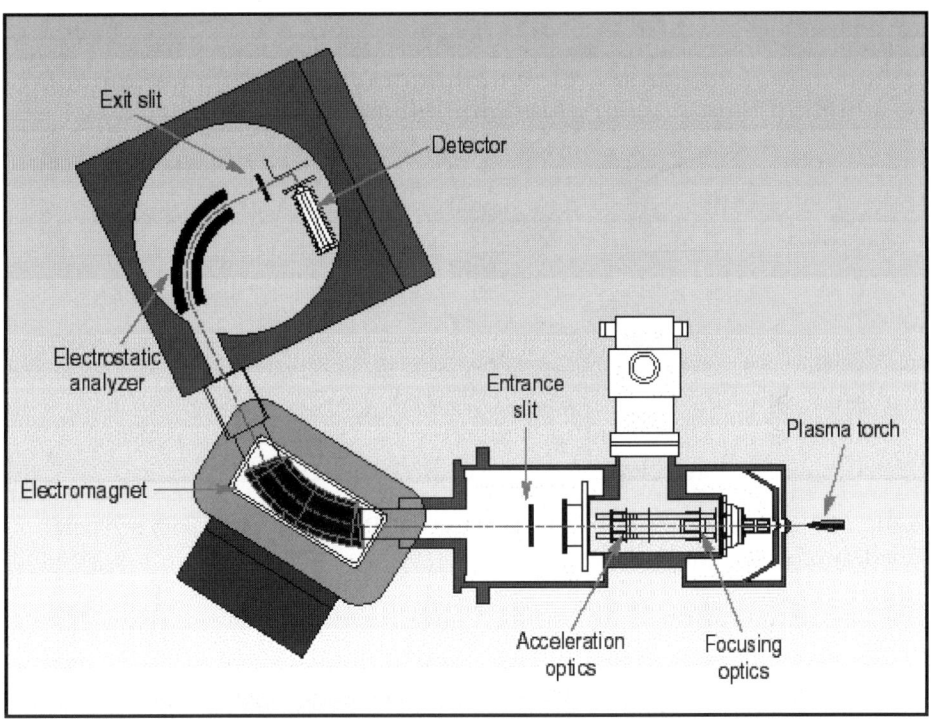

Figure 14.2 Schematic of a reverse Nier–Johnson double-focusing magnetic-sector mass spectrometer. (Courtesy of Thermo Finnegan.)

with m referring to the nominal mass at which the peak occurs and Δm being the mass difference between two resolved peaks [2]. In double-focusing sector mass analyzers different entrance and exit slit widths are employed to achieve desired resolution. However, the narrower slit widths used to yield higher resolutions also limit the ions transmitted to the detector, subsequently reducing sensitivity. For that reason, it is necessary to use the minimal resolution required to resolve the analyte of interest from the interfering complex.

Despite the obvious advantages of HR-ICP-MS instruments, they are significantly more expensive than the quadrupole mass analyzers commonly employed. They also require longer analysis times, which make them less than ideal for high-throughput applications. However, these instruments have dramatically enhanced ICP-MS analysis by increasing the number of elements that can effectively be measured.

14.2.1.3 Collision/reaction cell instruments

Analytes that suffer from major spectral interferences, such as those generated from the plasma or sample matrix, may require a collision or reaction cell instru-

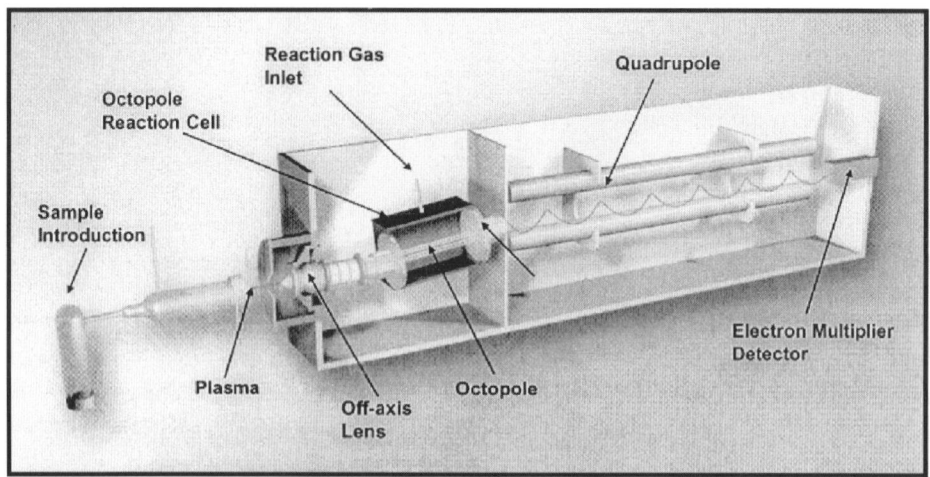

Figure 14.3 Schematic of a collision/reaction cell ICP-MS instrument with octopole. (Courtesy of Agilent Technologies.)

ment for reliable detection. These instruments, which were first introduced in the late 1990s as a tool by which to identify structures of organic molecules by MS, are actually able to alter the problematic species or even the analyte itself so that they no longer interfere with each other when they are introduced to the mass analyzer and detector.

A collision/reaction cell instrument, which is depicted in Fig. 14.3, is laid out in a similar manner as a traditional ICP-MS instrument except that a collision/reaction cell with gas inlet is positioned just before the quadrupole mass analyzer. The cell itself consists of another quadrupole, hexapole or octapole to which only rf-voltages are applied. When subjected to this rf-field, incoming ions are focused prior to colliding or reacting with the collision/reaction gas that is pumped through the gas inlet. The collisions and reactions that occur between the analyte ions and the gas molecules can eliminate problematic interferences by: (1) transforming the analyte ion into one that is not interfered with, or (2) converting the interfering species into ones that will not hinder the detection of the desired analyte.

Unfortunately, rather complex secondary collisions and reactions can occur within the cell to produce other interfering species, which must also be eliminated before they reach the detector. These secondary interferences are rejected in different ways, depending on the nature of the cell. Quadrupoles act through mass discrimination, whereas higher order multipoles, such as hexapoles and octapoles can discriminate analytes from interferences based on kinetic energy [36].

Quadrupoles can function as selective bandpass filters, which can be optimized and easily changed for various interferences. Also, in a practice known as dynamic reaction cell (DRC) technology, quadrupoles permit the use of highly reactive gases, including ammonia and methane, as the collision/reaction gas [37]. Such gases can act as catalysts for ion molecule chemistry, converting any interfering species into neutral or otherwise noninterfering species. Because the gases are highly reactive, the number of potential ion-molecule interactions is increased, as is the efficiency of interference elimination. Furthermore, by-products of these reactions can be prevented from forming other interfering species simply by scanning the bandpass of the cell's quadrupole.

Hexapoles, on the other hand, do not permit the use of highly reactive gases due to their inability to act as mass filters and control secondary reactions. Rather, these multipoles act as kinetic energy discriminators and require strictly low reactivity gases like helium, hydrogen, or xenon. Analyte ions can be distinguished from other ions by allowing the cell bias to be slightly less positive than the mass filter bias. This causes ions with the same energy as the cell bias to be rejected, whereas the analyte ions, which have energy higher than the cell bias, can be transmitted. Octapoles operate in a similar manner, but offer higher transmission characteristics, particularly at the lower end of the mass range.

Collision/reaction cell instruments have allowed for the analysis of more varied and novel samples, while still utilizing the advantages associated with the traditional quadrupole mass analyzers. As will be demonstrated in the following sections, these instruments, and the use of isotope dilution and HR-ICP-MS, have created a new dimension of versatility in the area of analytical chemistry, and have given ICP-MS the potential to become more than just a tool for trace metal analysis.

14.2.2 Applications

14.2.2.1 Halogens

As alluded to previously, the determination of halogens by ICP-MS has been somewhat difficult, owing to their high ionization energies and associated spectral interferences. The exception to this has been iodine, which, with a first ionization energy of 1.0084 MJ/mol [32], is amenable to ICP-MS analysis. In fact, the speciation of iodine complexes by LC-ICP-MS was documented as far back as 1992. In this initial work by Salov et al., IO_3^- and I^- were separated on a gel permeation column and could be detected at absolute amounts as low as 0.2 pmol when using a quadrupole mass analyzer [38].

Since iodine is a biologically essential element whose metabolism is highly dependent on chemical form, the applicability of LC-ICP-MS to the study of iodine-containing biological molecules is obvious. Hence, Taketera and Watanabe chose to use LC-ICP-MS for the speciation of iodo amino acids [39]. By enzymatically digesting the protein thyroglobin and employing reverse phase chromatography, monoiodotyrosine (MIT), diiodotyrosine (DIT), 3,3', 5- and 3,3',

$5'$-triiodothyronine (T_3 and reverse T_3) and thyroxine could be identified. Detection limits in the range of 35 to 130 pg iodine were reported, representing an order of magnitude improvement over previous methods. Consequently, zebrafish tissue [40] as well as human urine [41] and serum [41] have also been subjected to RP-HPLC-ICP-MS, which typically consists of a C_{18} column and TRIS buffer mobile phase, in order to detect these same iodine compounds. Capillary electrophoresis (CE) has also been used for these purposes [42].

Ion exchange chromatography (IEC), as well as size exclusion chromatography (SEC), has also proved useful in iodine analysis with ICP-MS. Studies with IEC have dealt mainly with inorganic species, but have been used to illustrate the decomposition and interconversion of many iodine complexes as a result of various sample pretreatment methods [43]. SEC, on the other hand, is able to separate larger compounds and has been used to determine the molecular weight distribution of iodine compounds bound to humic substances [44–46] in addition to detecting the nutritionally relevant iodine species in milk [47–50].

Many of these same techniques have also been applied to bromine analysis [46,48]. Despite a first ionization energy of 11.8 eV, bromine analysis by ICP-MS is a viable option. However, it is often necessary to increase the forward power of the ICP-MS from a normal value of about 1350 W to one in the range of 1550–1700 W in order to achieve a sufficient ionization efficiency [51,52].

Like iodine, bromine detection by ICP-MS is gaining popularity. With the realization that bromate, a disinfection by-product of the ozonation of drinking water, is a potential cancer causing agent associated with a lifetime cancer risk of 10^{-5} at levels of just 3 µg/L in drinking water [53], the need for a highly sensitive bromine detection system has become imperative. Several authors have reported using ion chromatography (IC) in conjunction with quadrupole ICP-MS instruments to specifically detect bromate in water samples [51,54–59]. While EPA method 300.0 [60] calls for the use of IC with conductivity detection for these purposes, it has been found that the high chloride matrix of the water samples can significantly hinder bromate detection. This is reflected in the 20 µg/L detection limit associated with this technique. Although, ICP-MS can also be inhibited by chloride related interferences, they are not an issue when monitoring the m/z of the bromine isotopes. Consequently, ICP-MS can be used to measure bromate despite its co-elution with chlorine by IC. Detection limits on the order of 1.6 µg/L have been reported for bromate by IC-ICP-MS, with even lower detection limits being obtained when preconcentration [55] or post-column derivatization methods [54,56] are employed. This method has even been deemed suitable for various food samples [61], which represent more complex matrices.

While chlorine may not be an issue, several other interferences, including the formation of $ArArH^+$ polyatomic ions, can hamper the detection of both bromine isotopes. Although, the signal-to-noise ratio at m/z 79, which is superior to that at

m/z 81, can permit quantification, it has been determined that accurate quantification is dependent on the use of isotope dilution [59,62].

Electrothermal vaporization (ETV) has been implemented, in place of conventional pneumatic nebulization, in an effort to minimize potential interferences. ETV involves placing the sample, which can be in either liquid or solid form, on a tungsten filament or in a graphite furnace. By using an appropriate temperature program and/or chemical modifier the analyte can be separated from the matrix and introduced to the ICP-MS, thereby reducing solvent and matrix related interferences. It has been reported to be beneficial in the analysis of organic chlorine [63] as well as iodine species in biological samples [64]. Yan et al. have even investigated the use of ETV and reduced pressure (RP) ICP-MS in the determination of halogens. By lowering the plasma pressure, signal intensity increased, thereby improving the detection limits of bromine and chlorine by roughly ten times in comparison with those at atmospheric pressure [65].

Aside from interferences, the volatility of halogenated species must also be considered when conducting ICP-MS analysis. Because species such as I_2, Br_2, and HBr are highly volatile, significant losses of analyte can occur when digestion and extraction methods cause the halogen species in a sample to be converted into these forms. For iodine, this can be avoided by employing methods that transform the iodine into nonvolatile species, such as iodide (I^-) or iodate (IO_3^-). Dry ashing, wet ashing and combustion with oxygen are examples of ways in which this can be accomplished [66]. Bromine loss can be minimized through the addition of $AgNO_3$ to a typical acid digestion [67]. The silver ions cause bromides to precipitate as the sparingly soluble AgBr, which can be separated from the digest and redissolved in ammonia. However, the removal of $[Ag(NH_3)_2]^+$ is necessary prior to reacidification for ICP-MS analysis.

In some situations, the potential volatility of halogenated compounds can be advantageous in that it may make them amenable to speciation by GC-ICP-MS. Such is the case for a class of compounds known as brominated flame retardants (BFRs), which were designed to stifle combustion by depriving flames of oxygen. Because they are now believed to be a health risk, the assessment of these chemicals in environmental and biological samples has become especially important. However, they are extremely insoluble in many of the LC mobile phases commonly used with ICP MS detection. Luckily, several congeners are able to be volatilized at temperatures below 325 °C, and GC-ICP-MS can, therefore, be used to separate and detect these compounds in less than 10 min [68]. Despite the highly organic nature of the samples analyzed, detection limits in the low to sub-ppb range were obtained by monitoring m/z 81 and employing a conventional quadrupole mass analyzer.

The detection of numerous iodine containing compounds has also benefited from GC-ICP-MS [69–73]. Most recently, iodophenols have been studied by this method. These compounds, which like bromate, are by-products of the disinfection of drinking water, could be seen at extremely low levels, with detection limits ranging from 0.07 to 0.12 ng/L for 2-iodophenol, 4-iodophenol, and 2,4,6-iodophenol. These detection limits, however, were obtained following sensitive solid-phase microextractions and were dependent on the use of oxygen as an optional gas [69].

14.2.2.2 Sulfur and phosphorus

The analysis of sulfur and phosphorus containing compounds by ICP-MS can be rather challenging. Phosphorus is monoisoptopic and suffers from intereferences with $^{15}N^{16}O^+$ and $^{14}N^{16}OH^+$. Sulfur, which has four stable isotopes at m/z 32, 33, 34, and 36, is interfered with by $^{16}O^{16}O^+$, $^{16}O^{16}O^1H^+$, $^{16}O^{18}O^+$, and $^{36}Ar^+$. The high ionization energies of these elements have also contributed to poor detection limits. Yet, because interest in these elements as well as the environmental and biological compounds that contain them has increased in recent years, methods have been developed to make their analysis by ICP-MS more feasible.

For example, Hanada and co-workers have proposed circumventing the problematic interferences by using ICP-MS to indirectly determine phosphorus in iron and steel samples [74]. This was achieved by converting all phosphorus species into molybdophosphate, $[P_2Mo_{18}O_{62}]^{6-}$ through the addition of ammonium molybdate and then monitoring for molybdenum rather than for phosphorus. Because molybdenum is relatively free of interferences, detection by ICP-MS is quite sensitive. Consequently, the determination of phosphorus by this method was effective at levels as low as 0.4 µg/g and was not inhibited by the presence of Al, Co, Cr, Cu, Mn, Ti, Ni, or Si.

Hughes et al. have also investigated the phosphorus content in iron samples, but have done so by direct determination using ETV [75], a technique that can also be used for sulfur determination. The use of ETV for sample introduction generates a dry aerosol, which drastically minimizes interferences from oxygen-containing polyatomic ions and provides more energy for analyte ionization. Both phosphorus and sulfur can form very volatile compounds, which allows low vaporization temperatures to be employed, but which can also result in analyte loss when higher temperatures must be used to adequately remove matrix ions. In order to combat potential analyte losses, chemical modifiers are added. Additionally, these modifiers can help increase the transport efficiency of the analyte and prevent co-volatilization of other elements, giving way to improved sensitivity. Hughes found that zirconium was an effective modifier in the ETV-ICP-MS analysis of phosphorus [75,76], allowing an absolute detection limit of 4 pg to be attained. Palladium has also been used as a chemical modifier for phosphorus [77] and even sulfur analyses [78], in conjunction with either ascorbic or nitric acid to further enhance matrix removal. Additionally, isotope dilution can be performed in conjunction with ETV-ICP-MS [79], yielding the potential for accurate quantitation of sulfur and phosphorus even at trace levels.

In spite of this, the biggest breakthrough in the ICP-MS analysis of sulfur and phosphorus has come with the introduction of high resolution and collision/reaction cell instruments. This technology facilitates the detection of these elements by either distinguishing them from interferences, or by significantly removing the interfering species.

Collision/reaction cell instruments use ion-molecule reactions between sulfur or phosphorus and O_2 to produce SO^+ and PO^+, which enjoy detection limits in the sub ng/g range. HR-ICP-MS instruments can provide resolution of 4000 or more, allowing the major sulfur and phosphorus isotopes at m/z 32 and 31, respectively, to be resolved from $^{16}O^{16}O^+$, $^{15}N^{16}O^+$, and $^{14}N^{16}OH^+$ interferences. As a result, detection limits on the order of 50 pg/g have been achieved for both elements. Consequently, low levels of sulfur in diesel fuel [80] and low levels of both sulfur and phosphorus in proteins [81–84] and high-purity process chemicals [85] have been detected by ICP-MS.

Isotope dilution has been used in conjunction with HR-ICP-MS to determine the sulfur content of geological materials. The information from these studies can be used to elucidate the Earth's sulfur cycle, but the accuracy of these measurements in rocks, soils, and sediments is crucial. Besides the interferences associated with sulfur detection, the plural oxidation states of sulfur have also presented problems when determining the total sulfur content of such samples. For that reason, Makishima and Nakamura have proposed oxidizing sulfur into sulfate, a highly stable form, with bromine generated in situ from HBr prior to the application of isotope dilution and HR-ICP-MS [86]. Not only has this technique yielded highly accurate sulfur measurements, but it requires only small sample amounts, even for materials with very low or high sulfur concentrations.

Organophosphorus pesticides were introduced in the 1970s as a replacement for organochlorine pesticides and have been used as insecticides, herbicides and fungicides. Since they could potentially pose a threat to human health, detection of these compounds in environmental samples has become pertinent. Recently, GC-ICP-MS with a collision/reaction cell was utilized for this purpose. By ramping the temperature of the GC oven to 250 °C, four different organophosphorus pesticides were separated before being carried to the ICP-MS by way of a 275 °C transfer line. While the use of nitrogen as an optional gas helped reduce background signal, it was the use of helium gas in the collision cell that, by energy discrimination, was able to minimize interferences at m/z 31. Consequently, the phosphorus species were easily identified by ICP-MS, with instrument detection limits in the high ng/L range achieved for each of the four organophosphorus pesticides.

The advances made in the detection of sulfur and phosphorus with high resolution and collision/reaction cell instruments have also been responsible for the emerging role of ICP-MS in biological and biochemical studies. While this section has highlighted some applications of ICP-MS to these elements, the vast majority pertains to the study of biological and clinical samples. These applications, as well as their implications and impact, will be the focus of the next section.

14.3 ICP-MS applications in studies of biological and clinical samples

Scientific progress in life sciences relies to some extent on the capabilities of current analytical technology. The common tasks include characterization of the composition of complex biological systems, elucidation of the biological pathways in living organisms, and understanding the pathological processes, toxicological aspects, etc. In many cases, the shortcomings of available analytical techniques have been limiting factors that hindered further development. An analytical approach is to use coupled techniques, based on combining the separation and detection step into one operating on-line system. A clear tendency can be observed toward using element specific detectors and, in particular, inductively coupled plasma— mass spectrometry. In this section, recent applications of ICP-MS in studies of biological and clinical significance are reviewed. The first part focuses on species-dependent bioavailability, uptake and distribution of elements in the living organisms, including plants, yeast, and animals. Several ICP-MS applications in studies on the pharmacokinetics of metal-based drugs are reviewed. Then, some examples are given on the role of this detection tool in studying the biological pathways of elements. Finally, applications concerning metal/metalloid interactions with biological molecules and protein analysis are reviewed.

14.3.1 Species dependent bioavailability

In the direct applications, ICP-MS has been used to assess a dose-dependent element uptake by different plant and/or animal tissues. One representative example is the work by Helgesen et al. [87], reporting the soil-to-carrot uptake rate of arsenic and copper. The increased accumulation of inorganic arsenic (As(III) and As(V)) was observed in carrots that were grown in highly contaminated soils compared with the accumulation from low contaminated soil. On the contrary, the accumulation of copper was not dose-dependent. The retention of cadmium, its distribution among different fish organs and its elimination were studied in a controlled pollution exposure experiment [88]. The element content in kidney and liver increased with the increment of cadmium concentration in polluted water. During the depuration step, the loss of accumulated cadmium was rapid in muscle but not in kidney and liver. The use of stable isotopes and ICP-MS to determine the fate of Mg absorption in rats was reported by Coudray et al. [89]. After oral and intravenous administration of different magnesium doses in form of two isotopes (^{25}Mg and ^{26}Mg), blood, feces, and urine were collected daily and the two isotopes measured by ICP-MS. The results indicated that with ^{25}Mg and ^{26}Mg having high natural abundance, the described methodology enables meaningful investigations of Mg bioavailability and metabolism. In order to evaluate cadmium absorption from wheat flour (porridge), other authors intrinsically labeled wheat with the stable ^{106}Cd isotope, keeping total element content similar to that expected in a normal diet [90]. The porridge was eaten at breakfast by adult and infant volunteers and bulk fecal collections were analyzed for unabsorbed Cd by ICP-MS. The data suggested that the apparent absorption of Cd might be higher than the 5% figure that is commonly quoted.

ICP-MS analyses of total elements in animal tissues were also carried out to assess possible relations between element status and various pathological disorders. As an example, elemental analysis of human brain tissues was performed in two regions of the brain in Alzheimer disease (AD) and in the age-matched control. The results showed significantly higher levels of Si, Sn, Al, and Mn in the parietal cortex of the AD brain than in the control [91].

It now is widely accepted that the bioavailability of elements as well as their pathways in living organisms is species dependent. Consequently, research interest has moved from total elemental analysis toward elemental speciation. For assessing a health risk related with soil contamination by arsenic (up to 56% of As(III) with respect to total element in soils), the leaching test was carried out. Using a rat model, the bioavailability of As(III) and As(V) from soils was evaluated by ICP-MS measurements that ranged from 1.02 to 9.87% and 0.26 to 2.98%, respectively, [92]. Species-dependent bioavailability of arsenic was studied by exposing mussels (*Mytilus edulis*) to trimethyl(carboxymethyl)arsonium bromide (arsenobetaine, C-1 betaine), trimethyl(2-carboxyethyl)arsonium bromide (C-2 betaine), or trimethyl(3-carboxypropyl)arsonium bromide (C-3 betaine). Arsenic was accumulated by the mussels in all cases but the efficiency of uptake decreased with the number of methylene units in the carboxyalkyl group. Chromatographic analysis (HPLC-ICP-MS) of the mussel extracts demonstrated that the arsenic compounds were accumulated unchanged [93].

In another work on arsenic uptake and distribution in animal organisms, speciation of As(III), As(V), dimethylarsinic acid (DMAs(V)), monomethylarsonic acid (MMAs(V)), trimethylarsine oxide (TMAsO), tetramethylarsonium ion (TMAs), arsenobetaine (AsB), arsenocholine (AsC), dimethylarsinoyl ethanol (DMAsE) and four common dimethylarsinoylribosides was performed by cation and anion exchange HPLC-ICP-MS. These species were determined in seaweed extracts and in the urine, blood serum, kidney, liver, and muscle of seaweed-eating sheep from Northern Scotland. The major arsenic species found in the diet were arsenoribosides. The major metabolite excreted into urine and blood was DMAs(V) $(95 +/- 4.1\%)$ with minor amounts of MMAs, riboside X, TMAs and an unidentified species. The analysis of tissue samples and wool from exposed sheep revealed the elevated levels of arsenic as compared with those in the control group [94].

Studies on bioavailability are also needed to evaluate the efficiency of supplementation with micronutrients. Thus, ICP-MS was used to determine an iodine uptake and excretion from different milk and dairy products [95]. By coupling HPLC and ICP-MS, Bratter et al. [96] compared the binding patterns of several elements (among them Ca, Mg, Co, Cu, Fe, Se, and Zn) to proteins in breast milk and various infant formulas. The results obtained for the two types of sample were significantly different, depending on the main component (human/cow milk or soy), its processing (hydrolysis) and the chemical form of the element added [96]. In recent decades, a number of studies have focused on selenium supplementation, because in many geographical regions the daily intake of this important antioxidant is below the recommended amount [97]. One approach is the characterization of element forms in different dietary products rich in selenium in order to identify species that could be responsible for the observed anticarcinogenic potential of these products. The common source of selenium is selenized yeast, containing protein-bound selenomethionine as the main species [98,99]. Enhanced anticarcinogenic activity was also observed for Se-enriched garlic and other vegetables from the Allium family. A few selenium analogs of organosulfur compounds were identified using hyphenated analytical techniques with ICP-MS and ESI-MS detection [100,101]. The primary species found in allium vegetables was Se-methylselenocysteine, reported as the most effective anticarcinogenic selenocompound. For selenized ramps, the species-dependent bioavailability of the element and tissue deposition in the rat model was evaluated by ICP-MS, showing different levels of selenium in blood plasma, kidney, muscle, and mammary gland as related to the animals fed with selenate or selenomethionine [102]. Owing to their high natural selenium levels, Brazil nuts can also be considered as an important dietary source of this element. Recently HPLC-ICP-MS determination of selenium distribution and speciation in different types of nuts was carried out in our laboratory [103,104]. The characterization of unidentified selenium-containing peaks was attempted by the employment of several procedures, including electrospray-mass spectrometry (ESI-MS) [105]. The results obtained showed about 65% of selenium in the form of selenomethionine.

Another ICP-MS approach to the selenium supplementation is to study the species-dependent efficiency of the fortification process. Lintschinger et al. used selenate solution to enrich wheat, alfalfa (*Medicago sativa*), and sunflower (*Helianthus annuus*) seeds during the sprouting stage. Uptake rates were studied by determination of total selenium. Metabolism of the absorbed selenium was analyzed by determination of selenium species in extracts of the sprouts using anion exchange HPLC coupled to ICP-MS [97]. The feasibility of wheat enrichment by selenate addition to soil fertilizers was investigated by Stadlober et al. [106]. Anion exchange HPLC-ICP-MS was optimized for the separation of selenite, selenate, selenocysteine, and selenomethionine. Total selenium determination and speciation analysis were performed in water extracts and in enzymatic digests of the wheat samples. It was shown that a major part of the selenate taken up by cereals was converted to selenomethionine. The fortification of onion leaves grown hydroponically in selenate or selenite containing medium was done recently [107]. Using two different pretreatment procedures and ion-pairing HPLC-ICP-MS, the primary organic selenium species found in leaves extracts was Se-methylselenocysteine. Four enrichment protocols for selenized yeast were examined by Ponce de Leon et al. [108]. The authors concluded that the most efficient incorporation to selenomethionine is obtained during the growth phase using small concentrations of sodium selenite.

14.3.2 Metal-based drugs

Increasing applications of ICP-MS can be observed in studies on the pharmacokinetics of metal-based drugs, most commonly different platinum anticancer formulations. The usefulness of electrothermal vaporization and laser ablation techniques for studying spatial distribution of platinum in micro samples of tissues and tumors was demonstrated by McKay et al. [109]. Before reaching the target cells, the platinum complexes undergo biotransformations in the bloodstream. The resulting aquated forms are considered biologically active [110]. Owing to the haematological toxicity observed for platinum drugs, the analytical speciation was undertaken. The common scheme is to evaluate "free" Pt in the ultrafiltrable plasma and the element level in erythrocytes versus total blood Pt. In parallel, the body clearance is examined by the analysis of urine [111]. In the cited study, up to 40% of the 2-h infusion dose of oxaplatin (diaminocyclohexane Pt coordination complex) was found in erythrocytes with the half-life of 12 to 50 days. The total half-life for ultrafiltrated plasma platinum, which could be available for tissues, was determined by flame atomic absorpion spectrometry (26 h) and by ICP-MS (up to 270 h), clearly indicating the necessity for using high sensitivity detection in such studies. The analytical figures of merit for platinum aquated species in plasma were compared for HPLC-ICP-MS and HPLC hyphenation with triple quadrupole MS [110]. The MS-MS and MS-MS-MS measurements enabled species characterization. However, the advantage of HPLC-ICP-MS was its better sensitivity (detection limit lower by an order of magnitude with respect to that achieved by triple quadrupole). It should be mentioned that the pharmacokinetics results obtained using different infusional modalities might be useful for optimization of drug administration schedules. Similar analytical schemes to that described above were used to study different platinum formulas, including a triplatinum complex BBR3464 [112] and satraplatin [113]. The simultaneous administration of various drugs has been proposed to decrease the toxicity of platinum complexes. Thus, Liu et al. applied ICP-MS to investigate the pharmacokinetics and biological effects of oxaplatin co-administered with paclitaxel versus oxaplatin alone in rat model [114]. In each case, the platinum level in the DNA fraction of peripheral blood cells correlated well with platinum in ultrafiltrable plasma. On the other hand, the elimination rate (urine analysis) increased at all dose levels of oxaplatin when combined with the second drug. The effect of *para*-aminohippurate and *tetra*-ethylammonium on cellular accumulation of *cis*-platin (*cis*-dichlorodiammine platinum(II)) was studied by Kolb et al. [115]. ICP-MS was one of the tools used to investigate changes of membrane lipids in human lung adenocarcinoma cells and their resistance to *cis*-platin [116]. In particular, platinum ICP-MS determination was carried out to measure the *trans*-membrane influx of the platinum drug for assessing the resistance of cancer cells. The mechanism of thallium influx in tumor cells is believed to be similar to that of *cis*-platin and ^{201}Tl scintigraphy has been used to detect malignant pulmonary disease. The feasibility of ICP-MS for such measurements was studied by Hanada et al. [117]. The lung carcinoma cell lines were exposed to *cis*-platin and to thallium, and cellular accumulation of the two elements was determined by ICP-MS. The results suggested that the accumulation of Tl in cells could reflect the cytotoxicity of *cis*-platin.

The pharmacokinetics of metal-containing agents for magnetic resonance imaging (MRI) was also studied by ICP-MS [118]. Frame et al. compared the results obtained for Gd-based contrast agent in rat tissues by using ICP-AES and ICP-MS [119]. They concluded that high sensitivity of ICP-MS detection is required to measure Gd concentrations in low dose MRI studies. It was also suggested that ICP-MS could replace radiotracer methods currently used for the low doses, as the detection limit obtained in tissue (6 pg/g) was an order of magnitude lower than in studies using the radiotracer technique.

The treatment of different diseases relies on the element-containing drugs and a few studies on their pharmacokinetics were performed using ICP-MS detection. Thus, the determination of gold-based drugs, auranofin, myochrysine, and their metabolites in urine from arthritis patients was performed by ion-pairing HPLC-ICP-MS. Dicyanogold(I) anion, $[Au(CN)2]^-$, was detected in the urine of several patients [120]. In another study, the restricted permeability of orally administered selenide phenylaminoalkyl antihypertensive agents to the central nervous system was demonstrated by ICP-MS measurements in a rat model. An oxidative procedure for the digestion and processing of tissue samples was developed, enabling the ICP-MS determination of Se-containing metabolites after the administration of selenide compounds. The results obtained using different selenium compounds indicated that the hydroxylation of aromatic ring caused lower levels of Se-containing metabolites in brain, while maintaining antihypertensive activity [121].

14.3.3 Biological pathways

Understanding the element pathways in living organisms is a necessary step toward elucidation of its specific biological function. Biomethylation processes are recognized as the major metabolic route for a number of elements, including arsenic, antimony, tin, mercury, etc. For arsenic, biologically mediated methylation has long been considered a natural detoxification process, while for mercury or antimony the methylated species present higher toxicity with respect to inorganic forms [122]. Although, the bibliography on element transformations in different organisms is quite extended, relatively few ICP-MS applications in this field have been reported. To date, the great majority of studies have focused on arsenic and selenium. The methodological development has been the primary interest in analytical research [99,123–132]. On the other hand, the ICP-MS technique was used in studies on pathways of biologically mediated processes.

The route of arsenic biotransformation in marine systems has been described as arsenate uptake by algae and it subsequent conversion to arsenosugars. Other marine organisms were shown to accumulate arsenic in various organic forms. The application of analytical speciation schemes with ICP-MS or ESI-MS detection contributed both in the characterization of these compounds [133,134] and in their determination in marine organisms [135–139]. For example, biotransformation of As(V) to the TMAs in the marine polychaetes was studied by Geiszinger et al. [140]. After evaluation of As speciation, these organisms were exposed to different concentrations of As(v) in seawater. The dose-dependent accumulation of the element was observed, as well as its ready methylation. TMAs was the major metabolite found and smaller amounts of the intermediate MMAs(V) and TMAsO were also observed.

The pathway of inorganic As in terrestrial environment has not been ultimately elucidated. It is known that inorganic arsenic forms predominantly in abiotic compartments, while the increasing contribution of different methylated species has been observed in the presence of bacteria and fungi [141]. Bacterial degradation of arsenobetaine (AsB) to TMAsO, DMAs(V), and MMAs(V) was demonstrated by HPLC coupled to hydride generation and atomic fluorescence spectrometry [142]. The application of ICP-MS detection in this work enabled the detection of the intermediate compounds in the biotransformation of AsB. The initial cleavage of the methyl–arsenic bond in AsB to form dimethylarsinoylacetate and the subsequent breakdown of the carboxymethyl–arsenic bond to yield DMAs(V) was demonstrated. ICP-MS detection was also used for the identification/quantification of arsenic metabolites in mushrooms, lichens, and green plants [143,144]. In mushrooms and lichens, the accumulation of organic forms was observed and few arsenite hyperaccumulating green plants were identified [143–145]. As reviewed in a paper by Keuhnelt et al., various experimental models have been used to study the biotransformation of arsenic in animals including rats, mice, rabbits, and monkeys [146]. In each case (except monkeys), the conversion of inorganic arsenic to MMAs(V), DMAs(V), TMAs, TMAsO, AsB, and arsenocholine (AsC) was observed. However, the efficiency of this process as well as the relative contribution of methylated arsenic compounds was different, depending on the animal species used in the experiment. Significant differences were also detected among the individuals of this same biological species [146]. In parallel, different susceptibility to arsenic toxicity in those organisms was observed [147,148]. In studies on arsenic biotransformation in mammals, ICP-MS was used in speciation analysis performed on urine samples [94,123,130,149–151]. An interesting study was recently reported by Katayama et al. [152]. Considering rapid and efficient conversion of inorganic arsenic to nontoxic organic species in marine organisms, the authors investigated the possible effect of dietary intake of seaweed on the biotransformation of As in rats. After administration of soluble or insoluble fraction of seaweeds in the two experimental groups, a large dose of arsenite was given to rats and arsenic speciation in urine was determined. The patterns obtained for arsenic compounds suggested that some components of seaweed could accelerate arsenic metabolism in rats. The proposed methylation pathway of arsenic in mammals involves the following steps:

   (i)   reduction of As(V) to As(III);
  (ii)   oxidative addition of methyl group on the As atom to form MMAs(V);
 (iii)   reduction to monomethylarsonous acid (MMAs(III));
 (iv)   oxidative methylation to DMAs(V);
  (v)   reduction to dimethylarsinous acid (DMAs(III));
 (vi)   oxidative methylation to trimethylarsine oxide (TMAsO); and
(vii)   reduction to TMAs [153].

The two labile intermediate arsenic species (MMAs(III), DMAs(III)) were detected in human urine by ion-pairing HPLC with hydride generation and element-specific (but not ICP-MS) detection [154,155]. These results provided the direct evidence in support of the putative route of As methylation. Due to the high toxicity of organic As(III) species, the detoxifying character of arsenic biomethylation has become controversial [156,157]. The mechanism of biomethylation was described as the enzymatic process controlled by specific methyltransferases, involving glutathione as the reducing agent and S-adenosylmethionine (AdoMet) as the donor of methyl group [158]. On the other hand, nonenzymatic methylation of arsenite was observed in the presence of methylvitamin B_{12} and glutathione [159].

The feasibility of ion-pairing HPLC-ICP-MS for studying the methylation processes of As was studied by Kannamkumarath et al. [160]. For enzymatic experiments, the extract of rabbit liver was fractionated by SEC and each fraction was then assayed for As(III) or MMAs(V) methylation. The results obtained seem to confirm that the first step in methylation was the element binding to the thiol group in glutathione. In the nonenzymatic experiments, methylvitamin B_{12} was used and the methylation of MMAs(V) to DMAs(V) was demonstrated. In addition to the complex composition and low element level in the samples, the important limitation in such speciation studies is the lability of target compounds. Suzuki et al. assumed that in body fluids arsenite might form the conjugates with glutathione or proteins through thiol groups [161]. Using HPLC-ICP-MS the authors demonstrated that these conjugated species of As(III) eluted intact on an anion exchange column, but on a cation column decomposition to free As(III) was observed.

In the case of selenium, studies on biologically mediated transformations focus on the following three aspects:

(i) biomethylation of inorganic forms;
(ii) specific incorporation of selenocysteine into proteins;
(iii) biological routes that involve competitive substitution of sulfur compounds by selenium analogs.

The first on this list is considered the elimination process and the putative pathway of biomethylation strongly parallels that of arsenic [162]. The reported methylated species of selenium include dimethylselenide (DMSe), dimethyldiselenide (DMDSe) (volatile compounds), and trimethylselonium ion (TMSe). The possible analogs of MMAs, DMAs, and TMAsO were not detected, possibly due to the fact that selenium–oxygen bonds are more easily cleaved than arsenic–oxygen bonds [122]. ICP-MS was used for the speciation of methylated selenium species in different plants [163,164], microorganisms [165], and clinical materials [166–168]. There has been a particular interest in certain microbe organisms and plants able to convert Se into volatile species. Since DMSe and DMDSe are harmless to the environment, phytovolatilization has been accepted as a potential bioremediation strategy [165]. The use of a new commercially available GC/ICPMS interface in conjunction with solid-phase microextraction allowed the virtually simultaneous monitoring of S and Se species from the headspace of the *Brassica juncea* seedlings grown in closed vials and treated with Se [163]. The addition of optional gases (O_2 and N_2) to the argon discharge proved to increase the sensitivity for Se and S as well as for Xe, which, as a trace contaminant gas, was used for ICP-MS optimization studies.

The important biological route of selenium is the synthesis of selenocysteine and its specific incorporation into a number of selenoproteins (ii) [169]. Some of these proteins have vital enzymatic functions [170]. Analytical speciation by HPLC-ICP-MS with the use of stable isotopes was used to follow key chemical reactions in the metabolism of selenite injected intravenously to rats [171]. The results showed that selenite was first reduced to selenide in the red blood cells; then the element moved into blood plasma and bound to albumin. It was also suggested that Se-albumin was transported to liver and utilized to synthesize selenoprotein P through selenophosphate and selenocysteine intermediates. This same research group applied ICP-MS for assessing Se fate in rats after intravenous injection of selenite enriched with [82]Se [172]. The concentration of labeled [82]Se and naturally occurring [77]Se was quantified in serum and in different tissues. The distributions of both exogenous and endogenous Se in serum, and supernatant fractions of the liver and kidneys were determined on a gel filtration column by HPLC-ICP-MS. The results obtained confirmed the rapid and efficient incorporation of Se into selenoprotein P in liver. For yeast and plants, in our studies only non-protein bound selenocysteine (proteinogenic amino acid 21) was found, suggesting the lack of specific incorporation of Se-cys to proteins in these biological species [103,105,107,173].

Different biotransformations of selenium occur by simple competition between sulfur and selenium analogs in sulfur pathways (iii). Speciation schemes were used for the characterization and determination of selenium metabolites in plants, microorganisms and mammals. In biological materials, the target selenium species include the low molecular weight compounds, and the forms weakly or strongly associated with biomolecules. The spatial distribution of selenium species in cells, tissues, and organs can also provide biologically useful information. Thus, the sample pretreatment becomes a key step in the analytical procedure. On the other hand, special care is needed in the selection of chromatographic conditions in order to keep the labile species intact during their separation [99,104, 164,173,174]. The speciation results obtained for yeast exposed to selenium and for nonaccumulating plants showed nonspecific incorporation of selenomethionine into proteins as the major metabolic route [98,99,104]. In the ICP-MS applications for selenium speciation in accumulating plants, the nonproteinogenic Se-amino acids such as Se-methylselenocysteine, Se-cystathionine, Se-homocysteine, or Se-methylmethionine were identified [100,129,164,175]. These data seem to confirm the proposed protective mechanism against selenium incorporation into proteins in these plants [176]. The genetic modification of *Brassica juncea* for more efficient synthesis of Se-methylselenocysteine was carried out and HPLC-ICP-MS or ESI-MS was used to characterize selenium species [175].

Among various questions remaining to clarify the diverse biological functions of selenium, the elucidation of its possible role in biomethylation is still a challenge. In studies carried out in enriched yeast, Se-adenosylmethionine (AdoSeMet) and Se-adenosylhomocysteine (AdoSeHcy) were found in cell extracts by ion-pairing HPLC-ICP-MS and confirmed both by enzymatic assay and by ESI-MS [173,177]. These data suggest that selenomethionine may enter the metabolic pathway of methionine in transmethylation reactions occurring in these microorganisms. A significantly lower ratio between AdoSeMet and AdoSeHcy was observed compared with the ratio between sulfur analogs; this could be ascribed to the higher methylation capacity of AdoSeMet compared with that of AdoMet [173].

The characterization of selenium metabolites in urine has been a common approach toward better understanding of selenium metabolism in mammals. Depending on the selenium intake, its chemical form, total urine selenium, and the pretreatment procedure applied, different selenium compounds were observed and a number of species still remain unidentified. Inorganic selenium and TMSe were often detected, confirming the biomethylation route of element elimination [167,178–180]. In recent studies significant progress in species characterization has been achieved by the application of different clean-up and extraction procedures as well as by the use of collision and reaction cells in ICP-MS [174,181,182]. The selenoamino acids, including selenomethionine, methylselenomethionine, and selenocystamine, were identified and confirmed by ESI-MS [167,174,180,181, 183]. AdoSeMet was also reported as a selenium metabolite in urine, using solid phase extraction, ion-pairing HPLC-ICP-MS, and laboratory synthesized standard [174]. As to the biological significance, these results seem to confirm the participation of selenium in pathways of sulfur compounds. However, in spite of using advanced analytical technology, the main selenium peaks observed on human urine chromatograms remain unidentified. After a clean-up procedure for the elimination of chloride and urea, Ogra et al. carried out selenium speciation in rat urine by ICP-MS and electrospray tandem mass spectrometry ESI-MS-MS using both positive and negative ion modes [184]. From the results, the major urinary Se metabolite was assigned as a diastereomer of a selenosugar, Se-methyl-N-acetyl-selenohexosamine.

14.3.4 Element interactions with proteins

In the early work on ICP-MS coupling to size exclusion chromatography (SEC), the feasibility of this system was studied for the separation and subsequent elemental analysis of metalloproteins in biological samples. The elution profiles of eight elements were acquired simultaneously and compared with the protein profile. The versatility of the technique was demonstrated by the quantitative multielement analysis of cytosolic metal-binding proteins separated from the polychaete worm (*Neanthes arenaceodentata*) [185]. Later, on-line SEC-ICP-MS coupling was used to assess the element distribution patterns in different biological samples [186,187]. In the analysis of protein liver extract, the use of a magnetic sector mass analyzer at medium mass spectral resolution helped to remove polyatomic interferences observed for Fe, S, and P [188]. Batch treatment with chelating resin (Chelex-100) enabled the speciation of zinc and copper binding with proteins in human serum by SEC-ICP-MS [189]. In this work, the concentrations of α 2-macroglobulin-zinc, ceruloplasmin-copper and the element species with albumin were estimated.

Better chromatographic resolution of biomolecules can be achieved using ion exchange columns (both HPLC and fast protein liquid chromatography FPLC). Anion exchange chromatography coupled to double focusing ICP-MS (DF-ICP-MS) via postcolumn isotope dilution analysis mode was used for the quantitative speciation of Fe, Cu, and Zn in human serum [190]. A similar separation system with DF-ICP-MS was applied to multielemental speciation [191]. The use of variable resolution allowed the detection of Ca, Sr, Fe, Cu, Zn, Se, Mn, Cr, Pb, Al, and Sn in different serum samples.

The common practice in the fractionation of proteins is the use of sodium dodecylsulfate polyacrylamide gel electrophoresis (SDS-PAGE). In a recent report, electrothermal vaporization (ETV)-ICP-MS was applied for the detection and quantification of selenium in proteins directly from the bands of stained gel [192]. The method was optimized by using glutathione peroxidase and then applied to the determination of selenium in proteins fractionated from a selenium-yeast candidate reference material. In another study, Lustig et al. used two-dimensional electrophoretic separation (isoelectric focusing and native PAGE) combined with sector field ICP-MS for the speciation of Cr, Ga, In, Pt, and V in human and rabbit sera [193].

The great majority of ICP-MS applications in studies on protein interactions with metals focused on the characterization of metallothioneins (MT). The biological importance of these proteins is due to their role in the homeostatic regulation of essential heavy metals like Cu and Zn. On the other hand, MT protect the cells from harmful chemicals, like nonessential and excessive essential heavy metals, reactive oxygen species, radicals, and alkylating agents [194]. In a recent review, Prange et al. [195] presented the development in the coupling of capillary electrophoresis (CE) and HPLC to element-specific and molecule-specific detectors, including ICP-MS and ESI-MS. More specifically, Fararello et al. [196] reviewed different chromatographic approaches with ICP-MS detection for multielemental speciation in metallothioneins and metallothioneine-like proteins.

Several papers focused on the element binding to transferrin (Tf). As to the possible biological significance, the identification of this iron-transporting glycoprotein as a target for other elements could contribute in a better understanding of the pharmacokinetics of these elements in the human body. It should be stressed that the application of a highly sensitive ICP-MS detection system allowed element speciation in unspiked clinical samples to be performed [197]. Four molecular forms of Tf with different iron-binding states were separated on an ion exchange column and iron was detected in each chromatographic peak by on-line ICP-MS [198]. The chemical forms of Al and Fe bound to human serum Tf were investigated by anion exchange HPLC coupled to high resolution (HR) ICP-MS [199]. The levels of ^{27}Al, ^{56}Fe, and ^{32}S, which are interfered with by polyatomic ions such as $^{13}C^{14}N^+$, $^{12}C^{15}N^+$, $^{12}C^{14}N^1H^+$, $^{40}Ar^{16}O^+$, $^{40}Ca^{16}O^+$, and $^{16}O_2^+$ when using a quadrupole mass analyzer, were monitored simultaneously by HR-ICP-MS at a mass resolution of 3000. Monitoring of ^{32}S was used to assess the protein levels in the HPLC eluate. In the following study, the same analytical system was used to investigate the effect of structural changes in Tf on the affinity of Al and Fe to this protein [200]. It was shown that the loss of sialic acid in Tf caused enhanced binding of Al. The effect could contribute to a better understanding of aluminium accumulation in patients with renal insufficiency, especially in dialysis treatment. The same research group reported the binding patterns of V(III), V(IV), and V(V) to human serum Tf using similar analytical methodology [201]. The observed affinity decreased from the lowest to the highest oxidation state of the element in the presence of bicarbonate. This order was altered in the absence of bicarbonate, which generally is required for metal binding to Tf. Another element of interest is bismuth, which is widely used in clinical treatment of different gastrointestinal ulcers. FPLC-ICP-MS was the analytical tool for studying the competitive binding of Bi to human albumin and transferrin [202]. It was found that over 70% of bismuth binds to Tf, even in the presence of a large excess of albumin (albumin/transferrin = 13:1).

Another promising application of ICP-MS in studies on metal–protein interactions is related to the detoxifying effect of selenium observed for mercury, cadmium, arsenic, etc. Possible interaction of these elements with selenoproteins has been proposed as a mechanism responsible for the reduction of their toxicity [203]. After simultaneous administration of ^{82}Se-enriched selenite and mercuric chloride into a rat, the co-elution of two elements with a protein peak was observed on the SEC-ICP-MS chromatogram of the serum sample. Although Se and Hg could bind to many plasma proteins, the affinity to the specific protein was high, with an equimolar ratio between elements of (1:1) [203]. In a further development, these same authors suggested that the target protein could be selenoprotein P (Sel P) [204]. A similar HPLC-ICP-MS system was used to study the interaction of silver, cadmium, and mercury ions with selenium in vitro [205]. The results obtained indicated that the interaction between transition metals and Se might occur through the general mechanism, i.e., transition metal ions and selenide form the unit complex (metal-Se), which then binds to selenoprotein P to form the ternary complex (metal-Se)$_n$(Sel P)$_m$, where n is the number of Hg–Se equimolar complexes and m is the number of binding sites in Sel P.

A few reports on metal ion characterization in metalloenzymes by hyphenated techniques with ICP-MS detection should also be mentioned. Inhibition, reactivation, and determination of metal ions in membrane metalloproteases of bacterial origin were studied by Leopold et al. [206]. The results obtained by SEC-ICP-MS and the results of enzymological methods indicated that two different membrane proteases from *Bacillus cereus* and *Pseudomonas aeruginosa* were zinc metalloproteases. In another study, ICP-MS measurements helped in the characterization of fibrinolytic metalloprotease from the fruiting bodies of an edible mushroom [207]. Using HPLC-ICP-MS, Suzuki et al. detected a zinc-binding protein present specifically in the livers of male adult rats [208].

Finally, an interesting application of ICP-MS in studies of the mechanisms involved in different encephalopathies was reported by Jobling et al. [209]. It was shown that copper and zinc binding modulated the aggregation and neurotoxic properties of prion protein, important in fibrillogenesis.

14.3.5 Applications for proteomics

Without any doubt, proteomics is a primary focal area in life sciences. In early approaches, proteomics focused on cataloging levels of protein expression. Today this concept is much broader and can be defined as the effort to establish the identities, quantities, structures as well as the biochemical and cellular functions of all proteins in an organism, organ, or organelle and how these properties vary in space, time, and physiological state [210]. In other words, in addition to the identification and quantification of proteins, proteomics includes the determination of their localization, modifications, interactions, activities and, ultimately, their function. Major types of proteomics technologies include protein fractionation, enzymatic digestion, one- or two-dimensional electrophoresis, liquid chromatography, capillary electrophoresis, mass spectrometry, and protein informatics.

In the last several years, a number proteomics approaches involving ICP-MS detection have been reported. The unique feature of atomic mass spectrometry is its high selectivity and sensitivity, which enable obtaining accurate quantitative data over a wide dynamic rang, almost independent of the chemical form of the analyte. By taking advantage of these features, ICP-MS was introduced for the analysis of proteins. To do so, a certain form of metal is attached specifically to the target protein, which can be then distinguished from the background of similar molecules and quantified by ICP-MS measurement of the element. As summarized by Baranov et al., tagging with gold, different lanthanides, tungsten, silver, and platinum was proposed in different analytical schemes [84]. Nanoparticles of gold were used for the quantification of low-density lipoproteins and also for labeling antibodies in the application of ICP-MS to immunoassay [84,211–213]. Fluorescent chelates of Eu, Tb, Dy, and Sm labeled antibodies with up to 8 metal atoms resulted in enhanced sensitivity with ICP-MS detection [212,214]. It should be stressed that this detection technique could be competitive in terms of both sensitivity and selectivity with respect to molecular fluorescence used in automated immunoassay systems. Recently, the possibility of the use of ICP-MS for simultaneous determination of at least two proteins by element-tagged immunoassay was demonstrated [215].

As discussed above, the advance in instrumental development through the introduction of the collision/reaction cell and high resolution ICP-MS dramatically improved the sensitivity and selectivity for non-metals. The methodological progress caused an immediate expansion of ICP-MS biological applications. In particular, quantitative determination of sulfur and phosphorus in biological micro samples is possible at the ultratrace levels [81,182,216,217]. On the other hand, simultaneous speciation of phosphorus and trace elements in plant seeds was carried out by coupling size exclusion chromatography with ICP-MS detection [218]. In further developments, ICP-MS was hyphenated with different separation schemes for proteomics applications. Within this context, Wind et al. focused on protein phosphorylation studies [82,83,219]. Being the most abundant posttranslational covalent modification of proteins, phosphorylation is involved in the regulation of important cellular processes including stress response and carcinogenesis [220]. The analytical scheme, in which the capillary liquid chromatography (μLC) was coupled alternatively to HR-ICP-MS with ^{31}P detection or to ESI-MS, was proposed for the identification of phosphopeptides [83]. The feasibility of this system was demonstrated by analyzing a mixture of synthetic phosphopeptides and a set of tryptic digests of three phosphoproteins. A similar strategy was used to determine the phosphorylation degree in proteins and peptides containing cysteine and/or methionine residues [221]. The phosphorus to sulfur (^{31}P to [32]S) ratio was determined by μLC-ICP-MS and converted into the degree of phosphorylation based on protein/peptide sequence information. In the subsequent study, the identification of phosphorylation sites in the polo-like kinases was achieved [82]. In the proposed procedure, the peptides were generated by tryptic digestion of proteins and then separated by capillary liquid chromatography with ^{31}P detection by HR-ICP-MS. In the third step, the identification of peptides was carried out by ESI-MS and tandem MS-MS. While the identification of phosphorylation sites was approached by molecular mass spectrometry, elemental mass spectrometry provided the quantitative phosphorylation profile. Although liquid chromatography is a powerful separation tool, complete sequence coverage of proteins generally is obtained by gel electrophoresis. Laser ablation ICP-MS for ^{31}P was used for spotting and quantification of proteins purified by 1-D SDS-PAGE [219]. In this work, two phosphoproteins (bovine α-caseine, 24 kDa and human fibrinogen Aα, 66 kDa) (1), three nonphosphorylated proteins (whale myoglobin, 14 kDa, human fibrinogen γ, 48 kDa, and human fibrinogen Bβ, 54 kDa) (2), and their mixture (1 + 2) were separated by gel electrophoresis. Blotting was performed prior to LA-ICP-MS in order to increase the sensitivity and also to eliminate inorganic phosphate. The images obtained by LA-ICP-MS and by staining with naphthalene black were compared, confirming that only phosphoproteins were detected in the ^{31}P analysis. The quantitative aspect was also discussed.

On the other hand, reversed phase chromatography coupled to HR-ICP-MS with ^{31}P and to ESI-MS was proposed for the quantitative determination of nucleotides from chemically modified DNA [222]. In the recent study by Profrock et al., four deoxynucleotide monophosphates (dNMP) present in the DNA chain were separated by CE and by HPLC, both coupled to ICP-MS with an octapole reaction cell [223].

14.4 Conclusions

ICP-MS has become a useful analytical tool in studies carried out in different life sciences. The versatility of this technique relies on its multielemental capabilities, high selectivity, and sensitivity as well as its easy coupling with different separation techniques. The determination of total elements in small amounts of biological and clinical samples can provide information on uptake rates, accumulation, spatial distribution, and excretion of the elements in living organisms. Speciation studies are helpful in assessing pharmacokinetics aspects of elements, in elucidation of their biological pathways and in understanding possible cellular mechanisms responsible for the observed health effects. The methodological progress toward enhanced performance of non-metals determination caused the expansion of biological applications of ICP-MS. Today, ICP-MS can be considered a valuable, complementary tool in various proteomics approaches.

References

1. A. Montasser, J.A. McLean, H. Liu, and J.M. Mermet, In: A. Montasser (Ed.). *Inductively Coupled Plasma Mass Spectrometry.* Wiley, Berlin, 1998, p. 964.

2. F. Adams, R. Gijbels, and R. Van Grieken, In: *Inorganic Mass Spectrometry.* Wiley, New York, 1988.

3. R.S. Houk, V.A. Fassel, and H.J. Svec, Inductively coupled plasma-mass spectrometry: sample introduction, ionization, ion extraction and analytical results. *Dynamic Mass Spectrom.* **6** (1981) 234.

4. A.R. Date and A.L. Gray, Plasma source mass spectrometry using an inductively coupled plasma and high resolution quadrupole mass filter. *Analyst* **106** (1981) 1255.

5. D.J. Douglas and J.B. French, Elemental analysis with a microwave-induced plasma/quadrupole mass spectrometer system. *Anal. Chem.* **53** (1981) 37.

6. R. Thomas, A beginner's guide to ICP-MS: part I. *Spectroscopy* **16** (2001) 38.

7. S.J. Hill, M.J. Bloxham, and P.J. Worsfold, Chromatography coupled with inductively coupled plasma atomic emission spectrometry and inductively coupled plasma mass spectrometry. A review. *J. Anal. Atom. Spectrom.* **8** (1993) 499.

8. F.A. Byrdy and J.A. Caruso, Elemental analysis of environmental samples using chromatography coupled with plasma mass spectrometry. *Environ. Sci. Technol.* **28** (1994) 528A.

9. K. Sutton, R.M.C. Sutton, and J.A. Caruso, Inductively coupled plasma mass spectrometric detection for chromatography and capillary electrophoresis. *J. Chromatogr. A* **789** (1997) 85.

10. C.A. Ponce de Leon, M. Montes-Bayon, and J.A. Caruso, Elemental speciation by chromatographic separation with inductively coupled plasma mass spectrometric detection. *J. Chromatogr. A* **974** (2002) 1.

11. M. Montes-Bayon, K. DeNicola, and J.A. Caruso, Liquid chromatography-inductively coupled plasma mass spectrometry. *J. Chromatogr. A* **1000** (2003) 457.

12. S.J. Hill, L.J. Pitts, and A.S. Fisher, High-performance liquid chromatography-isotope dilution inductively coupled plasma mass spectrometry for speciation studies: an overview. *TrAC* **19** (2000) 120.

13. G.N. Bowers Jr., J.D. Fassett, and E. White, Isotope dilution mass spectrometry and the National Reference System. *Anal. Chem.* **65** (1993) 475r.

14. J.D. Fassett and P.J. Paulsen, Isotope dilution mass spectrometry for accurate elemental analysis. *Anal. Chem.* **61** (1989) 643a.

15. H.G. Heumann, *Inorganic Mass Spectrometry*. Wiley, New York, 1988, p. 301.

16. H.P. Longerich, The application of isotope dilution to inductively coupled plasma-mass spectrometry. *Atom. Spectrosc.* **10** (1989) 112.

17. "EPA S-. Method 6800: elemental and speciated isotope dilution mass spectrometry." In: *Test Methods for Evaluating Solid Waste*. Environmental Protection Agency, Washington, DC, 1998.

18. D. Huo, H.M. Kingston, and B. Larget, "Application of isotope dilution in elemental speciation: speciated isotope dilution mass spectrometry (SIDMS)." In: D. Barcelo (Ed.). *Comprehensive Analytical Chemistry*. Elsevier, Amsterdam, 2000, p. 277.

19. K.G. Heumann, *Metal Speciation in the Environment*, Springer, Berlin/Heidelberg/New York, 1990.

20. K.G. Heumann, Isotope dilution mass spectrometry. *Int. J. Mass Spectrom. Ion Process.* **118/119** (1992) 575.

21. D. Tanzer and K.G. Heumann, Determination of dissolved selenium species in environmental water samples using isotope dilution mass spectrometry. *Anal. Chem.* **63** (1991) 1984.

22. L. Rottmann and K.G. Heumann, Development of an online isotope dilution technique with HPLC/ICP-MS for the accurate determination of elemental species. *Fresenius J. Anal. Chem.* **350** (1994) 221.

23. R. Nusko and K.G. Heumann, Cr(III)/Cr(VI) speciation in aerosol particles by extractive separation and thermal ionization isotope dilution mass spectrometry. *Fresenius J. Anal. Chem.* **357** (1997) 1050.

24. R. Nusko and K.G. Heumann, Chromium speciation with isotope dilution mass spectrometry. *Anal. Chim. Acta* **286** (1994) 283.

25. K.G. Heumann, S.M. Gallus, G. Radlinger, and J. Vogl, Accurate determination of element species by online coupling of chromatographic systems with ICP-MS using isotope dilution technique. *Spectrochim. Acta Part B* **53** (1998) 273.

26. K.G. Heumann, L. Rottman, and J. Vogl, Elemental speciation with liquid chromatography–inductively coupled plasma isotope dilution mass spectrometry. *J. Anal. Atom. Spectrom.* **9** (1994) 1351.

27. A.A. Brown, L. Ebdon, and S.J. Hill, Development of a coupled liquid chromatography-isotope dilution inductively coupled plasma mass spectrometry method for lead speciation. *Anal. Chim. Acta* **286** (1994) 391.

28. J.R. Encinar, Isotope dilution analysis for speciation. *Anal. Bioanal. Chem.* **375** (2003) 41.

29. H.M. Kingston, D. Huo, Y. Lu, and S. Chalk, Accuracy in species analysis: speciated isotope dilution mass spectrometry (SIDMS) exemplified by the evaluation of chromium species. *Spectrochim. Acta Part B* **53** (1998) 299.

30. H.M. Kingston, Inventor of method of speciated isotope dilution mass spectrometry. USA patent 5,414,259, 1995.

31. R. Thomas, A beginner's guide to ICP-MS: part VII: mass separation devices—double-focusing magnetic-sector technology. *Spectroscopy* **16** (2001) 22.

32. K.L. Ackley, K.L. Sutton, and J.A. Caruso, "The use of ICP-MS as a detector for elemental speciation studies." In: D. Barcelo (Ed.). *Comprehensive Analytical Chemistry.* Elsevier, Amsterdam, 2000, p. 249.

33. G. O'Connor and E.H. Evans, In: S.J. Hill (Ed.). *Inductively Coupled Plasma Spectrometry and Its Applications.* Sheffield Academic Press, Sheffield, 1999, p. 370.

34. R. Hutton, A. Walsh, D. Milton, and J. Cantle, Ultratrace elemental analysis by plasma source high resolution mass spectrometry. *ChemSA.* **17** (1991) 213.

35. U. Geissmann and U. Greb, High resolution ICP-MS—a new concept for elemental mass spectrometry. *Fresenius J. Anal. Chem.* **350** (1994) 186.

36. R. Thomas, A beginner's guide to ICP-MS: part IX—mass analyzers: collision/reaction cell technology. *Spectroscopy* **17** (2002) 42.

37. S.D. Tanner and V.I. Baranov, Theory, design, and operation of a dynamic reaction cell for ICP-MS. *Atom. Spectrosc.* **20** (1999) 45.

38. V. Salov, J. Yoshinaga, Y. Shibata, and M. Morita, Determination of inorganic halogen species by liquid chromatography with inductively coupled argon plasma mass spectrometry. *Anal. Chem.* **64** (1992) 2425.

39. K. Takatera and T. Watanabe, Speciation of iodo amino acids by high-performance liquid chromatography with inductively coupled plasma mass spectrometric detection. *Anal. Chem.* **65** (1993) 759.

40. R. Simon, B. Michalke, and K.W. Schramm, Background levels of iodinated thy-ronines and thyrosines in whole body homogenates of adult male and female zebrafish (Danio rerio) via LC/ICP-MS. *Organohalogen Comp.* **52** (2001) 120.

41. B. Michalke, P. Schrammel, and H. Witte, Method developments for iodine spe-ciation by reversed-phase liquid chromatography-ICP-mass spectrometry. *Biol. Trace Elem. Res.* **78** (2000) 67.

42. B. Michalke, Potential and limitations of capillary electrophoresis inductively coupled plasma mass spectrometry. *J. Anal. Atom. Spectrom.* **14** (1999) 1297.

43. H.J. Stark, J. Mattusch, R. Wennrich, and A. Mroczek, Investigation of the IC-ICP-MS determination of iodine species with reference to sample digestion pro-cedures. *Fresenius J. Anal. Chem.* **359** (1997) 371.

44. G. Radlinger and K.G. Heumann, Transformation of iodine in natural and waste-water systems by fixation on humic substances. *Environ. Sci. Technol.* **34** (2000) 3932.

45. G. Radlinger and K.G. Heumann, Determination of halogen species of humic substances using HPLC/ICP-MS coupling. *Fresenius J. Anal. Chem.* **359** (1997) 430.

46. K.G. Heumann, G. Radlinger, M. Erbes, I. Heiber, U. Obst, Z. Filip, and H. Claus, Ageing of dissolved halogenated humic substances and the microbiological influ-ence on this process. *Acta Hydrochim. Hydrobiol.* **28** (2000) 193.

47. P. Bratter, I.N. Blasco, V.E.N. de Bratter, and A. Raab, Speciation as an analytical aid in trace element research in infant nutrition. *Analyst* **123** (1998) 821.

48. F.A.R. Martino, M.L.F. Sanchez, and A.S. Mendel, Multi-elemental fractiona-tion in milk whey by size exclusion chromatography coupled online to ICP-MS. *J. Anal. Atom. Spectrom.* **17** (2002) 1271.

49. M. Leiterer, D. Truckenbrodt, and K. Franke, Determination of iodine species in milk using ion chromatographic separation and ICP-MS detection. *Eur. Food Res. Technol.* **213** (2001) 150.

50. L.F. Sanchez and J. Szpunar, Speciation analysis for iodine in milk size exclusion chromatography with inductively coupled plasma mass spectrometric detection (SEC-ICP-MS). *J. Anal. Atom. Spectrom.* **14** (1999) 1697.

51. A. Dudoit and S.A. Pergantis, Ion chromatography in series with conductivity detection and inductively coupled plasma-mass spectrometry for the determina-tion of nine halogen, metalloid and non-metal species in drinking water. *J. Anal. Atom. Spectrom.* **16** (2001) 575.

52. M. Nowak and A. Seubert, Ultra-trace determination of bromate in drinking wa-ters by means of microbore column ion chromatography and online coupling with inductively coupled plasma mass spectrometry. *Anal. Chim. Acta* **359** (1998) 193.

53. WHO, *Guidelines for Drinking Water Quality*, second ed., Geneva, 1993.

54.   M. Yamanaka, T. Sakai, H. Kumagai, and Y. Inoue, Specfic determination of bromate and iodate in ozonated water by ion chromatography with postcolumn derivatization and inductively coupled plasma mass spectrometry. *J. Chromatogr. A* **789** (1997) 259.

55.   J.T. Creed, M.L. Magnuson, J.D. Pfaff, and C. Brockhoff, Determination of bromate in drinking waters by ion chromatography with inductively coupled plasma mass spectrometric detection. *J. Chromatogr. A* **753** (1996) 261.

56.   H.P. Wagner, B.V. Pepich, D.P. Hautman, and D.J. Munch, Analysis of 500 ng/L levels of bromate in drinking water by direct-injection suppressed ion chromatography coupled with a single, pneumatically delivered post-column reagent. *J. Chromatogr. A* **850** (1999) 119.

57.   J.T. Creed, M.L. Magnuson, and C. Brockhoff, Determination of bromate in the presence of brominated haloacetic acids by ion chromatography with inductively coupled plasma mass spectrometric detection. *Environ. Sci. Technol.* **31** (1997) 2059.

58.   J. Diemer and K.G. Heumann, Bromide/bromate speciation by NTI-IDMS and ICP-MS coupled with ion exchange chromatography. *Fresenius J. Anal. Chem.* **357** (1997) 74.

59.   J.T. Creed and C.A. Brockhoff, Isotope dilution analysis of bromate in drinking water matrixes by ion chromatography with inductively coupled plasma mass spectrometric detection. *Anal. Chem.* **71** (1999) 722.

60.   US EPA, *EPA Method 300.0, Methods for the Determination of Inorganic Substances in Environmental Samples*, US EPA, Washington, DC, 1993.

61.   D.T. Heitkemper, L.A. Kaine, D.S. Jackson, and K.A. Wolnik, Practical applications of element-specific detection by inductively coupled plasma atomic emission spectroscopy and inductively coupled plasma mass spectrometry to ion chromatography of foods. *J. Chromatogr. A* **671** (1994) 101.

62.   J. Dahmen, M. Pfluger, M. Martin, L. Rottman, and G. Weichbrodt, Trace element determination of high-purity chemicals for the processing of semiconductors with high-resolution ICP-mass spectrometry using stable isotope dilution analysis (IDA). *Fresenius J. Anal. Chem.* **359** (1997) 410.

63.   P.K.G. Manninen, Determination of extractable organic chlorine by electrothermal vaporization inductively coupled plasma mass spectrometry. *J. Anal. Atom. Spectrom.* **9** (1994) 209.

64.   P. Schramel and S. Hasse, Iodine determination in biological materials by ICP-MS. *Mikrochim. Acta* **116** (1994) 205.

65.   X. Yan, T. Tanaka, and H. Kawaguchi, Electrothermal vaporization for the determination of halogens by reduced pressure inductively coupled plasma mass spectrometry. *Spectrochim. Acta Part B* **51** (1996) 1345.

66. G. Knapp, B. Maichin, P. Fecher, S. Hasse, and P. Schramel, Iodine determination in biological materials: options for sample preparation and final determination. *Fresenius J. Anal. Chem.* **362** (1998) 508.

67. F. Di Narda, R. Toniolo, and G. Bontempelli, Improved microwave digestion procedure for inductively coupled plasma mass spectrometric determinations of inorganic bromide residues in foodstuffs fumigated with methyl bromide. *Anal. Chim. Acta* **436** (2001) 245.

68. A.P. Vonderheide, M. Montes-Bayon, and J.A. Caruso, Development and application of a method for the analysis of brominated flame retardants by fast gas chromatography with inductively coupled plasma mass spectrometric detection. *J. Anal. Atom. Spectrom.* **17** (2002) 1480.

69. R.G. Wuilloud, J.C.A. Wuilloud, A.P. Vonderheide, and J.A. Caruso, Determination of iodinated phenol species at parts-per-trillion concentration levels in different water samples by solid-phase microextraction/offline GC-ICP-MS. *J. Anal. Atom. Spectrom.* **18** (2003) 1119.

70. A. Schwarz and K.G. Heumann, Two-dimensional on-line detection of brominated and iodinated volatile organic compounds by ECD and ICP-MS after GC separation. *Anal. Bioanal. Chem.* **374** (2002) 212.

71. U.M. Gruter, J. Kresimon, and A.V. Hirner, A new HG/LT-GC/ICP-MS multi-element separation technique for real samples in different matrices. *Fresenius J. Anal. Chem.* **368** (2000) 67.

72. S. Slaets, E. Laturnus, and F.C. Adams, Microwave induced plasma atomic emission spectrometry. A suitable detection system for the determination of volatile halocarbons. *Fresenius J. Anal. Chem.* **364** (1999) 133.

73. C. Brede, E. Lundanes, Y. Greibrokk, and S. Pedersen-Bjergaard, Simultaneous element-selective detection of C, F, Cl, Br and I by capillary gas chromatography coupled with microplasma mass spectrometry. *J. High Res. Chromatogr.* **21** (1998) 633.

74. K. Hanada, K. Fujimoto, M. Shimura, and K. Yoshioka, Determination of trace amounts of Si and P in iron and steel using gel chromatographic separation followed by ICP-MS. *Phys. Status Solidi* **167** (1998) 383.

75. D.M. Hughes, D.C. Gregoire, H. Naka, and C.L. Chakrabarti, Determination of trace amounts of phosphorus in high purity iron by electrothermal vaporization inductively coupled plasma mass spectrometry. *Spectrochim. Acta Part B* **53** (1998) 1079.

76. D.M. Hughes, D.C. Gregoire, H. Naka, and C.L. Chakrabarti, The vaporization of phosphorus compounds and the use of chemical modifiers for the determination of phosphorus by electrothermal vaporization inductively coupled plasma mass spectrometry. *Spectrochim. Acta Part B* **52** (1997) 517.

77. M. Resano, M. Verstraete, F. Vanhaecke, L. Moens, A. van Alphen, and E.R. Denoyer, Simultaneous determination of Co, Mn, P and Ti in PET samples by

solid sampling electrothermal vaporization ICP-MS. *J. Anal. Atom. Spectrom.* **15** (2000) 389.

78.   M. Resano, M. Verstraete, F. Vanhaecke, L. Moens, and J. Claessens, Direct determination of sulfur in Bisphenol A at ultratrace levels by means of solid sampling-electrothermal vaporization-ICP-MS. *J. Anal. Atom. Spectrom.* **16** (2001) 793.

79.   L.L. Yu, W.R. Kelley, J.D. Fassett, and R.D. Vocke, Determination of sulfur in fossil fuels by isotope dilution electrothermal vaporization inductively coupled plasma mass spectrometry. *J. Anal. Atom. Spectrom.* **16** (2001) 140.

80.   P. Evans, C. Wolff-Briche, and B. Fairman, High accuracy analysis of low level sulfur in diesel fuel by isotope dilution high resolution ICP-MS, using silicon for mass bias correction of natural isotope ratios. *J. Anal. Atom. Spectrom.* **16** (2001) 964.

81.   D.R. Bandura, V.I. Baranov, and S.D. Tanner, Detection of ultratrace phosphorus and sulfur by quadrupole ICP-MS with dynamic reaction cell. *Anal. Chem.* **74** (2002) 1497.

82.   M. Wind, H. Wesch, and W.D. Lehmann, Protein phosphorylation degree: determination by capillary liquid chromatography and inductively coupled plasma mass spectrometry. *Anal. Chem.* **73** (2001) 3006.

83.   M. Wind, M. Elder, N. Jakubowski, M. Linscheid, H. Wesch, and W.D. Lehmann, Analysis of protein phosphorylation by capillary liquid chromatography coupled to element mass spectrometry with 31P detection and to electrospray mass spectrometry. *Anal. Chem.* **73** (2001) 29.

84.   V.I. Baranov, Z.A. Quinn, D.R. Bandura, and S.D. Tanner, The potential for elemental analysis in biotechnology. *J. Anal. Atom. Spectrom.* **17** (2002) 1148.

85.   H. Wildner, Application of inductively coupled plasma sector field mass spectrometry for the fast and sensitive determination at isotope ratio measurement of non-metals in high-purity process chemicals. *J. Anal. Atom. Spectrom.* **13** (1998) 573.

86.   A. Makishima and E. Nakamura, Determination of total sulfur at microgram per gram levels in geological materials by oxidation of sulfur into sulfate with in situ generation of bromine using isotope dilution high-resolution ICPMS. *Anal. Chem.* **73** (2001) 2547.

87.   H. Helgesen and E.H. Larsen, Bioavailability and speciation of arsenic in carrots grown in contaminated soil. *Analyst* **123** (1998) 791.

88.   C. de Conto Cinier, M. Petit-Ramel, R. Faure, D. Garin, and Y. Bouvet, Kinetics of cadmium accumulation and elimination in carp Cyprinus carpio tissues. *Comp. Biochem. Physiol. C, Pharmacol. Toxicol. Endocrinol.* **122** (1999) 345.

89.   C. Coudray, D. Pepin, J.C. Tressol, J. Bellanger, and Y. Rayssiguier, Study of magnesium bioavailability using stable isotopes and the inductively-coupled plasma mass spectrometry technique in the rat: single and double labelling approaches. *Br. J. Nutr.* **77** (1997) 957.

90. H.M. Crews, L.M. Owen, N. Langford, S.J. Fairweather-Tait, T.E. Fox, L. Hubbard, et al., Use of the stable isotope (106)Cd for studying dietary cadmium absorption in humans. *Toxicol. Lett.* **112–113** (2000) 201.

91. R.A. Srivastava and J.C. Jain, Scavenger receptor class B type I expression and elemental analysis in cerebellum and parietal cortex regions of the Alzheimer's disease brain. *J. Neurol. Sci.* **196** (2002) 45.

92. J.C. Ng, S.M. Kratzmann, L. Qi, H. Crawley, B. Chiswell, and M.R. Moore, Speciation and absolute bioavailability: risk assessment of arsenic-contaminated sites in a residential suburb in Canberra. *Analyst* **123** (1998) 889.

93. K.A. Francesconi, J. Gailer, J.S. Edmonds, W. Goessler, and K.J. Irgolic, Uptake of arsenic-betaines by the mussel Mytilus edulis. *Comp. Biochem. Physiol. C, Pharmacol. Toxicol. Endocrinol.* **122** (1999) 131.

94. J. Feldmann, K. John, and P. Pengprecha, Arsenic metabolism in seaweed-eating sheep from Northern Scotland. *Fresenius J. Anal. Chem.* **368** (2000) 116.

95. G. Jahreis, W. Hausmann, G. Kiessling, K. Franke, and M. Leiterer, Bioavailability of iodine from normal diets rich in dairy products—results of balance studies in women. *Exp. Clin. Endocrinol. Diabetes* **109** (2001) 163.

96. P. Bratter, I.N. Blasco, V.E. Negretti de Bratter, and A. Raab, Speciation as an analytical aid in trace element research in infant nutrition. *Analyst* **123** (1998) 821.

97. J. Lintschinger, N. Fuchs, J. Moser, D. Kuehnelt, and W. Goessler, Selenium-enriched sprouts. A raw material for fortified cereal-based diets. *J. Agric. Food Chem.* **48** (2000) 5362.

98. C. B'Hymer and J.A. Caruso, Evaluation of yeast based selenium food supplements using HPLC and ICP-MS. *J. Anal. Atom. Spectrom.* **15** (2000) 1531.

99. H. Chassaigne, C.C. Chery, G. Bordin, and A.R. Rodriguez, Development of new analytical methods for selenium speciation in selenium-enriched yeast material. *J. Chromatogr. A* **976** (2002) 409.

100. M. Kotrebai, J.F. Tyson, E. Block, and P.C. Uden, High-performance liquid chromatography of selenium compounds utilizing perfluorinated carboxylic acid ion-pairing agents and inductively coupled plasma and electrospray ionization mass spectrometric detection. *J. Chromatogr. A* **866** (2000) 51.

101. C. Ip, M. Birringer, E. Block, M. Kotrebai, J.F. Tyson, P.C. Uden, et al., Chemical speciation influences comparative activity of selenium-enriched garlic and yeast in mammary cancer prevention. *J. Agric. Food Chem.* **48** (2000) 2062.

102. P.D. Whanger, C. Ip, C.E. Polan, P.C. Uden, and G. Welbaum, Tumorgenesis, metabolism, speciation, bioavailability, and tissue deposition of selenium in selenium-enriched ramps (allium tricoccum). *J. Agric. Food Chem.* **48** (2000) 5723.

103.   S.S. Kannamkumarath, K. Wrobel, A. Vonderheide, and J.A. Caruso, HPLC-ICP-MS determination of selenium distribution and speciation in different types of nut. *Anal. Bioanal. Chem.* **373** (2002) 454.

104.   K. Wrobel, S.S. Kannamkumarath, and J.A. Caruso, Hydrolysis of proteins with methanesulfonic acid for improved HPLC-ICP-MS determination of seleno-methionine in yeast and nuts. *Anal. Bioanal. Chem.* **375** (2003) 133.

105.   A.P. Vonderheide, K. Wrobel, S.S. Kannamkumarath, C. B'Hymer, M. Montes-Bayon, C. Ponce De Leon, et al., Characterization of selenium species in Brazil nuts by HPLC-ICP-MS and ES-MS. *J. Agric. Food Chem.* **50** (2002) 5722.

106.   M. Stadlober, M. Sager, and K.J. Irgolic, Effects of selenate supplemented fer-tilisation on the selenium levels in cereals—identification and quantification of selenium compounds by HPLC-ICP-MS. *Food Chem.* **73** (2001) 357.

107.   K. Wrobel, K. Wrobel, S.S. Kannamkumarath, J.A. Caruso, A. Wysocka, E. Bul-ska, et al., HPLC-ICP-MS speciation of selenium in enriched onion leaves—a potential dietary source of Se-methylselenocysteine. *Food Chem.*, in press.

108.   C.A. Ponce de León, M.M. Bayón, C. Paquin, and J.A. Caruso, Selenium incor-poration into Saccharomyces cerevisiae cells: a study of different incorporation methods. *J. Appl. Microbiol.* **92** (2002) 602.

109.   K. McKay, New techniques in the pharmacokinetic analysis of cancer drugs. II. The ultratrace determination of platinum in biological samples by inductively coupled plasma-mass spectrometry. *Cancer Surv.* **17** (1993) 407.

110.   C.J. Smith, I.D. Wilson, F. Abou-Shakra, R. Payne, T.C. Parry, P. Sin-clair, et al., A comparison of the quantitative methods for the analy-sis of the platinum-containing anticancer drug [*cis*-[amminedichloro(2-methylpyridine)]platinum(II)] (ZD0473) by HPLC coupled to either a triple quadrupole mass spectrometer or an inductively coupled plasma mass spectrom-eter. *Anal. Chem.* **75** (2003) 1463.

111.   F. Levi, G. Metzger, C. Massari, and G. Milano, Oxaliplatin: pharmacokinetics and chronopharmacological aspects. *Clin. Pharmacokinet.* **38** (2000) 1.

112.   C. Sessa, G. Capri, L. Gianni, F. Peccatori, G. Grasselli, J. Bauer, et al., Clini-cal and pharmacological phase I study with accelerated titration design of a daily times five schedule of BBR3464, a novel cationic triplatinum complex. *Ann. On-col.* **11** (2000) 977.

113.   J.L. Carr, M.D. Tingle, and M.J. McKeage, Rapid biotransformation of satraplatin by human red blood cells in vitro. *Cancer Chemother. Pharmacol.* **50** (2002) 9.

114.   J. Liu, E.H. Kraut, S. Balcerzak, M. Grever, S. D'Ambrosio, and K.K. Chan, Dosing sequence-dependent pharmacokinetic interaction of oxaliplatin with pa-clitaxel in the rat. *Cancer Chemother. Pharmacol.* **50** (2002) 445.

115.   R.J. Kolb, A.M. Ghazi, and D.W. Barfuss, Inhibition of basolateral transport and cellular accumulation of cDDP and N-acetyl- L-cysteine-cDDP by TEA and PAH in the renal proximal tubule. *Cancer Chemother. Pharmacol.* **51** (2003) 132.

116.   X. Liang and Y. Huang, Physical state changes of membrane lipids in human lung adenocarcinoma A(549) cells and their resistance to cisplatin. *Int. J. Biochem. Cell Biol.* **34** (2002) 1248.

117.   T. Hanada, H. Isobe, T. Saito, S. Ogura, H. Takekawa, K. Yamazaki, et al., Intracellular accumulation of thallium as a marker of cisplatin cytotoxicity in non-small cell lung carcinoma: an application of inductively coupled plasma mass spectrometry. *Cancer* **83** (1998) 930.

118.   S. Gaillard, C. Kubiak, C. Stolz, B. Bonnemain, and D. Chassard, Safety and pharmacokinetics of p792, a new blood-pool agent: results of clinical testing in nonpatient volunteers. *Invest. Radiol.* **37** (2002) 161.

119.   E.M. Frame and E.E. Uzgiris, Gadolinium determination in tissue samples by inductively coupled plasma mass spectrometry and inductively coupled plasma atomic emission spectrometry in evaluation of the action of magnetic resonance imaging contrast agents. *Analyst* **123** (1998) 675.

120.   Z. Zhao, W.B. Jones, K. Tepperman, J.G. Dorsey, and R.C. Elder, Determination of gold-based antiarthritis drugs and their metabolites in urine by reversed-phase ion-pair chromatography with ICP-MS detection. *J. Pharm. Biomed. Anal.* **10** (1992) 279.

121.   S.W. May, L. Wang, M.M. Gill-Woznichak, R.F. Browner, A.A. Ogonowski, J.B. Smith, et al., An orally active selenium-based antihypertensive agent with restricted CNS permeability. *J. Pharmacol. Exp. Ther.* **283** (1997) 470.

122.   J.S. Thayer, Review: biological methylation of less studied elements. *J. Inorg. Biochem.* **16** (2002) 677.

123.   L.S. Milstein, A. Essader, E.D. Pellizzari, R.A. Fernando, J.H. Raymer, K.E. Levine, et al., Development and application of a robust speciation method for determination of six arsenic compounds present in human urine. *Environ. Health Perspect.* **111** (2003) 293.

124.   U. Kohlmeyer, E. Jantzen, J. Kuballa, and S. Jakubik, Benefits of high resolution IC-ICP-MS for the routine analysis of inorganic and organic arsenic species in food products of marine and terrestrial origin. *Anal. Bioanal. Chem.* **377** (1) (2003) 6.

125.   J.A. Brisbin and J.A. Caruso, Comparison of extraction procedures for the determination of arsenic and other elements in lobster tissue by inductively coupled plasma mass spectrometry. *Analyst* **127** (2002) 921.

126.   K. Wrobel, K. Wrobel, B. Parker, S.S. Kannamkumarath, and J.A. Caruso, Determination of As(III), As(V), monomethylarsonic acid, dimethylarsinic acid and arsenobetaine by HPLC ICP-MS: analysis of reference materials, fish tissues and urine. *Talanta* **58** (2002) 899.

127.   B. Michalke and P. Schramel, Application of capillary zone electrophoresis-inductively coupled plasma mass spectrometry and capillary isoelectric focusing-

inductively coupled plasma mass spectrometry for selenium speciation. *J. Chromatogr. A* **807** (1998) 71.

128. J. Zheng, M. Ohata, N. Furuta, and W. Kosmus, Speciation of selenium compounds with ion-pair reversed-phase liquid chromatography using inductively coupled plasma mass spectrometry as element-specific detection. *J. Chromatogr. A* **874** (2000) 55.

129. M. Montes-Bayon, E.G. Yanes, D. Ponce, K. Jayasimhulu, A. Stalcup, J. Shann, et al., Initial studies of selenium speciation in Brassica juncea by LC with ICPMS and ES-MS detection: an approach for phytoremediation studies. *Anal. Chem.* **74** (2002) 107.

130. X. Wei, C.A. Brockhoff-Schwegel, and J.T. Creed, Application of sample pre-oxidation of arsenite in human urine prior to speciation via on-line photo-oxidation with membrane hydride generation and ICP-MS detection. *Analyst* **125** (2000) 1215.

131. X. Qianli, R. Kerrich, E. Irving, K. Liber, and F. Abou-Shakra, Determination of five arsenic species in aqueous samples by HPLC coupled with a hexapole collision cell ICP-MS. *J. Anal. Atom. Spectrom.* **17** (2002) 1037.

132. A. Chatterjee, Y. Shibata, J. Yoshinaga, and M. Masatoshi, Application of a nitrogen microwave-induced plasma mass spectrometer as an element-specific detector for arsenic speciation analysis. *J. Anal. Atom. Spectrom.* **14** (1999) 1853.

133. M. Miguens-Rodriguez, R. Pickford, J.E. Thomas-Oates, and S.A. Pergantis, Arsenosugar identification in seaweed extracts using high-performance liquid chromatography/electrospray ion trap mass spectrometry. *Rapid Commun. Mass Spectrom.* **16** (2002) 323.

134. S. McSheehy, P. Pohl, D. Velez, and J. Szpunar, Multidimensional liquid chromatography with parallel ICP MS and electrospray MS/MS detection as a tool for the characterization of arsenic species in algae. *Anal. Bioanal. Chem.* **372** (2002) 457.

135. G. Raber, K.A. Francesconi, K.J. Irgolic, and W. Goessler, Determination of 'arsenosugars' in algae with anion-exchange chromatography and an inductively coupled plasma mass spectrometer as element-specific detector. *Fresenius J. Anal. Chem.* **367** (2000) 181.

136. P.A. Gallagher, J.A. Shoemaker, X. Wei, C.A. Brockhoff-Schwegel, and J.T. Creed, Extraction and detection of arsenicals in seaweed via accelerated solvent extraction with ion chromatographic separation and ICP-MS detection. *Fresenius J. Anal. Chem.* **369** (2001) 71.

137. W. Li, C. Wei, C. Zhang, M. Van Hulle, R. Cornelis, and X. Zhang, A survey of arsenic species in Chinese seafood. *Food Chem. Toxicol.* **41** (2003) 1103.

138. A. Chatterjee, Determination of total cationic and total anionic arsenic species in oyster tissue using microwave-assisted extraction followed by HPLC-ICP-MS. *Talanta* **51** (2000) 303.

139.   A.E. Geiszinger, W. Goessler, and K.A. Francesconi, The marine polychaete Arenicola marina: its unusual arsenic compound pattern and its uptake of arsenate from seawater. *Mar. Environ. Res.* **53** (2002) 37.

140.   A.E. Geiszinger, W. Goessler, and K.A. Francesconi, Biotransformation of arsenate to the tetramethylarsonium ion in the marine polychaetes Nereis diversicolor and Nereis virens. *Environ. Sci. Technol.* **36** (2002) 2905.

141.   R. Bentley and T.G. Chasteen, Microbial methylation of metalloids: arsenic, antimony, and bismuth. *Microbiol. Mol. Biol. Rev.* **66** (2002) 250.

142.   R.O. Jenkins, A.W. Ritchie, J.S. Edmonds, W. Goessler, N. Molenat, D. Kuehnelt, et al., Bacterial degradation of arsenobetaine via dimethylarsinoylacetate. *Arch. Microbiol.* **180** (2003) 142.

143.   D. Kuehnelt, J. Lintschinger, and W. Goessler, Arsenic compounds in terrestrial organisms. IV. Green plants and lichens from an old arsenic smelter site in Austria. *Appl. Organomet. Chem.* **14** (2000) 411.

144.   A. Geiszinger, W. Goessler, and W. Kosmus, Organoarsenic compounds in plants and soil on top of an ore Vein. *Appl. Organomet. Chem.* **16** (2002) 245.

145.   K. Francesconi, P. Visoottiviseth, W. Sridokchan, and W. Goessler, Arsenic species in an arsenic hyperaccumulating fern, Pityrogramma calomelanos: a potential phytoremediator of arsenic-contaminated soils. *Sci. Total Environ.* **284** (2002) 27.

146.   D. Kuehnelt, W. Goessler, C. Schlagenhaufen, and K.J. Irgolic, Arsenic compounds in terrestrial organisms III: arsenic compounds in formica sp. from an old arsenite smelter site. *Appl. Organomet. Chem.* **11** (1997) 859.

147.   T. Gebel, Confounding variables in the environmental toxicology of arsenic. *Toxicology* **144** (2000) 155.

148.   M. Vahter, Methylation of inorganic arsenic in different mammalian species and population groups. *Sci. Prog.* **82** (1999) 69.

149.   C. Hopenhayn Rich, M.L. Biggs, D.A. Kalman, L.E. Moore, and A.H. Smith, Arsenic methylation patterns before and after changing from high to lower concentrations of arsenic in drinking water. *Environ. Health Perspect.* **104** (1996) 1200.

150.   P. Apostoli, D. Bartoli, L. Alessio, and J.P. Buchet, Biological monitoring of occupational exposure to inorganic arsenic. *Occup. Environ. Med.* **56** (1999) 825.

151.   Z. Gong, G. Jiang, W.R. Cullen, H.V. Aposhian, and X.C. Le, Determination of arsenic metabolic complex excreted in human urine after administration of sodium 2,3-dimercapto-1-propane sulfonate. *Chem. Res. Toxicol.* **15** (2002) 1318.

152.   M. Katayama, Y. Kouya, C. Furusawa, C. Sakiyama, T. Kaise, and Y. Sugawa-Katayama, Effects of the fractionated components of the seaweed Hizikia Fusiforme Okam. on the arsenic metabolism in rats administered a large dose of arsenate. *Appl. Organomet. Chem.* **16** (2002) 427.

153.   W.R. Cullen and K.J. Reimer, Arsenic speciation in the environment. *Chem. Rev.* **89** (1989) 713.

154.   X.C. Le, X. Lu, M. Ma, W.R. Cullen, H.V. Aposhian, and B. Zheng, Speciation of key arsenic metabolic intermediates in human urine. *Anal. Chem.* **72** (2000) 5172.

155.   M. Alauddin, S.T. Alauddin, M. Bhattacharjee, S. Sultana, D. Chowdhury, H. Bibi, et al., Speciation of arsenic metabolite intermediates in human urine by ion-exchange chromatography and flow injection hydride generation atomic absorption spectrometry. *J. Environ. Sci. Health Part A, Tox. Hazard Subst. Environ. Eng.* **38** (2003) 115.

156.   M. Styblo, "Biological mechanisms and toxicological consequences of methylation of arsenic." In: R.A. Goyer and M.G. Cherian (Eds.), *Toxicology of Metals: Biochemical Aspects*, Springer-Verlag, Berlin, 1995, p. 407.

157.   M. Styblo, Z. Drobna, I. Jaspers, S. Lin, and D.J. Thomas, The role of biomethylation in toxicity and carcinogenicity of arsenic: a research update. *Environ. Health Perspect.* **110 Suppl 5** (2002) 767.

158.   H.V. Aposhian, Enzymatic methylation of arsenic species and other new approaches to arsenic toxicity. *Ann. Rev. Pharmacol. Toxicol.* **37** (1997) 397.

159.   R.A. Zakharyan and H.V. Aposhian, Arsenite methylation by methylvitamin B12 and glutathione does not require an enzyme. *Toxicol. Appl. Pharmacol.* **154** (1999) 287.

160.   S.S. Kannamkumarath, K. Wrobel, K. Wrobel, and J.A. Caruso, "Application of HPLC-ICP-MS for in-vitro studies on biomethylation of arsenic compounds." In: *The 29th Annual Meeting of Federation of Analytical Chemistry & Spectroscopy Societes—FACSS 2002*, Providence, Rhode Island, 2002.

161.   K.T. Suzuki, B.K. Mandal, and Y. Ogra, Speciation of arsenic in body fluids. *Talanta* **58** (2002) 111.

162.   J.S. Thayer, "Heavy-metal organo species, chemistry and remediation." In: R.A. Meyers (Ed.), *Encyclopedia of Environmental Analysis and Remediation*, Wiley, 1998, p. 2126.

163.   J. Meija, M. Montes-Bayon, D.L. Le Duc, N. Terry, and J.A. Caruso, Simultaneous monitoring of volatile selenium and sulfur species from se accumulating plants (wild type and genetically modified) by GC/MS and GC/ICPMS using solid-phase microextraction for sample introduction. *Anal. Chem.* **74** (2002) 5837.

164.   M. Montes Bayon, T.D. Grant, J. Meija, and J.A. Caruso, Selenium in plants by mass spectrometric techniques: developments in bio-analytical methods. *J. Anal. Atom. Spectrom.* **17** (2002) 1015.

165.   W.T. Frankenberger Jr. and M. Arshad, Bioremediation of selenium-contaminated sediments and water. *Biofactors* **14** (2001) 241.

166.   M. Itoh and K.T. Suzuki, Effect of dose on the methylation of selenium to monomethylselenol and trimethylselonium ion in rats. *Arch. Toxicol.* **71** (1997) 461.

167.   B. Gammelgaard, O. Jøns and L. Bendahl, Selenium speciation in pretreated human urine by ion-exchange chromatography and ICP-MS detection. *J. Anal. Atom. Spectrom.* **16** (2001) 339.

168.   M.A. Quijano, A.M. Gutierrez, M.C. Perez-Conde, and C. Camara, Determination of selenium species in human urine by high performance liquid chromatography and ICP-MS. *Talanta* **50** (1999) 165.

169.   T.C. Stadtman, Selenocysteine. *Ann. Rev. Biochem.* **65** (1996) 83.

170.   M.P. Rayman, The importance of selenium to human health. *Lancet* **356** (2000) 233241.

171.   K.T. Suzuki, K. Shiobara, K. Ishiwata, and M. Ohmichi, Speciation by HPLC/ICP-MS with use of stable isotopes: chemical reactions in the metabolism of selenium administred as selenite. *J. Inorg. Biochem.* **67** (1997) 21.

172.   K.T. Suzuki, K. Ishiwata, and Y. Ogra, Incorporation of selenium into selenoprotein P and extracellular glutathione peroxidase: HPLC-ICPMS data with enriched selenite. *Analyst* **124** (1999) 1749.

173.   K. Wrobel, K. Wrobel, and J.A. Caruso, Selenium speciation in low molecular weight fraction of Se-enriched yeast by HPLC-ICP-MS: detection of selenoadenosylmethionine. *J. Anal. Atom. Spectrom.* **17** (2002) 1048.

174.   K. Wrobel, K. Wrobel, S.S. Kannamkumarath, and J.A. Caruso, Identification of selenium species in urine by ion-pairing HPLC-ICP-MS using laboratory synthesized standards. *Anal. Bioanal. Chem.* **377** (2003) 670.

175.   M. Montes Bayon, D.D.L. Duc, N. Terry, and J.A. Caruso, Selenium speciation in wild-type and genetically modified selenium accumulating plants with HPLC separation and ICP-MS/SE-MS detection. *J. Anal. Atom. Spectrom.* **17** (2002) 872.

176.   B. Neuhierl, F. Thanbichler, F. Lottspeich, and A. Bock, A family of *S*-methylmethionine-dependent thiol/selenol methyltransferases: role in selenium tolerance and evolutionary relation *J. Biol. Chem.* **274** (1999) 5407.

177.   M. Kotrebai, M. Birringer, J.F. Tyson, E. Block, and P.C. Uden, Identification of the principal selenium compounds in selenium-enriched natural sample extracts by ion pair chromatography with ICP and ES-MS detection. *Anal. Commun.* **36** (1999) 249.

178.   M.A. Quijano, A.M. Gutierrez, M.C. Perez-Conde, and C. Camara, Determination of selenium species in human urine by HPLC and ICP-MS. *Talanta* **50** (1999) 165.

179.   B. Gammelgaard, K.D. Jessen, F.H. Kristensen, and O. Jons, Determination of trimethylselonium ion in urine by ion chromatography and ICP-MS detection. *Anal. Chim. Acta* **404** (2000) 47.

180. J. Zheng, M. Ohata, and N. Furuta, Reversed-phase liquid chromatography with mixed ion-pair reagents coupled with ICP-MS for the direct speciation analysis of selenium compounds in human urine. *J. Anal. Atom. Spectrom.* **17** (2002) 730.

181. B. Gammelgaard, L. Bendahl, U. Sidenius, and O. Jøns, Selenium speciation in urine by ion-pairing chromatography with perfluorinated carboxylic acids and ICP-MS detection. *J. Anal. Atom. Spectrom.* **17** (2002) 570.

182. J.M. Marchante Gayón, I. Feldmann, C. Thomas, and N. Jakubowski, Speciation of selenium in human urine by HPLC-ICP-MS with a collision and reaction cell. *J. Anal. Atom. Spectrom.* **16** (2001) 457.

183. T.H. Cao, R.A. Cooney, M.M. Woznichak, S.W. May, and R.F. Browner, Speciation and identification of organoselenium metabolites in human urine using inductively coupled plasma mass spectrometry and tandem mass spectrometry. *Anal. Chem.* **73** (2001) 2898.

184. Y. Ogra, K. Ishiwata, H. Takayama, N. Aimi, and K.T. Suzuki, Identification of a novel selenium metabolite, Se-methyl-N-acetylselenohexosamine, in rat urine by high-performance liquid chromatography—inductively coupled plasma mass spectrometry and—electrospray ionization tandem mass spectrometry. *J. Chromatogr. B, Analyt. Technol. Biomed. Life Sci.* **767** (2002) 301.

185. A.Z. Mason, S.D. Storms, and K.D. Jenkins, Metalloprotein separation and analysis by directly coupled size exclusion high-performance liquid chromatography inductively coupled plasma mass spectroscopy. *Anal. Biochem.* **186** (1990) 187.

186. C.N. Ferrarello, M.R. Fernandez de la Campa, C. Sariego Muniz, and A. Sanz-Medel, Metal distribution patterns in the mussel Mytilus edulis cytosols using size-exclusion chromatography and double focusing ICP-MS detection. *Analyst* **125** (2000) 2223.

187. O. Mestek, J. Kominkova, R. Koplik, T. Zima, M. Miskusova, and P. Stern, Speciation of Cu, Se, Zn and Fe in blood serum of hemodialysed patients. *Sb. Lek.* **103** (2002) 23.

188. J. Wang, D. Dreessen, D.R. Wiederin, and R.S. Houk, Measurement of trace elements in proteins extracted from liver by size exclusion chromatography-inductively coupled plasma-mass spectrometry with a magnetic sector mass spectrometer. *Anal. Biochem.* **288** (2001) 89.

189. K. Inagaki, N. Mikuriya, S. Morita, H. Haraguchi, Y. Nakahara, M. Hattori, et al., Speciation of protein-binding zinc and copper in human blood serum by chelating resin pre-treatment and inductively coupled plasma mass spectrometry. *Analyst* **125** (2000) 197.

190. C. Sariego Muñiz, J.M. Marchante Gayón, J.I. García Alonso, and A. Sanz-Medel, Speciation of essential elements in human serum using anion exchange chromatography coupled to postcolumn isotope dilution analysis with double focusing ICPMS. *J. Anal. Atom. Spectrom.* **16** (2001) 587.

191. M. Bayón Montes, A.B. Cabezuelo Soldado, E. González Blanco, and A. Sanz-Medel, Capabilities of fast protein liquid chromatography coupled to a double focusing inductively coupled plasma mass spectrometer for trace metal speciation in human serum. *J. Anal. Atom. Spectrom.* **14** (1999) 947.

192. C.C. Chery, H. Chassaigne, L. Verbeeck, R. Cornelis, F. Vanhaecke, and L. Moens, Detection and quantification of selenium in proteins by means of gel electrophoresis and electrothermal vaporization ICP-MS. *J. Anal. Atom. Spectrom.* **17** (2002) 576.

193. S. Lustig, D. Lampaert, K. De Cremer, J. De Kimpe, R. Cornelis, and P. Schramel, Capability of flatbed electrophoresis (IEF and native PAGE) combined with sector field ICP-MS and autoradiography for the speciation of Cr, Ga, In, Pt and V in incubated serum samples. *J. Anal. Atom. Spectrom.* **14** (1999) 1357.

194. Y. Ogra and K.T. Suzuki, Biological significance of non-acetylated metallothionein. *J. Chromatogr. B, Biomed. Sci. Appl.* **735** (1999) 17.

195. A. Prange and D. Schaumloffel, Hyphenated techniques for the characterization and quantification of metallothionein isoforms. *Anal. Bioanal. Chem.* **373** (2002) 441.

196. C.N. Ferrarello, M.R. Fernandez de la Campa, and A. Sanz-Medel, Multielement trace-element speciation in metal-biomolecules by chromatography coupled with ICP-MS. *Anal. Bioanal. Chem.* **373** (2002) 412.

197. A.B. Soldado Cabezuelo, M. Montes Bayón, E. Blancon González, J.I. García Alonso, and A. Sanz-Medel, Speciation of basal aluminium in human serum by fast protein liquid chromatography with inductively coupled plasma mass spectrometric detection. *Analyst* **123** (1998) 865.

198. K. Harad, A. Kuniyasu, H. Nakayama, M. Nakayama, T. Matsunaga, Y. Uji, et al., Separation of human serum transferrins with different iron-binding states by high-performance liquid chromatography using a pyridinium polymer column. *J. Chromatogr. B, Analyt. Technol. Biomed. Life Sci.* **767** (2002) 45.

199. M.H. Nagaoka and T. Maitani, Binding patterns of co-existing aluminium and iron to human serum transferrin studied by HPLC-high resolution ICP-MS. *Analyst* **125** (2000) 1962.

200. M.H. Nagaoka and T. Maitani, Effects of sialic acid residues of transferrin on the binding with aluminum and iron studied by HPLC/high-resolution ICP-MS. *Biochim. Biophys. Acta* **1526** (2001) 175.

201. M. Nagaoka, T. Yamazaki, and T. Maitani, Binding patterns of vanadium ions with different valence states to human serum transferrin studied by HPLC/high-resolution ICP-MS. *Biochem. Biophys. Res. Commun.* **296** (2002) 1207.

202. H. Sun and K.Y. Szeto, Binding of bismuth to serum proteins: implication for targets of Bi(III) in blood plasma. *J. Inorg. Biochem.* **94** (2003) 114.

203. S. Yoneda and K.T. Suzuki, Detoxification of mercury by selenium by binding of equimolar Hg–Se complex to a specific plasma protein. *Toxicol. Appl. Pharmacol.* **143** (1997) 274.

204. S. Yoneda and K.T. Suzuki, Equimolar Hg–Se complex binds to selenoprotein P. *Biochem. Biophys. Res. Commun.* **231** (1997) 7.

205. C. Sasakura and K.T. Suzuki, Biological interaction between transition metals (Ag, Cd and Hg), selenide/sulfide and selenoprotein P. *J. Inorg. Biochem.* **71** (1998) 159.

206. I. Leopold and B. Fricke, Inhibition, reactivation, and determination of metal ions in membrane metalloproteases of bacterial origin using high-performance liquid chromatography coupled on-line with inductively coupled plasma mass spectrometry. *Anal. Biochem.* **252** (1997) 277.

207. J.H. Kim and Y.S. Kim, A fibrinolytic metalloprotease from the fruiting bodies of an edible mushroom, armillariella mellea. *Biosci. Biotechnol. Biochem.* **63** (1999) 2130.

208. K.T. Suzuki, J. Takenaka, and Y. Ogra, Identification of the zinc-binding protein specifically present in male rat liver as carbonic anhydrase III. *Chem. Biol. Interact.* **122** (1999) 185.

209. M.F. Jobling, X. Huang, L.R. Steward, K.J. Barnham, and C. Curtain, Copper and zinc binding modulates the aggregation and neurotoxic properties of the prion peptide PrP106-126. *Biochemistry* **40** (2001) 8073.

210. *Defining the Mandate of Proteomics in the Post-Genomics Era: Workshop Report.* The National Academic Press, Washington, 2002, p. 1. http:/www.nap.edu/books/NI000479.

211. C. Zhang, Z. Zhang, B. Yu, J. Shi, and X. Zhang, Application of the biological conjugate between antibody and colloid Au nanoparticles as analyte to inductively coupled plasma mass spectrometry. *Anal. Chem.* **74** (2002) 96.

212. V.I. Baranov, Z. Quinn, D.R. Bandura, and S.D. Tanner, A sensitive and quantitative element-tagged immunoassay with ICPMS detection. *Anal. Chem.* **74** (2002) 1629.

213. E.J. Andreu, J.J. Martin de Llano, I. Moreno, and E. Knecht, A rapid procedure suitable to assess quantitatively the endocytosis of colloidal gold and its conjugates in cultured cells. *J. Histochem. Cytochem.* **46** (1998) 1199.

214. C. Zhang, F.B. Wu, Y.Y. Zhang, X. Wang, and X.R. Zhang, A novel combination of immunoreaction and ICP-MS as a hyphenated technique for the determination of thyroid-stimulated hormone in human serum. *J. Anal. Atom. Spectrom.* **16** (2001) 1393.

215. Z. Quinn, V.I. Baranov, S.D. Tanner, and J.L. Wrana, Simultaneous determination of proteins using an element-tagged immunoassay coupled with ICP-MS detection. *J. Anal. Atom. Spectrom.* **17** (2002) 892.

216. P.R.D. Mason, K. Kaspers, and M.J. van Bergen, Determination of sulfur isotope ratios and concentrations in water samples using ICP-MS incorporating hexapole ion optics. *J. Anal. Atom. Spectrom.* **14** (1999) 1067.

217. J.S. Becker, S.F. Boulyga, C. Pickhardt, J. Becker, S. Buddrus, and M. Przybylski, Determination of phosphorus in small amounts of protein samples by ICP-MS. *Anal. Bioanal. Chem.* **375** (2003) 561.

218. R. Koplik, H. Pavelkova, J. Cincibuchova, O. Mestek, F. Kvasnicka, and M. Suchanek, Fractionation of phosphorus and trace elements species in soybean flour and common white bean seeds by size exclusion chromatography-inductively coupled plasma mass spectrometry. *J. Chromatogr. B, Analyt. Technol. Biomed. Life Sci.* **770** (2002) 261.

219. M. Wind, I. Feldmann, N. Jakubowski, and W.D. Lehmann, Spotting and quantification of phosphoproteins purified by gel electrophoresis and laser ablation-element mass spectrometry with phosphorus-31 detection. *Electrophoresis* **24** (2003) 1276.

220. T. Hunter, Signaling-2000 and beyond. *Cell* **100** (2000) 113.

221. M. Wind, O. Kelm, E.A. Nigg, and W.D. Lehmann, Identification of phosphorylation sites in the Polo-like Kinases Plx1 and Plk1 by a novel strategy based on element and electrospray high resolution mass spectrometry. *Proteomics* **2** (2002) 1516.

222. C. Siethoff, I. Feldmann, N. Jakubowski, and M. Linscheid, Quantitative determination of DNA adducts using liquid chromatography/electrospray ionization mass spectrometry and liquid chromatography/high-resolution ICP-MS. *J. Mass Spectrom.* **34** (1999) 421.

223. D. Profrock, P. Leonhard, and A. Prange, Determination of phosphorus in phosphorylated deoxyribonucleotides using capillary electrophoresis and high performance liquid chromatography hyphenated to ICP-MS with an octopole reaction cell. *J. Anal. Atom. Spectrom.* **18** (2003) 708.

Advances in Mass Spectrometry, Volume 16
A.E. Ashcroft, G. Brenton and J.J. Monaghan (Editors)

CHAPTER 15

Some Experiments and Thoughts on Atmospheric Pressure Photoionization

Pietro Traldi and Ester Marotta

CNR, Istituto di Scienze e Tecnologie Molecolari, Corso Stati Uniti 4, 35127 Padova, Italy

15.1 Introduction

Photoionization has been considered, from the beginning of analytical mass spectrometry, highly attractive; it exhibits some theoretical advantages with respect to electron ionization, but also some severe limitations [1].

The energy transfer involved in ionization of atoms and molecules must be enough to excite one electron from a bond to an unquantized orbital. The ionization energy is just the lowest energy value required for the occurrence of this phenomenon. For electron ionization, this process can be written as

$$M + e^- \rightarrow [M^*]^- \rightarrow M^{+\bullet} + 2e^-$$

in which three events occur, according to Wigner [2]: (i) approach of an electron to a neutral molecule, (ii) formation of a collisional complex, (iii) dissociation of the complex in a positive ion and two electrons. The probability of ionization is critically dependent on step (iii). Wannier [3] has shown that this probability depends on the number of freedom degrees n for sharing the energy excess between the electrons. Defining E_c the minimum energy, the ionization probability can be defined as

$$P(E - E_c) = k(E - E_c)^n \qquad (15.1)$$

and, in the case of emission of two electrons from the collision complex, $n = 1$.

For photoionization, the basic reaction becomes

$$M + h\nu \rightarrow [M^*] \rightarrow M^{+\bullet} + e^-$$

and, by using the Wigner and Wannier arguments, in this case the probability expressed by Eq. (15.1) will have n less by 1 than the value for the EI-induced process. This implies that the ionization probability as a function of photon energy will be zero until the ionization energy is reached. When the IE is reached, the probability will rise immediately to the value determined by the electronic transition probability for the process. In other words, the necessary condition to obtain the photoionization of a molecule M is that $IE_M \leqslant h\nu$.

The main limitation to the extensive use of photoionization in mass spectrometry was that at the light frequencies suitable to produce ionization of most organic compounds (IE ranging up to 13 eV) it is not possible to use windows in the path of the light beam. All the window materials are essentially opaque at this photon energy. Consequently, the light source, usually involving a gas-discharge, must be mounted inside the ion source housing. A further aspect that in the past limited the common use of photoionization was surely the low sensitivity of the method. When operating in high vacuum conditions, typical of classical ion sources, the formation of ions is some orders of magnitude lower than is observed with the same sample density in EI conditions, and this can be related to the photon cross section.

However, it should be emphasized that photoionization has been used since 1976 as a detection method in gas chromatography, proving that, when the sample density is high enough, good sensitivity can be achieved, together with the specificity related to the wavelength employed.

Only a few papers appeared in the past on analytical applications of photoionization in mass spectrometry. Among them, that by Chen showed the analytical power of the method [4] by employing an argon resonance lamp emitting photons with energies of 11.6 and 11.8 eV with an intensity of 3×10^{12} photons s^{-1}. The interaction of a mixture of alkanes in nitrogen (at a pressure of 10^{-2} Torr) with the light beam leads to good quality mass spectra, with a detection limit of about 10 ppb. Analogous results were achieved by Revel'skii et al. [5].

Of course, with high-power lasers, the photoionization is no longer limited to photons whose energy exceeds that of the ionization energy, as multiphoton processes [1] now become operative. However, the use of "conventional" light beam with high density vapours coming from the vaporization of sample solution has been considered of interest and the Bruins's group has developed and tested a first experimental apparatus [6], devoted to LC/MS experiments. Considering the analogies with the well-established atmospheric pressure chemical ionization (APCI) technique, this new method has been called atmospheric pressure photoionization (APPI).

In an APPI source, a series of different processes can be activated by photon irradiation. Calling ABC the analyte molecule, S the solvent and G other gaseous species present in the source (N_2, O_2, and H_2O at trace level), the first step can be considered their photoexcitation:

$$ABC + h\nu \rightarrow ABC^*,$$
$$S + h\nu \rightarrow S^*,$$
$$G + h\nu \rightarrow G^*.$$

At this stage, inside the source, a collection of excited and nonexcited species is present and a series of further processes can occur, as:

$$ABC^* \rightarrow ABC + h\nu \quad \text{radiative decay,}$$
$$ABC^* \rightarrow AB^{\boldsymbol{\cdot}} + B^{\boldsymbol{\cdot}} \quad \text{photodissociation,}$$
$$ABC^* \rightarrow ACB^* \quad \text{isomerization,}$$

$$\left. \begin{array}{l} ABC^* + S \rightarrow ABC + S^* \\ ABC^* + G \rightarrow ABC + G^* \\ ABC + S^* \rightarrow ABC^* + S \end{array} \right\} \quad \text{collisional quenching,}$$

$\ldots$

Only when $h\nu \geqslant IE$, can the ionization take place

$$ABC^* \rightarrow ABC^{+\boldsymbol{\cdot}} + e^-,$$
$$S^* \rightarrow S^{+\boldsymbol{\cdot}} + e^-,$$
$$G^* \rightarrow G^{+\boldsymbol{\cdot}} + e^-.$$

Recombination processes can also occur. Hence, inside an ion source at atmospheric pressure, a highly complex mixture of ions, neutrals at ground and excited states, radicals and electrons, is photo-generated.

To put some order in this complex environment, the photon energy can be chosen in order to avoid the ionization processes of the S and G species. The plot of intensity vs frequency of the most commonly employed lamps is reported in Fig. 15.1.

By this data it can be deducted that the Kr lamp is the most suitable one for analytical purposes. It shows an energy distribution from 8 to 10 eV and consequently it does not lead, in principle, to ionization of O_2 (IE = 12.07 eV), N_2 (IE = 15.58 eV), H_2O (IE = 12.62 eV), CH_3OH (IE = 10.86 eV), and CH_3CN (IE = 12.26 eV) [7], typical G and S species present inside the source. The negative counterpart is that, by its use, the direct photoionization of organic compounds with IE > 10 eV cannot be obtained and in these cases the use of a dopant (D) has been proposed. A suitable substance, added in relatively large amount (with

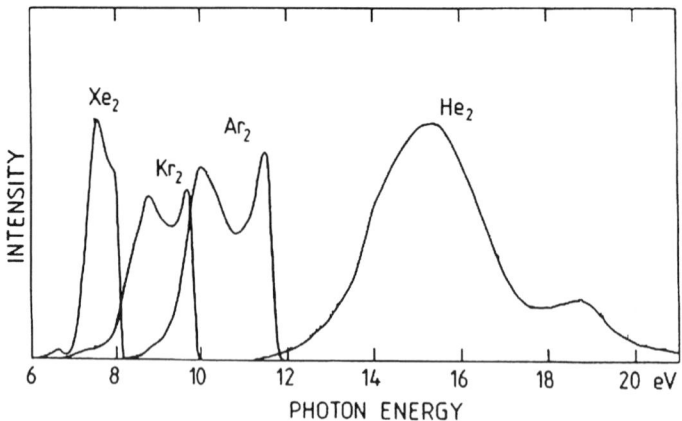

Figure 15.1 Plot of intensity vs frequency of the most commonly employed lamps.

IE value $\leqslant 10$ eV), leads to the production of a large number of analyte ions, reasonably through charge-exchange (electron transfer) and/or proton transfer [8]. In the former case the related mechanism can be simply described as

$$D + h\nu \rightarrow D^{+\bullet},$$

$$D^{+\bullet} + ABC \rightarrow D + ABC^{+\bullet}$$

and the internal energy of $ABC^{+\bullet}$ can be calculated by

$$E_{int} = RE(D^{+\bullet}) - IE(ABC),$$

where $RE(D^{+\bullet})$ is the recombination energy of the dopant ion and $IE(ABC)$ is the ionization energy of the sample molecule. Relatively intense molecular ions are expected if $RE(D^{+\bullet})$ is greater than $IE(ABC)$. Considering that benzene and toluene, often used as dopants, exhibit RE values of about 9 eV [9] the simple dopant mechanism above described seemed to us quite improbable. For these reasons, a series of experiments was performed to investigate the ionization mechanisms operating in APPI conditions.

First of all, it is important to describe the instrumental configuration employed for the experiments. Two geometries are usually employed in APPI sources. The in-line geometry, due to Robb et al. [6], has been derived from the standard heated nebulizer of the PE/Sciex 300 and 3000 (MDS Sciex, Concord, Ontario, Canada) series triple quadrupole mass spectrometers; in this case the lamp is mounted perpendicular to the ion guide tube (Fig. 15.2). In the orthogonal geometry source, developed by Syagen Technology starting from the scheme of the Agilent Technologies APCI source (Fig. 15.3), the heated nebulizer and the Kr lamp are, re-

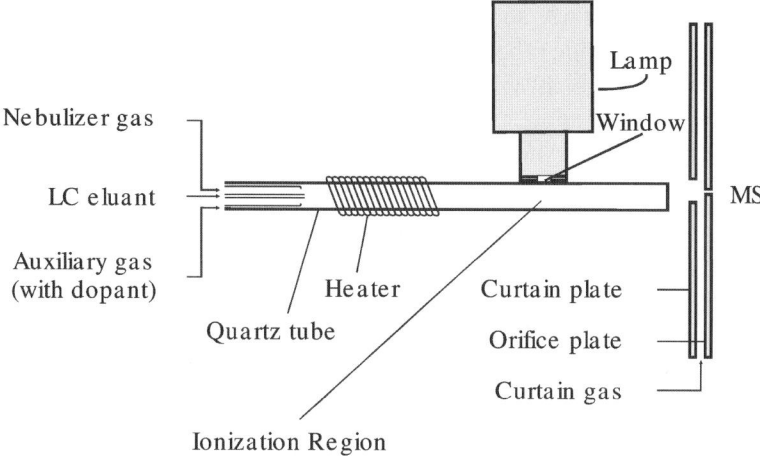

Figure 15.2 APPI source with in-line geometry, due to Robb et al. [6].

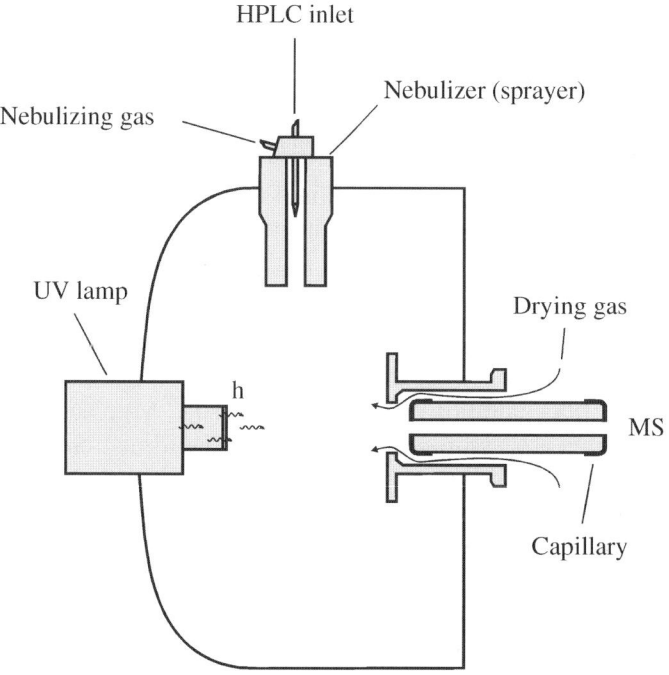

Figure 15.3 APPI source with orthogonal geometry (Syagen Technologies Inc., Tustin, CA, USA).

spectively, perpendicularly and in-line with respect to the mass spectrometer ion path and there is no guide tube present.

The experiments which will be described here have been carried out with the second source configuration, available on the Agilent 1100 Series MSD Trap (Agilent Technologies, Palo Alto, CA, USA) instrument.

15.2 The case of acetonitrile

The first set of experiments was performed by using, as analyte and solvent, chemical species which, in principle, would lead to unequivocal APPI results [10]. As analytes some trimethylfurocumarins, exhibiting an ionization energy < 10 eV were chosen, while acetonitrile (IE $= 12.2$ eV) was used as the solvent (10^{-6} M solutions). In these conditions the unique formation of $M^{+\bullet}$ of furocumarins would be expected and the participation of acetonitrile in M ionization processes would be excluded.

Compound **1**

On the contrary, as shown by the spectrum of compound **1** reported in Fig. 15.4a, the formation of $[M+H]^+$ ions is predominant. At first sight it could be ascribed to some "self-chemical ionization" reaction, i.e., by the reaction of $M^{+\bullet}$ with neutral molecules of the analyte

$$M^{+\bullet} + M \rightarrow [M-H]^{\bullet} + [M+H]^{+}.$$

However, the low concentration of furocumarins (10^{-6} M) places some doubts on the feasibility of this process in such a high yield. The formation of $[M+H]^+$ cannot be, in principle, ascribed to acetonitrile which, because of its high ionization energy, cannot directly generate reactant ions under APPI conditions. Considering the high-pressure regime typical of APPI, $[M+H]^+$ could be generated by complex mechanisms, involving other acidic ionic species present inside the source. The implication of H_3O^+ can be invoked, considering that water is present, even if at trace level, inside the source. However, the formation of H_3O^+ would require, as a first step, photoionization of water and this, due to the high ionization energy of water (12.62 eV) [7], cannot take place in the experimental setup employed. Also the implication of other molecular species (N_2 and O_2) present

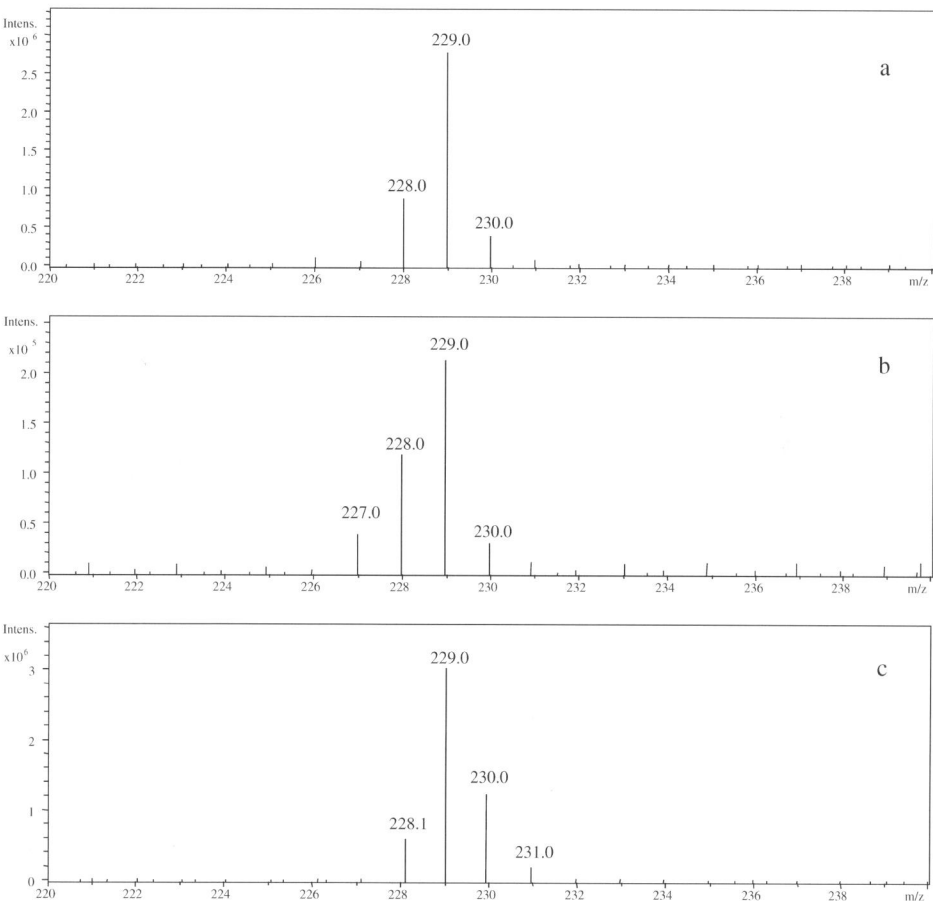

Figure 15.4 APPI mass spectrum of compound **1** dissolved in CH_3CN (a), in CCl_4 (b), and in CD_3CN (c).

inside the source must be excluded, due either to their high ionization energies (IE = 15.58 eV and 12.13 eV, respectively) [7], or to their capability to activate charge-exchange processes, leading to $M^{+\bullet}$ and not to $[M + H]^+$.

Further experiments were performed by using CCl_4 as solvent. In these conditions a clear increase in $M^{+\bullet}$ abundance is observed, even if $[M + H]^+$ still remains the most abundant species (see Fig. 15.4b). This result proves that, at least partially, the $[M + H]^+$ formation is due to self-chemical ionization processes. However, the higher $[M + H]^+/[M]^{+\bullet}$ ratio obtained by using acetonitrile as solvent (Fig. 15.4a) suggests that acetonitrile participates in the $[M + H]^+$ formation. In order to investigate the possible role of acetonitrile in $[M + H]^+$ formation, the same experiment was performed by using CD_3CN and, quite surprisingly, as

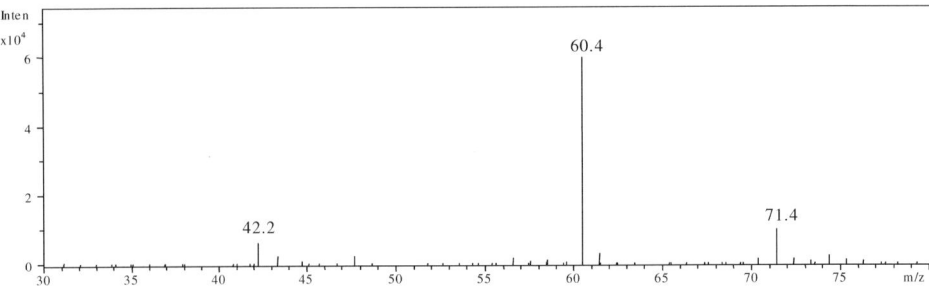

Figure 15.5 APPI spectrum of acetonitrile.

shown by Fig. 15.4c, the formation of $[M + D]^+$ ions was observed. This unequivocally proves the involvement of acetonitrile in the formation of $[M + H]^+$, even if its direct photoionization cannot take place. Consequently the photoionization reactions of acetonitrile itself were studied.

The gas-phase reactivity of acetonitrile has been widely studied, due to its implication in high stratosphere chemistry [11] and to its effectiveness as a reactant in chemical ionization experiments [12]. In those studies highly ionizing media were present while in the present case the photon energy (10 eV) is lower than the acetonitrile IE value (12.2 eV) and, hence, direct photoionization cannot occur.

However, by vaporizing only acetonitrile in the APPI source, ions are generated and the spectrum shown in Fig. 15.5 (with intensity of the order of 10^4 DAC units) is obtained. The most abundant ion is present at m/z 60, while the weak signal detected at m/z 42 is reasonably due to protonated acetonitrile. Interestingly the $[M − H]^+$ (m/z 40) and $C_3H_4N^+$ (m/z 54) ions, typical of the CI spectra of acetonitrile [12], are completely absent, proving that either the photon-irradiation or the highest sample density present in APPI experiments activate different reaction channels.

The formation of the ions at m/z 60 and 42 necessarily implies the occurrence of photoionization phenomena: in fact, by switching off the Kr lamp (but leaving constant all the voltages applied to the source) these ions completely disappear, proving that they are photochemical products.

Three different mechanisms have been considered for the formation of these ions:

(i) photoionization of a complex of acetonitrile with some neutral species present in the condensed phase and/or in the source environment, exhibiting an IE value < 10 eV;

(ii) photo-induced isomerization of acetonitrile, leading to neutral species exhibiting an IE < 10 eV;

(iii) photon-activated formation of an ion-pair.

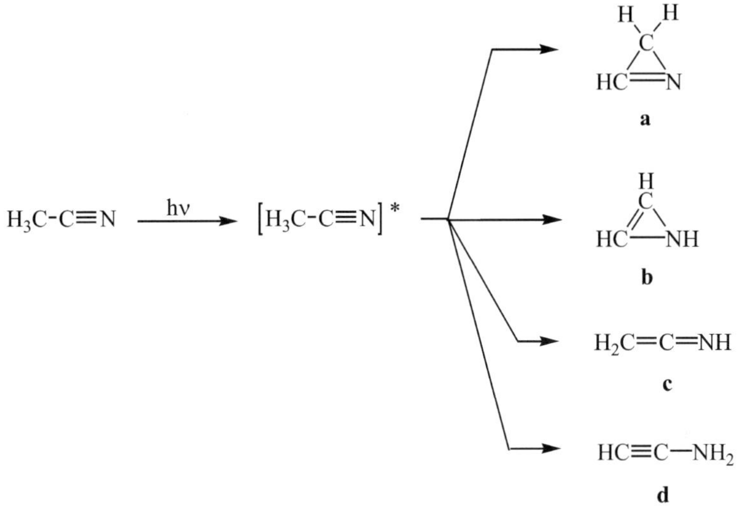

$$H_3C\text{-}C\equiv N \xrightarrow{h\nu} \left[H_3C\text{-}C\equiv N\right]^*$$

a

$$\begin{array}{c} H \quad H \\ \diagdown C \diagup \\ HC = N \end{array}$$

b

$$\begin{array}{c} H \\ C \\ HC \underline{\quad} NH \end{array}$$

c

$$H_2C = C = NH$$

d

$$HC \equiv C\text{-}NH_2$$

Scheme 15.1

Table 15.1: Experimental and calculated ionization energies of acetonitrile and its isomers

Structures	IE measured (eV)[3]	IE (eV) calculated by AM1	IE (eV) calculated by MP2/6-31G*
CH_3CN	12.2	12.5	12.3
a	10.1	11.1	11.0
b		9.2	8.9
c		9.5	9.5
d		9.5	9.2

The third point can be immediately excluded, considering that in negative ion mode no ions were detected.

Point (i) must be taken in consideration. In fact, the ion at m/z 60 just corresponds to the protonated complex of CH_3CN and H_2O. It could be hypothesized that the [$CH_3CN\bullet H_2O$] complex is already present in the solution and/or is easily formed inside the ion source. If its IE value is < 10 eV it could be ionized and the ion so formed could protonate the neutral complex, leading to the ion at m/z 60. Unfortunately, IE data on the [$CH_3CN\bullet H_2O$] complex are not available and consequently any experimental evidence on the validity of point (i) cannot be given.

The photon-induced isomerization of acetonitrile (M $\xrightarrow{h\nu}$ M* → M$_i^*$) could lead to different structures, as reported in Scheme 15.1. Interestingly, the experimental IE value of structure **a**, available in the literature [7], is 10.1 eV, very close to the

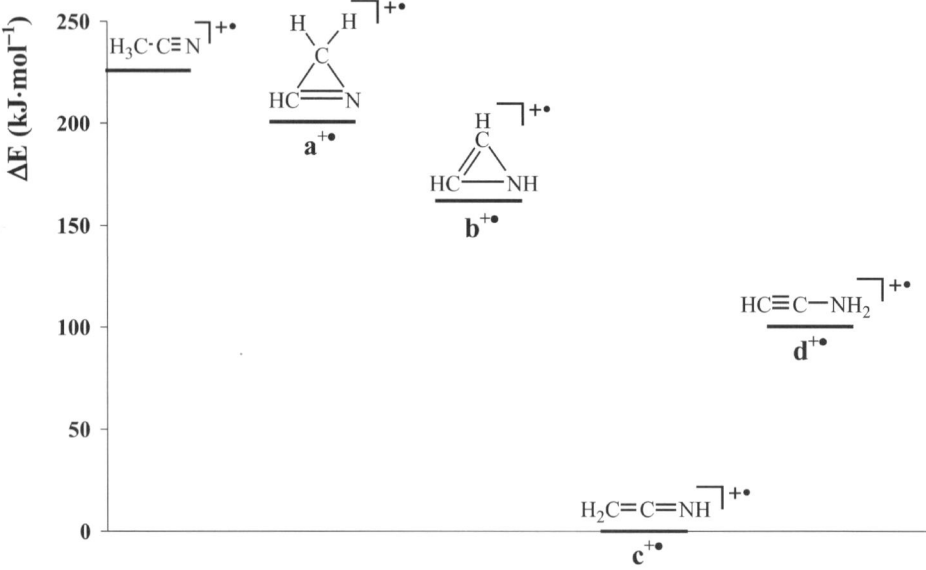

Figure 15.6 Relative electronic energy (ΔE, kJ mol^{-1}) of molecular ion of acetonitrile and its isomers with respect to the energy of structure **c**, representing the lowest determined value, calculated by the MP2/6-31G* method.

photon energy employed in the APPI source. Hence, structure **a** could lead, by irradiation, to the production of ionic species. Unfortunately, no experimental data for IE of **b**–**d** structures are available, and for this reason some theoretical calculations were undertaken by both AM1 and MP2/6-31G* methods [13]. The IE data so obtained, on the basis of Koopmans' theorem [14], are reported in Table 15.1. Good agreement has been found between the values obtained by the two methods, showing that structures **b**, **c**, and **d** all exhibit IE values lower than 10 eV. In our opinion structure **c** is to be assumed as the most stable photoionization product of acetonitrile: the cumulated double bonds account for its high stability and the relative total electron energies of the different isomers, reported in Fig. 15.6, fully confirm this hypothesis. Furthermore, it should be emphasized that the isomerization of acetonitrile to ketene imine **c** has been already described, as being produced by heating mixtures of acetonitrile in argon by reflected shock waves [15].

For the ion at *m/z* 42, present in the spectrum of Fig. 15.5, the reaction of M$^{+•}$ of **c** with neutral acetonitrile molecules can be invoked, with the formation of the highly stable ion **e**

$$H_2C=C=NH^{+•} \ + \ H_3C\text{-}C\equiv N \ \longrightarrow \ H_2C=C=\overset{+}{N}H_2 \ + \ H_2\overset{•}{C}\text{-}C\equiv N$$

        c   **e**

This ionic species can be considered the precursor of the abundant ion at m/z 60, justified by the structure **f**, generated by a complex between **e** ions and water molecules:

$$H_2C=C=\overset{+}{N}\cdots\begin{smallmatrix}H\\H\end{smallmatrix}$$

f

Further experiments were undertaken by collisional activation and ^{13}C and D isotope labelling confirming the structures proposed above and indicating ions **c** and **f** as protonating agents [10].

15.3 The case of benzene and toluene

By analyzing, in APPI conditions, hexadeuterobenzene solutions of compound **1**, the production of $[M + D]^+$ ions was again observed, proving the involvement of benzene as a protonating medium. Worth noting, in this context, is the mechanism proposed by Koster and Bruins [16] on the action of toluene (T) as dopant in other solvents (CH_3OH and CH_3CN), to rationalize the co-existence of $M^{+\bullet}$ and $[M + H]^+$ ions of the analytes. The results obtained with a triple quadrupole instrument (in which Q2 works as reaction chamber) suggest the occurrence of the following reactions:

$$T + h\nu \rightarrow T^{+\bullet},$$

$$T^{+\bullet} + S \rightarrow [T - H]^\bullet + [S + H]^+,$$

$$[S + H]^+ + M \rightarrow S + [M + H]^+.$$

However, in the present case, the above mechanism cannot be invoked. In fact, in this case the solvent is the dopant itself and no trace of protonated benzene was found in the spectrum. To investigate this unexpected behavior, APPI experiments were performed on benzene and toluene [17].

The APCI mass spectra of benzene and hexadeuterobenzene show the predominant formation of the odd-electron molecular ions at m/z 78 and 84, respectively, together with $[M - H]^+$ (m/z 77) and $[M - D]^+$ (m/z 82) ions with a relative abundance of $\approx 50\%$. When connecting the head space of a vial containing benzene to the APPI source, the formation of $M^{+\bullet}$ is highly favored. Due to the low internal energy deposition typical of APPI, the $H^\bullet$ loss observed in the APCI spectrum is practically suppressed and the phenyl ion exhibits a relative abundance lower than 5%.

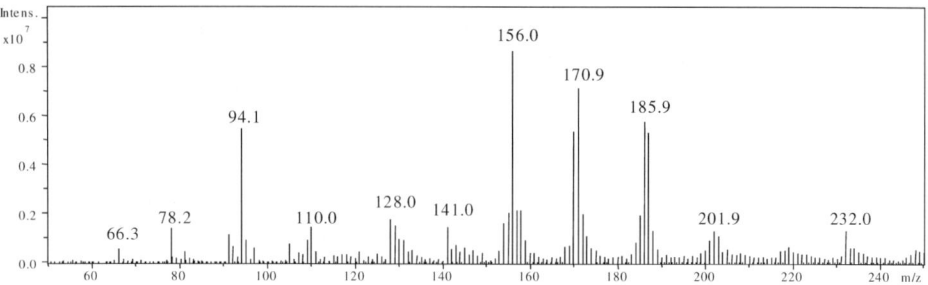

Figure 15.7 APPI spectrum of benzene.

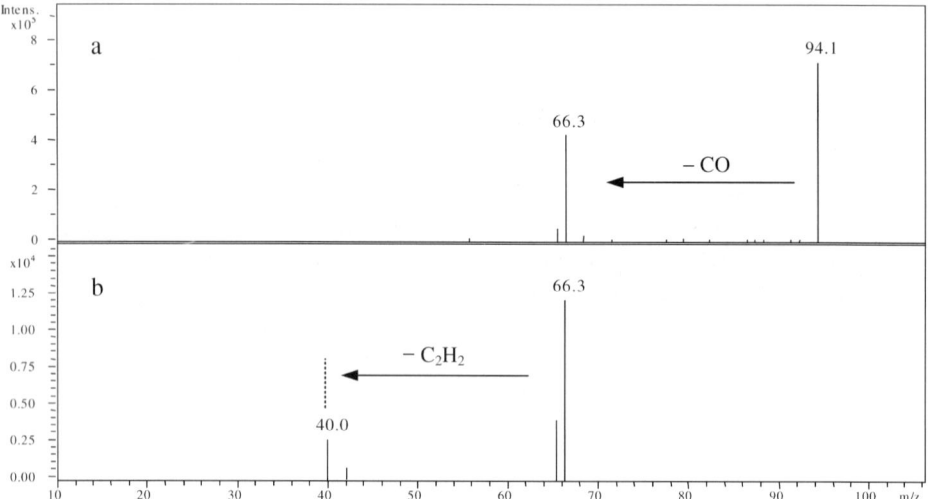

Figure 15.8 (a) MS^2 and (b) MS^3 spectrum of m/z 94 generated in APPI of benzene.

When benzene is nebulized inside the APPI source (i.e., when the benzene density is particularly high) the situation shows a deep change. As can be seen in Fig. 15.7, M$^{+\bullet}$ practically disappears and abundant ionic species at m/z 94, 156, 170, 171, 186, and 187 become present. These ions are surely generated by photoinitiated reactions: in fact, by switching off the Kr lamp, no signal can any longer be detected. The ion at m/z 156 can be easily attributed to an ion cluster constituted by C$_6$H$_6$$^{+\bullet}$ and neutral benzene. The ion at m/z 94 would correspond to a species originating from oxygen atom addition to the benzene molecule. MS^2 and MS^3 spectra of this ion, reported in Fig. 15.8, show an easy primary loss of a neutral species of 28 Da (reasonably a CO molecule) leading to the ion at m/z 66, followed by a C$_2$H$_2$ loss, giving rise to the ion at m/z 40. These results suggest for the ions at m/z 94 the structure of the phenol molecular ion.

Considering the ion source environment, the ions at m/z 94 could originate from two possible reactions:

(i) $\quad C_6H_6^{+\bullet} + O_2 \longrightarrow [C_6H_6-O_2]^{+\bullet} \overset{-O}{\longrightarrow} C_6H_6O^{+\bullet},$

$\qquad\qquad m/z\ 78 \qquad\qquad m/z\ 110 \qquad\qquad m/z\ 94$

(ii) $\quad C_6H_6^{+\bullet} + H_2O \longrightarrow [C_6H_6-OH_2]^{+\bullet} \overset{-H_2}{\longrightarrow} C_6H_6O^{+\bullet}.$

$\qquad\qquad m/z\ 78 \qquad\qquad m/z\ 96 \qquad\qquad m/z\ 94$

The latter hypothesis is highly attractive, when looking at the recent literature data on the $^\bullet$OH-iniziated oxidation of benzene studied in two simulation chambers, i.e., the outdoor European Photo-Reactor (EUPHORE) and the indoor photoreactor at the NIES [18]. The formation of phenol starting from benzene was studied using many different reactants (HONO, H_2O_2, HCHO + NO, $CH_3NO + NO$), all leading to the photolytic production of $^\bullet$OH, prone to radical attack to the benzene molecule, leading to neutral phenol. In our case, the experimental set-up is substantially different, as the reactants described above are completely absent from inside the reaction environment while the $M^{+\bullet}$ of benzene is present in quite high abundance. However, the feasibility of reaction (ii) has been investigated by adding 50 μL of water or deuterated water to a 3 mL benzene sample solution. The solution was injected, immediately after extensive sonication, by a syringe pump. In both cases the spectra are identical (from both the qualitative and the quantitative point of view) to that reported in Fig. 15.7, proving the lack of influence of water on the generation of the ions at m/z 94 (as well as of ions at m/z 170, 171, 186, and 187). Furthermore, experiments carried out by injecting deuterated benzene show that the ions at m/z 94 are shifted at m/z 100, proving that the hydrogen atoms present in the species are those originally present in the benzene molecule.

Looking at the spectrum of Fig. 15.7, the presence of an ion at m/z 110 should be emphasized. Its mass value just corresponds to that of the "activated complex" of reaction (i), giving a strong support to the related hypothesis. For this reaction the mechanism reported in Scheme 15.2 can be proposed.

This mechanism is reminiscent, to some extent, of the reactivity exhibited in the gas-phase by some alkyl and aryl radicals, which under tropospheric conditions react rapidly and solely with O_2 to form peroxyl radicals [19,20]

$$R^\bullet + O_2 \to RO_2^\bullet.$$

In particular, the reaction of the phenyl radical with O_2 has been described either in solution [21,22] or in the gas phase [22,23] and the related rate constants were experimentally determined. Hence, the occurrence of the reaction reported

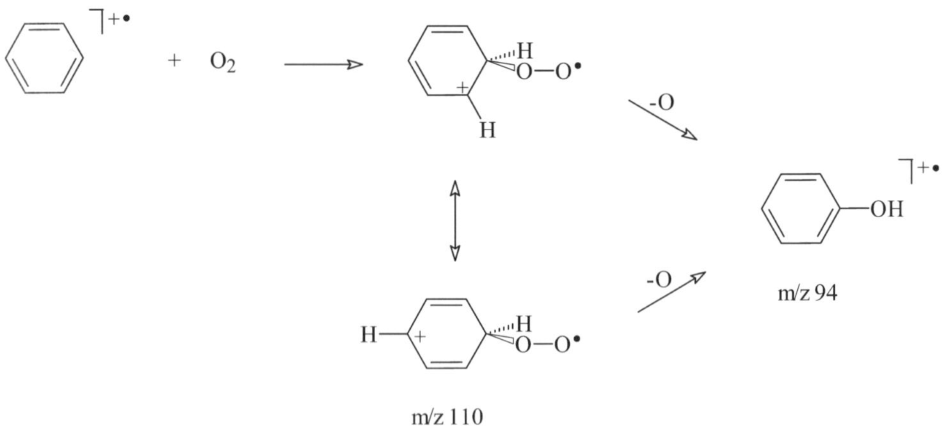

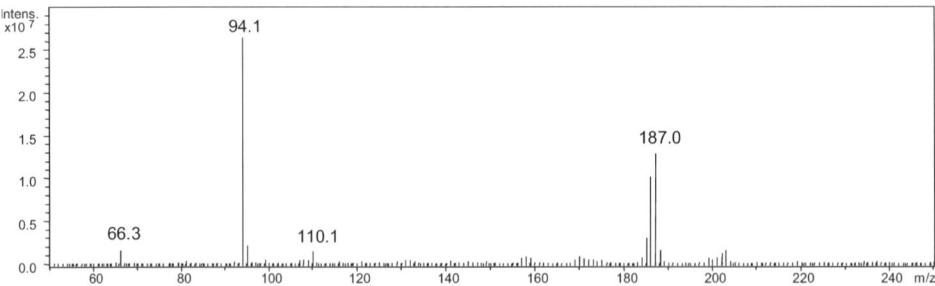

Scheme 15.2

Figure 15.9 APPI spectrum of a mixture of phenol 10% and benzene 90%.

in Scheme 15.2 can be rationalized by a mechanism not governed by the positive charge, but mainly by the radical character of $M^{+\bullet}$. Considering $M^{+\bullet}$ in distonic form, the positive charge and the radical could be placed in the *para*-position and the reactivity of benzene could be thought to be analogous to that proposed by Preidel and Zellner [23]. The high stability of the phenyl peroxyl radical cation, well detectable in the spectrum of Fig. 15.7 at m/z 110, should be emphasized and can be justified by the distonic nature of this intermediate.

A series of experiments were devoted to the investigation of the origin of the species at m/z 170, 171, 186, and 187. The injection of a sample of benzene (90%) and phenol (10%) leads to the spectrum reported in Fig. 15.9, showing the presence of abundant ions at m/z 94 ($M^{+\bullet}$ of phenol), 186 and 187 only. This result indicates that the production of the ions at m/z 186 and 187 is related to the phenol molecular ion concentration, i.e., that this last species must be considered as a synthon in their formation. The related mechanism (postulated only), shown in Scheme 15.3, can be described as an electrophilic attack of the phenol molecu-

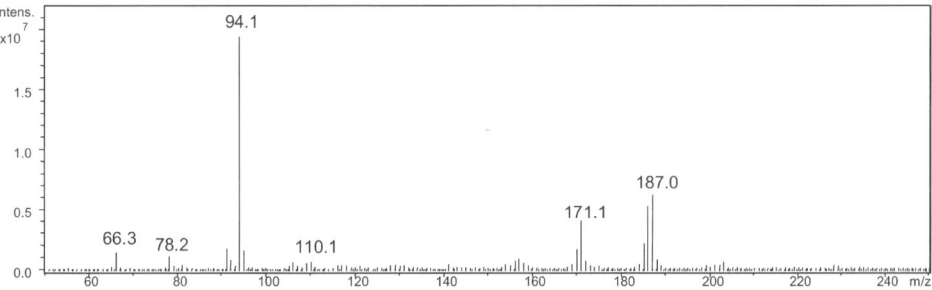

Scheme 15.3

Figure 15.10 APPI spectrum of a mixture of phenol 0.1% in benzene.

lar ion on the aromatic ring of the neutral phenol molecule. Due to the electron-donating nature of the –OH substituent, the electrophilic attack would take place preferentially in the *para-* and *ortho-*positions. In the second reaction step, H$_2$ or H$^•$ is ejected from the intermediate, leading to the formation of phenoxyphenol molecular ions **g** (m/z 186) or protonated phenoxyphenol molecules **h** (m/z 187). This mechanism is fully supported by deuterium labelling experiments. By injecting hexadeuterobenzene (90%) and phenol (10%), the same spectra as in Fig. 15.9 has been obtained, proving that benzene does not participate to the formation of the ions at m/z 186 and 187.

Reducing the phenol concentration to 0.1%, the spectrum reported in Fig. 15.10 has been obtained. In these conditions, even if M$^{+•}$ of phenol still represents the most abundant species, the ions at m/z 170 and 171 are readily detected. For their formation the reaction of phenol M$^{+•}$ with neutral benzene can be proposed, with a mechanism analogous to that reported in Scheme 15.3, leading to structures **i** and **l**. The absence of this reaction for high phenol concentration (Fig. 15.9) is at first sight surprising, but this behavior can be rationalized by considering the

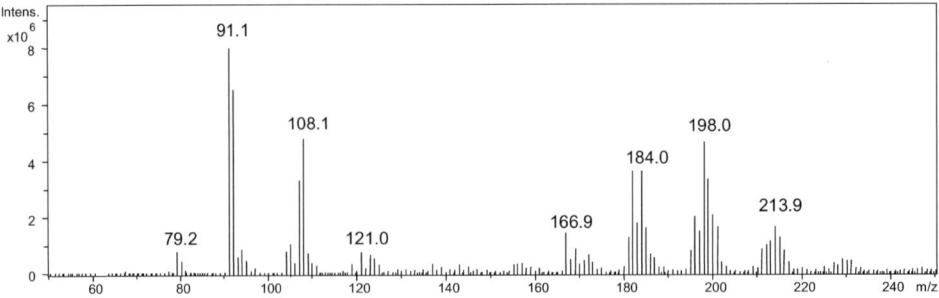

Figure 15.11 APPI spectrum of toluene.

lower reactivity of benzene in electrophilic attack with respect to phenol, where the aromatic ring is activated by the electron-donating substituent. Collisional spectra of the ions at m/z 170 and 171, compared with those of $M^{+\cdot}$

i

l

and $[M + H]^+$ of diphenylether and 4-phenylphenol, confirm for them the structures **i** and **l** reported above.

In the case of toluene, when injected at the same flow rate employed for benzene experiments (3 µL min^{-1}), the APPI spectra show behaviour very similar to that described above for benzene. In this case the base peak (see Fig. 15.11) is due to the tropylium ion (m/z 91) and the odd-electron molecular ion (m/z 92) is of high abundance. However, further ions are present at higher m/z values. In particular, the ion at m/z 108 is analogous to that at m/z 94 detected in the case of benzene and discussed above. Its MS^2 spectrum shows the easy loss of CO, leading to the ion at m/z 80, giving evidence for its "phenolic" nature. In other words, it reasonably originates from reaction of $M^{+\cdot}$ ions of toluene with molecular oxygen, and the oxygen atom is reasonably bonded to one of the benzene carbon atoms and not to the methyl group.

Other abundant ions are present at m/z 182, 184, 198, 199, and 214. While the ion at m/z 182 can be considered a real toluene dimer (as proved by the related MS^2 spectrum), the ion at m/z 184 can be considered as a cluster originating from the interaction of a neutral toluene molecule with the toluene molecular ion. The mass of the ion at m/z 199 corresponds to a structure constituted by the species at m/z 107 with neutral toluene. The MS/MS spectrum of this species, reported in Fig. 15.12, shows primary losses of $^\cdot$CH$_3$ (m/z 184), H$_2$O (m/z 181), and CO

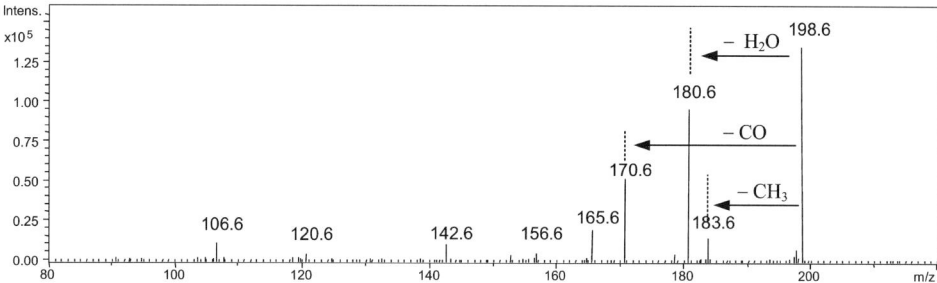

Figure 15.12 MS^2 spectrum of m/z 199 generated in APPI of toluene.

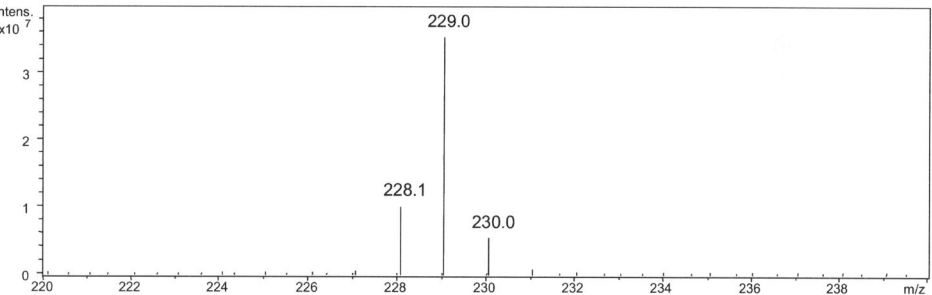

Figure 15.13 APPI mass spectrum of compound **1** dissolved in a mixture of benzene 90% and phenol 10%.

(m/z 171). Further product ions are detected at m/z 165, 143, and 107. Finally, the ion at m/z 215 can be reasonably considered as generated by the reaction of the neutral molecules of 108 Da with the ions at m/z 107. It should be emphasized that in the case of toluene, for which different possible reactant ions are produced (e.g., those at m/z 91, 92, 107, and 108), more complex ion clusters are generated at higher mass values, due to the wider number of available reactive channels. In the case of benzene, for which only M$^{+\bullet}$ ions are generated by irradiation, a simple reaction pattern is obtained.

In conclusion, the data discussed above show that benzene and toluene, when interacting with 10.2 eV photons at atmospheric pressure, undergo a series of reactions, mainly with oxygen molecules, giving rise to the formation of phenol and cresol, respectively. These species, reasonably originating through radical mechanism(s), are the reactants for further reactions, in this case following electrophilic mechanisms, leading, in the case of benzene, to diphenylether and phenoxyphenol, either as odd-electron cations or as protonated molecules. In the case of toluene analogous species are generated.

Some of the ions identified can act as effective protonating media; however, we introduced compound **1** into the solution of phenol (10%) and benzene (90%)

to verify the data just described at the beginning of this section. As can be seen in Fig. 15.13, the presence of phenol leads to a dramatic increase of protonated molecules of **1**, proving that phenol plays an important role in the $[M + H]^+$ production even if the other protonating species (m/z 171 and 187) are absent (or present in low abundance). This result suggests that the use of phenol as a dopant in APPI experiments would lead to effective data and work is now in progress in this direction.

15.4 Conclusions

The data described above show that interesting results can be obtained in APPI conditions either from the analytical or from the fundamental ion chemistry points of view. The possibility to photo-induce reactions leading to excited molecules and/or ions with low internal energy content gives to the analyst a highly effective method to investigate on ion structure and to obtain valid quantitative data. On the other hand, the APPI method, when coupled with a sufficiently specific mass analyzer, capable of MS^n experiments, allows us to carry out fundamental studies on reactions involving photons, neutral molecules (at fundamental and/or excited states), radicals, slow electrons, and ions, which represent the species present inside the source.

References

1. J.D. Morrison, In: J.H. Futtrell (Ed.), *Gaseous Ion Chemistry and Mass Spectrometry*, Wiley, New York, 1986, pp. 95–106.

2. E.P. Wigner, *Phys. Rev.* **73** (1968) 1002.

3. G.H. Wannier, *Phys. Rev.* **90** (1953) 817.

4. H.N. Chen, W. Genuit, A.J.H. Boerboom, and J. Los, *Int. J. Mass Spectrom. Ion Phys.* **51** (1983) 207.

5. I.A. Revel'skii, Y.S. Yashin, V.K. Kurochkin, and R.G. Kostyanovskii, Patent SU 1159412, 1985.

6. D.B. Robb, T.R. Covey, and A.P. Bruins, *Anal. Chem.* **72** (2000) 653.

7. R.D. Levin and S.G. Lias, *Ionization Potential and Appearance Measurements, 1971–1981, Issue October 1982*, U.S. Department of Commerce and National Bureau of Standards, Washington, DC 20234; http://webbook.nist.gov.

8. J.R. Chapman, *Practical Organic Mass Spectrometry*, second ed., Wiley, Chichester, 1993, pp. 74–82.

9. N. Einolf and B. Murson, *Int. J. Mass Spectrom. Ion Phys.* **9** (1972) 141.

10. E. Marotta, R. Seraglia, F. Fabris, and P. Traldi, *Int. J. Mass Spectrom.* **228** (2003) 841.

11. H. Bohringer and F. Arnold, *Nature* **290** (1981) 321.

12. (a) G. Moneti, G. Pieraccini, D. Favretto, and P. Traldi, *J. Mass Spectrom.* **33** (1998) 1148;
     (b) G. Moneti, G. Pieraccini, F. Dani, S. Turillazzi, D. Favretto, and P. Traldi, *J. Mass Spectrom.* **32** (1997) 1371.

13. E. Marotta and P. Traldi, *Rapid Commun. Mass Spectrom.* **17** (2003) 2846.

14. T. Koopmans, *Physica* **1** (1933) 104.

15. A. Doughty, G.B. Bacskay, and J.C. Mackie, *J. Phys. Chem.* **98** (1994) 13546.

16. G. Koster and A.P. Bruins, In: *Proceedings of the 49th ASMS Conference on Mass Spectrometry and Allied Topics, Chicago IL*, May 17–31, 2001.

17. M. Tubaro, E. Marotta, R. Seraglia, and P. Traldi, *Rapid Commun. Mass Spectrom.* **17** (2003) 2423.

18. R. Volkamer, B. Klotz, I. Barnes, T. Imamura, K. Wirtz, N. Washida, K.H. Becker, and U. Platt, *Phys. Chem. Chem. Phys.* **4** (2002) 1598.

19. R. Atkinson, *J. Phys. Chem. Ref. Data, Monograph* **2** (1994) 1.

20. R. Atkinson, *J. Phys. Chem. Ref. Data, Monograph* **1** (1989) 1.

21. G.A. Russell and R.F. Bridger, *J. Amer. Chem. Soc.* **85** (1963) 3765.

22. P.M. Sommeling, P. Mulder, R. Louw, D.V. Avila, J. Lusztyk, and K.U. Ingold, *Phys. Chem.* **97** (1993) 8361.

23. M. Preidel and R. Zellner, *Ber. Bunsen-Ges. Phys. Chem.* **105** (1989) 3609.

Advances in Mass Spectrometry, Volume 16
A.E. Ashcroft, G. Brenton and J.J. Monaghan (Editors)
© 2004 Published by Elsevier B.V.

CHAPTER 16

Electrospray Mass Spectrometry of Gas Phase Macromolecular Complexes

Margaret M. Sheil [1], Jennifer L. Beck [1], Raj Gupta [1], Stephen Watt [1],
Susan E. Brown [2], and Nicholas E. Dixon [2]

[1] Institute of Biomolecular Science and Department of Chemistry, University of Wollongong,
Wollongong, NSW 2522, Australia

[2] Research School of Chemistry, Australian National University, Canberra, ACT 0200, Australia

Abstract

Recent work involving the use of electrospray ionization mass spectrometry to study protein–DNA complexes is reviewed. In particular, our on-going work involving complexes in the bacterial replisome, a giant nucleoprotein complex that faithfully copies DNA during replication of the chromosome, is presented. We have examined different complexes important in DNA replication in *E. coli*, each with different binding modes. Results for two complexes studied to date (i) Tus protein (35,652 Da), with its double-stranded 21 mer DNA recognition sequence, *Ter B*; and (ii) the protein–protein complex formed between the θ and ε subunits of DNA polymerase III (i.e., a multisubunit enzyme that is the major replicative polymerase of *E. coli*) are presented here. ESI-MS studies of both these complexes have provided new information concerning the nature of the binding and enabled us to investigate the limitations and advantages of studying such complexes in the gas phase.

16.1 Introduction

Elucidation of the genome and proteome of an organism are the first steps in understanding biological processes at the molecular level. Equally as important as the determination of the primary structure of the protein or DNA, however, is characterization of the higher-order structures and the way in which biopolymers come together to form large macromolecular assemblies (e.g., protein-receptor complexes, ribosomes, and viruses). Interactions of proteins with the genome of an organism are important throughout all stages of its life cycle. DNA-binding proteins are involved in gene expression, i.e., replication, transcription, translation, and repair. DNA–protein interactions are also important in cancer, aging and in the success of pathogens such as viruses. The formation of ternary complexes between a protein, DNA, and metal ions or drugs is also important in many biological or drug-mediated processes. From a chemical viewpoint, specific interactions between a protein and DNA are highly complex since they require that the protein distinguish base sequences or local structural variations in DNA in genes that contain hundreds of base pairs. The binding of proteins to DNA may involve either *direct recognition* via selective hydrogen bonding or hydrophobic interactions between amino acid side chains and the DNA bases; and/or *indirect recognition* of local variations in DNA conformation. Nonspecific electrostatic interactions between phosphate groups of DNA and basic amino acid residues of proteins and water-mediated contacts are also important in conferring stability.

DNA–protein complexes are very stable (K_ds $\sim 10^{-9}$–10^{-12} M) and a number have been characterized by either X-ray crystallography or NMR spectroscopy [1]. These complexes can also be studied via indirect "footprinting" methods, either as a prelude to structural studies, or in cases where no detailed structural information is available. The central importance and extraordinary diversity of DNA–protein complexes, however, mean that efforts to elucidate the principles governing DNA–protein recognition are continuing to receive widespread attention [1].

16.2 Examples of DNA-binding proteins

An increasing number of DNA-binding proteins have been identified and characterized [2]. In particular, a variety of proteins involved in *E. coli* replication have been studied extensively [3]. These include: an initiator protein, dnaA [4], which binds at the origin of replication, a helicase (for relaxing supercoiled DNA) [5], a primase (for synthesizing primer RNA) [6], ten subunits of DNA polymerase III which include a polymerase, ATPases, a $3' \rightarrow 5'$ exonuclease (proofreading), and a sliding clamp, which confers processivity to the process [3]. Analogous proteins are present in viruses and eukaryotes; however, replication is even more complex in the latter because the chromosomes are larger and associated with histone proteins [7].

DNA-binding proteins that regulate transcription, generally termed transcription factors, include: the *E. coli* trp repressor [8]; Fos and Jun, which regulate growth of normal and transformed cells [9]; the vitamin D [10] and estrogen receptors [11], and the homeobox transcription factors important in developmental control [12]. Architectural proteins such as the mobility group (HMG) 1 proteins and HMG-box-containing proteins alter the shape of DNA to which they are bound [13]. Recently, interactions with HMG proteins have been proposed to play a role in the mode of action of some DNA-binding drugs such as cisplatin [14].

DNA-binding proteins are also directly or indirectly involved in disease processes. For example, certain tumor lines may develop following mutation of a stress-activated transcription factor; the tumor suppressor protein p53 [15], and regulation of telomerase has been implicated in both cancer and aging [16]. Precise knowledge of molecular interactions between transcription factors and target sequences, for example in oncogenes, is of interest for the development of new drugs [17].

16.3 Characterization of protein–DNA complexes

X-ray crystallography, which allows visualization of biological molecules at near-atomic resolution, remains the most powerful method for obtaining higher order structural information on individual proteins and DNA–protein complexes. Nuclear magnetic resonance (NMR) also provides detailed structural information and has the advantage of being able to analyze dynamic interactions in solution (in contrast to X-ray crystallography which is a solid phase technique). Without underestimating the enormous power of these two techniques, each has some limitations. In X-ray crystallography there may be problems with preparation of suitable crystals, whereas not all complexes are amenable to study by NMR because of size and solubility requirements. Further, these techniques are relatively insensitive, requiring preparation of relatively large amounts of recombinant proteins and DNA [18]. Several other biophysical techniques are also frequently used to obtain gross structural information and/or kinetic and thermodynamic data on DNA–protein interactions. These include: circular dichroism (CD) spectroscopy, fluorescence spectroscopy and more recently, surface plasmon resonance (SPR) [19].

DNA–protein complexes may also be examined using a variety of biochemical assays [19]. Footprinting methods rely on the bound protein or drug blocking the cleavage of DNA by a nonspecific enzyme such as DNase I. The resulting pieces of DNA are then analyzed by gel electrophoresis, which reveals gaps in the DNA sequences where the ligand or protein is bound. DNA–protein binding may also be detected by filter-binding assays and electrophoretic mobility "gel-shift" assays, which in combination with titration experiments enable stoichiometry and binding constants to be determined.

Noncovalent complexes are also commonly stabilized by UV [20] or chemical [21] crosslinking prior to proteolytic digestion and analysis, thereby allowing more precise determination of DNA–protein contact points. Mass spectrometry (either matrix assisted laser desorption (MALDI) [22,23] or ESI-MS) is frequently used to characterize peptides resulting from proteolytic digestion of the complexes [24]. ESI-MS allows the option of online chromatographic separation of reaction products via LC-MS. However, MALDI-MS has an advantage in that proteolysis reaction mixtures can be directly loaded onto MALDI targets and analyzed *in situ*.

16.4 Mass spectrometry of noncovalent macromolecular complexes

The rapid growth in the technology and applications of mass spectrometry (MS) over the past ten years has been such that it is now a well-established tool for the determination of *primary* structure of biomolecules. This has followed the introduction and development of the ionization techniques, electrospray (ESI) [25], and matrix-assisted laser desorption (MALDI) [26]. Recent improvements in instrumentation for tandem mass spectrometry, especially ion traps [27] and time-of-flight analyzers [28], have greatly expanded the application of mass spectrometry in biochemistry. In addition to important applications in the analysis of primary structures of biomolecules, ESI-MS has also been shown to play a role in the determination of *higher order structure* of proteins [29] and in the study of *noncovalent complexes*, important in molecular recognition. For example, protein–protein, DNA–protein, DNA–drug, protein–ligand, and antibody–antigen complexes have been studied by ESI-MS. Some of the reports, particularly from Robinson and co-workers, have featured very large complexes such as an 800 kDa complex of the chaperone GroEL [30], the 30S subunit of the *E. coli* ribosome ($\sim$ 850 kDa) [31] and the 2.5 MDa complex of the bacteriophage MS2 virus capsid [32]. Recently, the same group also demonstrated real time monitoring of subunit exchange between small heat shock proteins [33]. Another spectacular example is the detection of the whole intact rice yellow mottle virus (6.5 MDa) and tobacco mosaic virus consisting of 2140 identical protein subunits and an RNA molecule (40.5 MDa) [34].

A number of recent reviews have provided overviews of the application of ESI-MS for the analysis of noncovalent complexes [35–38] and protein folding and dynamics [29,39,40]. These have highlighted the potential advantages of ESI-MS for the study of noncovalent complexes. For example, ESI-MS can provide:

- Direct evidence for the formation of the complex since mass is an intrinsic property of all molecules;

- Precise information on all species present in a mixture in contrast to most other spectroscopic techniques which provide information on the *average* of all species present;

- Accurate measurement of the stoichiometry of binding of metal ions and ligands to large biopolymers and of the stoichiometry of macromolecular complexes;

- The ability to monitor transient structures or intermediates which are not accessible by other techniques [39,41];

- A highly sensitive analytical method (µmolar concentrations with 1–50 µL required for each analysis so total sample consumption is of the order of pmols or less);

- The capability for rapid screening of large numbers of variants so ESI-MS can either be used for prescreening prior to undertaking more detailed structural studies or may find application in diagnostic tests.

In summary, mass spectrometry is especially suited for confirmation of binding and for determination of stoichiometries, affording superior speed and sensitivity in comparison with other biophysical techniques. Furthermore, it is possible to observe components of equilibrium mixtures without removal of a component that might perturb the equilibrium position. Some caution must be exercised in interpretation of the data so that relative comparisons are only drawn between species of similar composition and, therefore (presumably), ionization efficiencies. It should also be noted that the ionization process itself might perturb equilibria and there is relatively little information available concerning changes in the strength or specificity of noncovalent interactions that occur on transfer from the condensed to the gas phase during the ionization.

Finally, while outside the scope of this paper, it should be noted, that MALDI mass spectrometry has also been used to characterize DNA– and RNA–protein complexes. MALDI is generally less amenable to the study of noncovalently bound complexes (because ions are not formed from solution) but offers particular advantages for studies of protein–DNA contacts as noted above.

16.5 ESI-MS of DNA–protein complexes

To fully exploit the advantages ESI-MS offers for studying DNA–protein complexes, there are a number of experimental details to consider. The first is the size of the DNA–protein complex under study. To the best of our knowledge, the largest ds DNA–protein complex so far detected had a mass of $\sim$ 134 kDa [42]. Ions of interest were observed in the range m/z 4000–7000 using a time-of-flight instrument. This and earlier studies (more below) highlight the need for a mass analyzer with extended m/z range in order to detect large noncovalent complexes. In cases where the size of the complex remains a limiting factor, DNA-binding domains of proteins prepared by proteolytic or chemical cleavage of native proteins, or via recombinant preparation of specific protein domains, can be used to derive information on DNA binding [43–45].

A limitation of ESI-MS lies in its incompatibility with the traditional buffers used for studying biomolecular complexes. Thus, the design of most ESI-MS experiments involves a compromise between the best ESI-MS solvents and the best solvents for maintenance of the complex. First, involatile salts in samples for ESI-MS analysis may interfere with the ionization process and salt adducts can contribute to peak broadening and limit the accuracy of mass measurement. Therefore, solutions of volatile salts such as ammonium acetate or ammonium bicarbonate (often 10 mM, but up to 50 mM concentration) are commonly used to acquire ESI mass spectra of oligonucleotides (DNA) and proteins but these raise issues from the biochemical standpoint. To overcome this problem, protein samples or DNA–protein complexes must be desalted just prior to analysis. We have explored a variety of techniques such as dialysis, ultrafiltration, and gel filtration using spin columns. We have found that simple dilution of the higher salt concentrations immediately prior to analysis can also be particularly effective. Typically, ESI-MS spectra of proteins are obtained under conditions of low pH (often 0.1–1% formic or acetic acid) and high capillary temperatures. For DNA–protein complexes, protonation of charged residues on the protein may disrupt hydrogen bonding within the protein and alter its conformation, affecting its DNA recognition surface. Release of metal ions at low pH may also perturb the DNA binding sites (e.g., in enzymes or zinc finger proteins). The use of high capillary temperatures has been shown to cause dissociation of a DNA–protein complex [43]; however, another study showed that a DNA–protein complex only dissociated under conditions where covalent bonds were broken [46]. For DNA–protein complexes the choice of ionization mode is also an important experimental consideration since proteins are usually detected as positive ions whereas oligonucleotides are analyzed as negative ions. Of the reports to date of observation of noncovalent DNA–protein complexes using ESI-MS, in some cases the complexes were detected as negative ions and in others as positive ions. Thus both ionization modes should be investigated to determine which gives the most complete information.

The contribution of various factors to gas phase stability of noncovalent complexes has been reviewed [29,35–38]. Electrostatic interactions are thought to be strengthened *in vacuo*. In contrast, in complexes where hydrophobic interactions are important, solution phase abundances were not mirrored in the gas phase. These results suggest that gas phase stability of complexes depends on the relative contributions of the different physical forces holding them together. One of the most important criteria to be satisfied concerns whether noncovalent complexes observed in the gas phase reflect solution behavior or are the result of nonspecific associations that occur during ionization. Consequently, many ESI-MS studies of noncovalent complexes to date have involved systems which have been well characterized in solution, although there are examples where ESI-MS has been used to derive new biological information [47]. Accumulation of information concerning a range of different binding partners prepared and analyzed under a range of conditions will shed light on whether mass spectrometry can be used to study noncovalent interactions.

The first ESI-MS study of a DNA–protein complex involved a complex between bovine serum albumin (BSA) and single-stranded (ss) 20-mer phosphorothioate oligonucleotides [48]. The interaction between serum albumin and oligonucleotides has attracted attention with respect to delivery of antisense therapeutics; however, binding is likely to occur through nonspecific electrostatic interactions. This was supported by estimates of dissociation constants (K_d) which showed a 100-fold increase when the solvent was changed from water to 100 mM ammonium actetate/30 mM imidazole. A 1:1 complex ($\sim$ 73 kDa) was detected in 10 mM ammonium acetate/33 mM imidazole (pH 7.5) for a mixture of 0.5 μM BSA/1 μM oligonucleotide. A 2:1 (oligonucleotide : protein) complex was observed when the concentration of oligonucleotide was raised to 3 μM, with all other conditions the same.

Smith and co-workers studied the stoichiometry of binding of several 13–16-mer ss oligonucleotides to the gene V protein of bacteriophage f1, a protein that stabilizes ss DNA during phage replication [49]. Complexes of protein with DNA < 15 bases had 2:1 (oligonucleotide : protein) stoichiometries, while a complex with a 16-mer bearing the sequence of its natural binding partner (an operator of gene II mRNA) had a 4:1 stoichiometry, yielding a mass of $\sim$ 43 kDa.

The first study in which ds DNA was complexed with its natural binding partner involved the DNA binding domain of the eukaryotic transcription factor, PU1 [43]. A 1:1 complex was observed in 10 mM ammonium acetate (pH 7.0). Their results were consistent with observations made using gel mobility shift assays. The fact that the complex showed fewer ions than in the ESI mass spectra of the protein alone was proposed to support the view that the complex is more compact than the protein alone. In this study, the effect of capillary heating was also investigated, highlighting the need for careful control of capillary temperature when analyzing these fragile complexes.

The trp operator/repressor is one of the most studied transcription systems in biochemistry [50]. Binding of tryptophan to the repressor protein changes its conformational shape and allows it to bind to its operator DNA sequence preventing transcription of genes by RNA polymerase. A time-of-flight instrument was used to study interactions between the trp repressor, tryptophan, and its specific operator DNA sequence [46]. This investigation highlighted the importance of sample preparation. No complex with the consensus DNA sequence was observed when attempts were made to prepare the complex in 5 mM ammonium acetate (pH 6.0) whereas a homodimer protein-ds DNA complex in 1:1 ratio was observed when the complex was prepared in 20 mM sodium phosphate buffer (pH 6.0), 90 mM in sodium chloride, followed by dialysis against 10 mM ammonium acetate (pH 6.0). Competition experiments demonstrated that the complex was observed only with the specific DNA operator sequence. In a more recent study from the same group, a 29 bp synthetic DNA was designed for binding to the proteins GclR and IclR. The IclR family of proteins prevents constitutive expression of genes involved in acetate usage in *E. coli*. Both the proteins formed noncovalent complexes with dsDNA, and the stoichiometries determined were 4:1 (protein monomer : dsDNA) [42].

Other examples of DNA–protein detected by ESI-MS include a ~ 118 kDa complex corresponding to a 4:1:1:1 complex of Zn^{2+} : vitamin D receptor : retinoid X receptor : vitamin D response element [52]; a ternary complex between Zn^{2+}, the DNA binding domain of the vitamin D receptor, and its response element (ds 34-mer DNA) [45]; and the complex of peptides of GCN4 with ds DNA [44]. A very recent study of the catalytic domain of bacteriophage λ integrase (λ-Int) enzyme by ESI-MS showed ions arising from three different species (folded, unfolded, and dimer). In the presence of DNA, ions from the complex and folded form only were detected, suggesting that binding of hairpin DNA (40 mer) stabilizes the fold of the protein [52].

There have also been reports of RNA-protein/peptide complexes studied by ESI-MS [53,54]. The use of ESI-MS for studying DNA or RNA-protein interactions has yet to reach its full potential, however, as more biologically relevant binding partners which have been well characterized by other methods need to be studied. As confidence in the technique grows with time, more examples of the use of ESI-MS to derive important new information on protein–DNA complexes will become available.

16.6 Selected results: complexes important in DNA replication

Much that is known today about DNA replication is largely a result of studies involving *Escherichia coli*. DNA replication, a three-step process consisting of initiation, elongation, and termination, occurs in a bidirectional manner along the bacterial chromosome at a rate of about 1000 base pairs/second [3]. The major replicative polymerase in *E. coli* is DNA polymerase III. DNA polymerase III consists of 10 different subunits and contains a polymerase core that is capable of simultaneous replication in leading and lagging strands [55]. The catalytic core of pol III consists of three subunits: α, ε, and θ. The α subunit (130 kDa) is responsible for the $5' \rightarrow 3'$ polymerase activity, and ε (27.5 kDa) is a $3' \rightarrow 5'$ proofreading exonuclease. The function of smaller subunit, θ (9 kDa), is still undetermined.

Replication terminates in a region opposite the origin. Two main components dictate replication arrest. First, there are 6 termination DNA sequences, termed *Ter* (*A–F*), that are about 20 base pairs long. Second, Tus protein binds as a monomer to termination sequences, halting replication [57,58].

We have an on-going program aimed at studying protein–protein and protein–DNA complexes involved in the bacterial replisome. Thus, to study these complexes, in particular, and in order to investigate the limitations and advantages of studying such complexes in the gas phase, we have examined different complexes important in DNA replication in *E. coli*, each with different binding modes. We present here some results from two examples of complexes studied to date, namely, the (i) Tus protein (35,652 Da), with its double-stranded (ds) 21 mer DNA recognition sequence, *Ter B* (mass of complex, 48,599 Da) [56] and (ii) the protein–protein complex formed between the θ and ε subunits of DNA polymerase III (i.e., a multisubunit enzyme that is the major replicative polymerase of *E. coli*; mass of complex, 29,434 Da).

16.7 The Tus–*TerB* complex

The availability of the X-ray crystal structure [59], combined with *in vivo* and *in vitro* binding studies of native Tus compared with mutant proteins, has enabled analysis of the relative contributions of various polar and hydrophobic interactions to binding. In addition, variant *Ter* sequences have been studied [60]. In the Tus–*Ter* complex, the DNA lies in a positively charged cleft between amino- and carboxy-terminal domains that are joined by interdomain β strands [59]. Fourteen Tus residues make sequence-specific contacts with *Ter* DNA, and there are numerous polar contacts between Tus and the phosphate backbone. On the side of the complex that allows the replication fork to proceed, one DNA strand makes extensive contact with Tus, whereas the other strand is predominantly exposed to solvent. In contrast, on the side of the complex where replication is halted, Tus makes extensive contacts with both DNA strands [59].

In the cleft, the side chain of Ala 173 is involved in a hydrophobic interaction with the methyl group of a thymine base. When this alanine residue is changed to the more bulky threonine (A173T), Tus binds ~ 4000-fold less tightly to *TerB* and is unable to halt replication in vivo [61,62]. Arg 198 lies just outside the core DNA binding region on the side of the complex where replication is halted and makes sequence-specific contacts with DNA.

Equilibrium dissociation constants (K_D) for the Tus–*TerB* complex have been measured using gel mobility shift and filter binding assays [60,61,63] and in surface plasmon resonance (SPR) experiments [62]. The binding is very tight: K_D values are 3.3×10^{-13} M in buffer (pH 7.5) containing 150 mM glutamate [63] and 0.5×10^{-9} M in buffer (pH 7.6) containing 250 mM KCl [62]. In the latter study, a range of KCl concentrations was investigated, and extrapolation of data to [KCl] = 150 mM gave $K_D \sim 1 \times 10^{-12}$ M. We studied the Tus–*TerB* interaction and used ESI-MS to compare the relative strengths of binding of native and mutant proteins with specific DNA sequences.

16.8 ESI-MS of Tus–*TerB*

ESI mass spectra of Tus–*TerB* complex were obtained using a Micromass QTOF2 mass spectrometer with a Z-spray source. The complex was prepared in 0.01 M ammonium acetate (pH 8). Interestingly, when ESI mass spectra of Tus protein alone were obtained using this solvent, the peaks observed were broad with a peak width at half height of ~ 115 mass to charge units. The quality of the spectra of the complex was sensitive to the desolvation temperature used [56]. In our earlier ESI-MS studies of noncovalent complexes, desolvation temperatures were kept low, as we reasoned that this would maintain the integrity of these interactions [64]. In our experiments using Tus–*TerB*, high quality ESI mass spectra were obtained using a desolvation temperature of 240 °C. The interaction between Tus and *TerB* is one of the tightest noncovalent interactions known. Thus gentle interface conditions may not be required in cases where the binding between partners in the noncovalent complexes is relatively strong. The quality of ESI mass spectra of the Tus–*TerB* complex was also affected by the integrity of the capillary needle. In some experiments, it was impossible to obtain mass spectra until a new needle was installed.

Our experiments were aimed at distinguishing the binding between Tus–*TerB* and Tus-variant DNA sequences, and the binding between Tus–*TerB* and mutant Tus–*TerB* complexes. The variant DNA sequences are shown in Table 16.1. Only the sequence of one strand is shown, all experiments, however, involved complementary pairs of DNA.

Coskun-Ari et al. [60] have measured equilbrium dissociation constants (K_{obs}) for complexes of Tus with *TerB*, posn5*TerB*, pos7*TerB*, and pos10*TerB*, respectively. *TerH* was identified as a possible binding site for Tus by searching the *E. coli* genomic DNA sequence, but had not been previously examined by experiment [60].

In preliminary experiments we attempted to use ESI-MS to distinguish between the binding of Tus and these dsDNA sequences, with a view to determining dissociation constants from titrations monitored by ESI-MS. When the complexes were

Table 16.1: DNA sequences used in this work, equilibrium dissociation constants and concentrations of ammonium acetate causing dissociation of 50% of the complex

DNA	Sequence	$K_{(obs)} \times 10^{13}$ [60]	[NH$_4$OAc] (mM)
TerB	5'-ATAAGTATGTTGTAACTAAAG-3'	9	1960
Posn5TerB	5'-ATAA**C**TATGTTGTAACTAAAG-3'	16	1780
Posn7TerB	5'-ATAAGT**G**TGTTGTAACTAAAG-3'	139	1180
Posn10TerB	5'-ATAAGTATG**G**TGTAACTAAAG-3'	1204	1150
TerH	5'-**GATC**GTATGTTGTAACTAAAG-3'	–	1490
Random	5'-CTTATGAGCTTATAAGCTCATTAAG-3'	–	

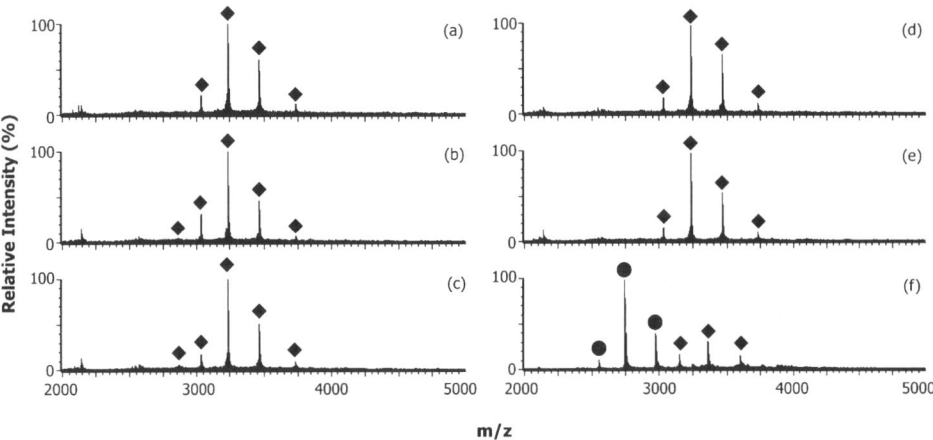

Figure 16.1 Positive ion ESI mass spectra of complexes of Tus with different *Ter* sequences: (a) *TerB*, (b) *TerH*, (c) posn5*TerB*, (d) posn7*TerB*, (e) posn10*TerB*, and (f) nonspecific 24 mer self-complementary DNA. (●) Ions from free Tus protein; (◆) Ions from the Tus–*TerB* complex.

prepared in 0.02 M ammonium acetate (pH 8), we could not distinguish between the binding of Tus with *TerB* and that with any of the *TerB* variants. These ESI mass spectra are shown in Fig. 16.1. The only significant ions in the spectra were from 1:1 Tus–DNA complexes ([M + 14H]$^{14+}$, [M + 15H]$^{15+}$, [M + 16H]$^{16+}$ ions at m/z 3465.5, 3224.5, and 3032.4, respectively). Importantly, the predominant ions were from free Tus in the spectrum of Tus with a random nonspecific DNA sequence obtained under the same conditions. This observation is strong evidence that the specific complexes detected in the gas phase are not the result of nonspecific associations in the ionization source or in solution.

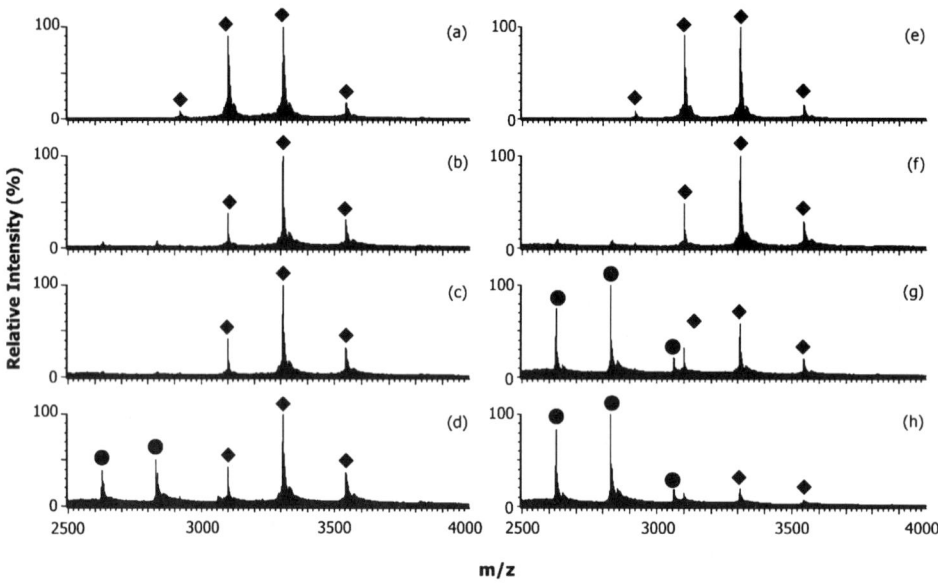

Figure 16.2 Positive ion ESI mass spectra of the Tus–dsDNA complexes his$_6$Tus–*TerB* in: (a) 10 mM NH$_4$OAc, (b) 800 mM NH$_4$OAc, (c) 1000 mM NH$_4$OAc, and (d) 1400 mM NH$_4$OAc and his$_6$Tus-posn10Tus–*TerB* at: (e) 10 mM NH$_4$OAc, (f) 800 mM NH$_4$OAc, (g) 1000 mM NH$_4$OAc, and (h) 1400 mM NH$_4$OAc. (●) Ions from free protein; (◆) Ions from the Tus–*TerB* complex.

In surface plasmon resonance experiments with Tus or Tus carrying a hexahistidine tag (his$_6$Tus), it was possible to determine equilibrium dissociation constants for interactions with dsDNA only when the binding interaction was weakened by increasing the KCl concentration of the analysis buffer [62]. We reasoned that increasing the ionic strength of the spray solvent might allow distinction between complexes of Tus and mutants with DNA (and variant sequences) in ESI mass spectra. Fig. 16.2 shows the effect of increasing concentrations of ammonium acetate (pH 8) used as spray solvent for acquisition of spectra of the complexes his$_6$Tus–*TerB* and his$_6$Tus–posn10*TerB*. In the case of the his$_6$Tus–*TerB* complex, no dissociation was observed until the complex was treated with 1400 mM ammonium acetate (pH 8) solution. Even at this salt concentration, ions from the complex predominate. On the other hand, the his$_6$Tus–posn10*TerB* complex shows a greater degree of dissociation with increasing ammonium acetate concentration. Almost all the complex has dissociated at 1400 mM ammonium acetate. This stands in contrast to earlier experiments where the spray solvent was 10 mM ammonium acetate (data not shown). The ammonium acetate concentrations at which all the Tus–dsDNA complexes dissociated are also shown in Table 16.1.

These results are consistent with values of K_{obs} measured in solution studies [60] for *TerB*, posn5*TerB*, pos7*TerB*, and pos10*TerB*, although it is difficult to ascertain from the data the relative binding order of pos10*TerB* relative to pos7*TerB*. In all experiments, as ions corresponding to free Tus increased, ions appeared that corresponded to both single strands and to ds*Ter* DNA (data not shown). These additional ions were at m/z 2119.7 and 1590.4 ($[M + 3H]^{3+}$ and $[M+4H]^{4+}$ of one *Ter* strand), at m/z 1624.7 ($[M+4H]^{4+}$ of the other *Ter* strand), and at m/z 2142.6 and 1836.9 ($[M + 6H]^{6+}$ and $[M + 7H]^{7+}$ of ds*Ter*). The appearance of the DNA as single strands when it dissociates from Tus is expected, based on other work in this laboratory showing that dsDNA denatures as the desolvation temperature is increased above 60 °C. It seems likely that the response factor for DNA would be different from that for free Tus protein or the Tus–*Ter* complexes. Therefore, we have compared abundances of ions from free Tus with the abundances of ions from complexes in the following experiments.

Based on the result above, experiments aimed at determining the relative binding affinities of Tus or mutant Tus proteins with dsDNA were carried out using 0.8 M ammonium acetate (pH 8). In SPR studies, the equilibrium dissociation constants for Tus and his$_6$Tus were indistinguishable [62]. In all the experiments (below), his$_6$Tus (rather than unmodified Tus) was used to enable more direct comparisons with mutant Tus proteins (A173T and R198A), which both carried a hexa-histidine tag.

An experiment was carried out over the same range of salt concentrations used to obtain the data in Fig. 16.2, but comparing Tus–*TerB* with his$_6$Tus–*TerB*, A173T–*TerB*, and R198A–*TerB*. No significant differences could be detected between ESI mass spectra of Tus–*TerB* and of his$_6$Tus–*TerB* at any salt concentration in the range 10 to 2200 mM (data not shown). This is in agreement with the SPR studies [62]. In contrast, the complexes R198A–*TerB* and A173T–*TerB* were observed to dissociate over this concentration range. The relative order of binding affinities was determined by comparing the concentrations of ammonium acetate at which each complex is 50% dissociated. This gave a relative order of binding affinity for *TerB* of his$_6$Tus (> 2200 mM) > R198A (1530 mM) > A173T (775 mM). The solution K_D values measured in 250 mM potassium chloride using SPR for the his$_6$Tus–*TerB*, R198A–*TerB*, and A173T–*TerB* complexes are 0.5×10^{-9}, 130×10^{-9}, and 2000×10^{-9} M, respectively [62]. The present ESI-MS results are in agreement with that data.

The Tus–*Ter* interaction involves many electrostatic contacts, so increasing salt concentration markedly decreases the stability of the Tus–*TerB* complex in solution studies. This general observation has been confirmed in the present ESI mass spectra. The solvent used in the SPR study was 50 mM Tris·HCl, pH 7.5, 0.1 mM EDTA, 0.1 mM DTT, 0.005% Nonidet P-20, 250 mM potassium chloride, which is markedly different from ammonium acetate used for ESI-MS. To enable further comment on whether the solution and gas phase complexes are the same, values of K_D need to be measured by titrating Tus with *TerB* and comparing these values with those determined in solution at similar ionic strength. The delineation of solvent and ESI-MS instrumental conditions that allow discrimination among the relative stabilities of various complexes is a starting point for being able to measure K_D values by ESI-MS. A complication is that electrostatic interactions are thought to be strengthened *in vacuo*. Therefore, the relative contributions of electrostatic interactions, hydrogen bonding, hydrophobic, and van der Waals interactions to the free energy of binding will influence stabilities of noncovalent complexes in the ESI source [35]. An example of where this may impact on values of dissociation constants estimated in the gas phase would be in comparisons of Tus with R198A. Arg 198 is involved in interactions with the negatively charged DNA backbone. If this (and other electrostatic interactions) were strengthened in the mass spectrometer, then the difference in the strength of *TerB* binding by Tus and R198A would be greater in the gas phase than in solution.

16.9 Interactions between the $\varepsilon 186$ and θ subunits of DNA polymerase III

The ε subunit of DNA polymerase III contains two domains. The domains have distinct functions; the N-terminal domain containing the amino acids 2–186 ($\varepsilon 186$) harbors the exonuclease active site and the binding site for θ whereas the C-terminal domain contains the residues 187–243 that include the α binding site [65]. Exonucleolytic proofreading of misincorporated nucleotides by ε contributes to the high fidelity of DNA replication (4.6 million base pairs in *E. coli* are replicated with an error frequency $\sim 10^{-10}$ per base pair). In addition to the proofreading function, ε plays an important role in the structure of pol III core as it interacts tightly with both α and θ [65].

The two subunits, θ and $\varepsilon 186$, form a tight complex [65]. Although, an X-ray crystal structure is not available for the θ–$\varepsilon 186$ complex, details of the interface between the two proteins in the complex and structural information on the individual subunits have become available from NMR experiments [67,68].

The ESI-MS experiments outlined above showed that relative binding affinities of mutant Tus proteins for *Ter* DNA were the same in the gas phase as in solution. The X-ray structure of the complex shows there are substantial polar and electrostatic contacts between binding partners [57]. Consistent with this, ESI mass spectra showed that the complex dissociated when treated with ammonium acetate at concentrations over the range 1–2 M [56].

We have carried out similar experiments with the θ–$\varepsilon 186$ complex. Solution conditions were varied and ESI-MS was used to provide "snapshots" of the equi-

Table 16.2: Dissociation $\varepsilon 186$–θ in different organic solvents observed by ESI-MS

	Concentration (M) at which 50% dissociation of the complex was observed	Dielectric constant at 25 °C
Methanol	8.9	32.6
Ethanol	4.8	24.3
1-Propanol	1.7	20.1
1-Butanol	1.1	17.1
Acetonitrile	4.3	38.8
Acetone	4.2	20.7
Isopropanol	3.1	18.3

librium positions (relative amounts of complex and free binding partners). In contrast to the Tus–*Ter* complex [56], $\varepsilon 186$–θ remained intact up to a concentration of at least 9 M ammonium acetate. This suggests that electrostatic interactions do not play a major role in stabilizing $\varepsilon 186$–θ and is consistent with NMR chemical shift mapping experiments that suggested the hydrophobic residues 21–27 of θ (AAAGVAF) are involved in its association with ε [67].

The hydrophobicity of the $\varepsilon 186$–θ interaction was probed further by treating the complex in 0.1 M ammonium acetate (pH 8) with the organic solvents methanol, ethanol, 1-propanol, 1-butanol, isopropanol, acetone, and acetonitrile. The total abundance of ions from $\varepsilon 186$–θ was expressed as a percentage of the total abundances of all the ions in the ESI mass spectra (θ, $\varepsilon 186$, and $\varepsilon 186$–θ) and these data are summarized in Table 16.2.

For the series of primary alkanols the concentrations of alkanol at which 50% of the complex decreased with decreasing dielectric constant. In other words, the propensity of the complex to dissociate increased with decreasing dielectric constant [69] of the primary alkanol. In a separate experiment, the concentration at which 50% dissociation occurred was measured for isopropanol, acetone, and acetonitrile. Since acetonitrile has the highest dielectric constant, it would have been expected to be the least effective solvent in dissociating the complex if the dielectric constant were the only factor involved. This clearly is not the case, suggesting that other intrinsic physical and structural features of the solvents (e.g., the ability to H-bond) may also be important. These data, however, do indicate that hydrophobic interactions contribute to the stability of the $\varepsilon 186$–θ complex. When an X-ray structure of the complex is available, mutational analyzes of residues at the interface should enable detailed analysis of the relative contributions of polar and nonpolar interactions to the overall stability of the complex in the condensed and gas phases. We are also exploring other factors that may influence the stability of these complexes in the gas phase.

16.10 Conclusions

The study of protein–DNA interactions is somewhat more challenging than examining protein–protein binding by ESI-MS. This follows from differences in the conditions required for optimum ionization of DNA and proteins and the need for an extended m/z range analyzer in many cases. Nonetheless, the relatively small numbers of studies in this area to date demonstrate that these challenges can be met in order to improve our understanding of cellular processes involving protein–DNA interactions at the molecular level. Clearly, ESI-MS has enormous potential to contribute to the study of such interactions. Based on our work, the very different gas phase behavior of the Tus–*Ter* and $\varepsilon 186$–θ complexes, for example, suggests that ESI-MS can be used to probe the nature of noncovalent interactions that hold assemblies of biological macromolecules together. Detailed studies of a range of biological complexes that have been characterized in solution, however, will be required to further understand the processes that influence the gas phase stability of large noncovalent complexes.

References

1. C.L. Larson and G.L. Verdine, In: S.M. Hecht (Ed.), *Bioinorganic Chemistry: Nucleic Acids*, Oxford, New York, 1996, pp. 324–346.

2. K. Frech, K. Quandt, and T. Werner, *Trends Biochem. Sci.* **22** (1997) 103–104.

3. Z. Kelman and M. O'Donnell, *Ann. Rev. Biochem.* **64** (1995) 171–200.

4. C. Speck, C. Weigel, and W. Messer, *EMBO J.* **18** (1999) 6169–6176.

5. J. Weigelt, S.E. Brown, C.S. Miles, N.E. Dixon, and G. Otting, *Structure* **7** (1999) 681–690.

6. A.A. Mustaev and G.N. Godson, *J. Biol. Chem.* **270** (1995) 15711–15718.

7. S. Waga and B. Stillman, *Ann. Rev. Biochem.* **67** (1998) 721–751.

8. A.N. Lane, J.F. Lefevre, and O. Jardetzky, *Biochim. Biophys. Acta* **909** (1987) 58–70.

9. G. McGill and D.E. Fisher, *Chem. Biol.* **5** (1998) R29–R38.

10. J.C. Hsieh, G.K. Whitfield, A.K. Oza, H.T.L. Dang, J.N. Price, M.A. Galligan, P.W. Jurutka, P.D. Thompson, C.A. Haussler, and M.R. Haussler, *Biochemistry* **38** (1999) 16347–16358.

11. M.R. Yudt, D. Vorojeikina, L. Zhong, D.F. Skafar, S. Sasson, T.A. Gasiewicz, and A.C. Notides, *Biochemistry* **38** (1999) 14146–14156.

12. W.J. Gehring, M. Affolter, and T. Burglin, *Ann. Rev. Biochem.* **63** (1994) 487–526.

13. C.A. Bewley, A.M. Gronenborn, and G.M. Clore, *Ann. Rev. Biophys. Biomol. Struct.* **27** (1998) 105–131.

14. U.M. Ohndorf, M.A. Rould, Q. He, C.O. Pabo, and S.J. Lippard, *Nature* **399** (1999) 708–712.

15. K. Roemer, *Biol. Chem.* **380** (1999) 879–887.

16. T.R. Hughes and V. Lundblad, *Front. Mol. Biol.* **20** (1998) 164–202.

17. M. Zewail-Foote and L.H. Hurley, *Anti-Cancer Drug Design* **14** (1999) 1–9.

18. J. Beachamp and N.W. Issacs, *Curr. Opin. Chem. Biol.* **3** (1999) 525–529.

19. M.J. Guille and G.G. Kneale, *Mol. Biotechnol.* **8** (1997) 35–52.

20. D.A. Connor, A.M. Falick, M.C. Young, and M.D. Shetlar, *Photochem. Photobiol.* **68** (1998) 299–308.

21. D. Bastia, *Structure* **4** (1996) 661–664.

22. S.L. Cohen, A.R. Ferre-D'Amare, S.K. Burley, and B.T. Chait, *Protein Sci.* **4** (1995) 1088–1099.

23. M.C. Golden, K.A. Resing, B.D. Collins, M.C. Willis, and T.H. Koch, *Protein Sci.* **8** (1999) 2806–2812.

24. D.L. Wong, J.G. Pavlovich, and N.O. Reich, *Nucleic Acids Res.* **26** (1998) 645–649.

25. J.B. Fenn, M. Mann, C.K. Meng, S.F. Wong, and C.M. Whitehouse, *Science* **246** (1989) 64–71.

26. M. Karas and F. Hillenkamp, *Anal. Chem.* 60 (1988) 2299–2301.

27. K.R. Jonscher and J.R. Yates III, *Anal. Chem.* **68** (1996) 659–667.

28. M. Guilhaus, D. Selby, and V. Mlynski, *Mass Spectrom. Rev.* **19** (2000) 65–107.

29. A.A. Rostom and C.V. Robinson, *Curr. Opin. Struct. Biol.* **9** (1999) 135–141.

30. A.A. Rostom and C.V. Robinson, *J. Amer. Chem. Soc.* **121** (1999) 4718.

31. A.A. Rostom, P. Fucini, D.R. Benjamin, R. Juenemann, K.H. Nierhaus, F.U. Hartl, C.M. Dobson, and C.V. Robinson, *Proc. Natl. Acad. Sci. USA* **97** (2000) 5185–5190.

32. M.A. Tito, K. Tars, K. Valegard, J. Hajdu, and C.V. Robinson, *J. Amer. Chem. Soc.* **122** (2000) 3550–3551.

33. F. Sobott. J.L.P. Benesch, E. Vierling, and C.V. Robinson, *J. Biol. Chem.* **277** (2002) 38921–38929.

34. S.D. Fuerstenau, W.H. Benner, J.J. Thomas, C. Brugidou, B. Bothner, and G. Siuzdak, *Angew. Chem., Int. Ed. Engl.* **40** (2001) 541–544.

35. J.A. Loo, *Mass Spectrom. Rev.* **16** (1997) 1–23.

36. J.L. Beck, M.L. Colgrave, S.F. Ralph, and M.M. Sheil, *Mass Spectrom. Rev.* **20** (2001) 61–87.

37. J.A. Loo, *Int. J. Mass Spectrom.* **200** (2000) 175–186.

38. S.A. Hofstadler and R.H. Griffey, *Chem. Rev.* **101** (2001) 377–390.

39. H. Hernandez and C.V. Robinson, *J. Biol. Chem.* **276** (2001) 46685–46688.

40. A.M. Last and C.V. Robinson, *Curr. Opin. Struct. Biol.* **3** (1999) 564–570.

41. A. Miranker, C.V. Robinson, S.E. Radford, R.T. Aplin, and C.M. Dobson, *Science* **262** (1993) 848–849.

42. L.J. Donald, D.J. Hosfield, S.L. Cuvelier, W. Ens, K.G. Standing, and H.W. Duckworth, *Protein Sci.* **10** (2001) 1370–1380.

43. X.H. Cheng, P.E. Morin, A.C. Harms, J.E. Bruce, Y. Bendavid, and R.D. Smith, *Anal. Biochem.* **239** (1996) 35–40.

44. L.J. Deterding, J. Kast, M. Przybylski, and K.B. Tomer, *Bioconj. Chem.* **11** (2000) 335–344.

45. T.D. Veenstra, L.M. Benson, T.A. Craig, A.J. Tomlinson, R. Kumar, and S. Naylor, *Nat. Biotechnol.* **16** (1998) 262–266.

46. N. Potier, L.J. Donald, I. Chernushevich, A. Ayed, W. Ens, C.H. Arrowsmith, K.G. Standing, and H.W. Duckworth, *Protein Sci.* **7** (1998) 1388–1395.

47. J. Fändrich, J.A. Tito, M.R. Lerous, A.A. Rostom, F.U. Hartl, C.M. Dobson, and C.V. Robinson, *Biochemistry* **97** (2000) 14151–14155.

48. M.J. Greig, H. Gaus, L.L. Cummins, H. Sasmor, and R.H. Griffey, *J. Amer. Chem. Soc.* **117** (1995) 10765–10766.

49. X.H. Cheng, A.C. Harms, P.N. Goudreau, T.C. Terwilliger, and R.D. Smith, *Proc. Natl. Acad. Sci. USA* **93** (1996b) 7022–7027.

50. S. Bass, V. Sorrells, and P. Youderian, *Science* **242** (1988) 240–245.

51. T.A. Craig, L.M. Benson, A.J. Tomlinson, T.D. Veenstra, S. Naylor, and R. Kumar, *Nat. Biotechnol.* **17** (1999) 1215–1218.

52. H.B. Kamadurai, S. Subramaniam, R.B. Jones, K.B. Green-Church, and M.P. Foster, *Protein Sci.* **12** (2003) 620–626.

53. K.A. Sannes-Lowery, P. Hu, D.P. Mack, H-Y. Mei, and J. Loo, *Anal. Chem.* **69** (1997) 5130–5135.

54. C. Liu, L. Pasa-Toilc, S.A. Hofstadler, A.C. Harms, R.D. Smith, C.H. Kang, and N. Siha, *Anal. Biochem.* **262** (1998) 67–76.

55. D.R. Herendeen, T.J. Kelly, and J. Thomas, *Cell* **84** (1) (1996) 5–8.

56. A. Kapur, J.L. Beck, S.E. Brown, N.E. Dixon, and M.M. Sheil, *Protein Sci.* **11** (2002) 147–157.

57. D.E. Bussiere and D. Bastia, *Mol. Microbiol.* **31** (1999) 1611–1618.

58. R.G. Wake and G.F. King, *Structure* **5** (1997) 1–5.

59. K. Kamada, T. Horiuchi, K. Ohsumi, N. Shimamoto, and K. Morikawa, *Nature* **383** (1996) 598–603.

60. F.F. Coskun-Ari and T.M. Hill, *J. Biol. Chem.* **272** (1997) 26448–26456.

61. A. Skokotas, M. Wrobleski, and T.M. Hill, *J. Biol. Chem.* **269** (1994) 20446–20455.

62. C. Neylon, S.E. Brown, A.V. Kralicek, C.S. Miles, C.A. Love, and N.E. Dixon, *Biochemistry* **39** (2000) 11989–11999.

63. P.A. Gottlieb, S. Wu, X. Zhang, M. Tecklenburg, P. Kuempel, and T.M. Hill, *J. Biol. Chem.* **267** (1992) 7434–7443.

64. A. Kapur, J.L. Beck, and M.M. Sheil, *Rapid Commun. Mass Spectrom.* **13** (1999) 2489–2497.

65. F.W. Perrino, S. Harvey, and S.M. McNeill, *Biochemistry* **38** (48) (1999) 16001–16009.

66. S. Hamdan, E.M. Bulloch, P.R. Thompson, J.L. Beck, J.Y. Yang, J.A. Crowther, P.E. Lilley, P.D. Carr, D.L. Ollis, S.E. Brown, and N.E. Dixon, *Biochemistry* **41** (2002) 5266–5275.

67. M.A. Keniry, H.A. Berthon, J.Y. Yang, C.S. Miles, and N.E. Dixon, *Protein Sci.* **9** (2000) 721–733.

68. E.F. De Rose, T. Darden, S. Harvey, S. Gabel, F.W. Perrino, R.M. Schaaper, and R.E. London, *Biochemistry* **42** (2003).

69. R.C. Weast and M.J. Astle (Eds.), *CRC Handbook of Chemistry and Physics*, CRC Press, Boca Raton, FL, 1982–1983.

Advances in Mass Spectrometry, Volume 16
A.E. Ashcroft, G. Brenton and J.J. Monaghan (Editors)

CHAPTER 17

Clinical Applications of Mass Spectrometry

R.N. Dalton* and C. Turner

The WellChild Trust Laboratory, Department of Child Health, Guy's, King's & St Thomas'
Medical School, Guy's Hospital, London SE1 9RT, UK

The aim of this mini-review is to promote both the clinical analyst's view of mass spectrometry (MS) and the application of "gold standard" methods in the clinical laboratory. The intention is not to be comprehensive but reflect the authors' interests and views on MS in the overall context of clinical diagnostics.

Clinical diagnostics, therapeutics, and research rely on the accurate identification and quantitation of a multitude of metabolites and proteins. MS offers potential solutions to the complex analytical challenges faced in clinical diagnostics, but, until recently, in routine clinical laboratories, it has remained a technology shrouded in the mystique of alchemy. This misperception is, in part, a product of the undergraduate teaching of MS by chemists whose interests tend to focus on the importance of molecular weight determination, structural information, and the cost of state-of-the-art instrumentation. Additionally, the introduction of gas chromatography-MS into clinical laboratories for "complex" analyses has only served to reinforce the prejudice of MS as capital and labour intensive and technically and intellectually complex. However, the development of robust liquid MS interfaces, multiple MS techniques, and cheap computing have presented the dramatic potential for routine specific and accurate "gold standard" isotope-dilution measurement of many clinically relevant compounds.

To appreciate the potential role of MS in clinical diagnostics it is important to understand current clinical laboratory practice and how it has developed during the last 20 years. In routine clinical chemistry high throughput is the aim; hence, the research effort has been directed into automation and information flow. Many of the analytical methods are simple colorimetric assays that, although they can be applied with excellent precision, are inherently inaccurate and prone to significant

---

*Corresponding author. E-mail address: neil.dalton@kcl.ac.uk.

interferences. A prime example, emphasising the compromises between accuracy, automation, and cost, is the measurement of creatinine using a method that was first described in 1886: the problems of creatinine measurement have been excellently reviewed [1].

Immunoassay, a readily automated technology, has played a central role in the development of modern clinical diagnostics. Protein, peptide hormone, steroid, and therapeutic drug analysis are all areas where immunoassay is now the primary analytical method, but once again specificity and accuracy have been sacrificed for ease of automation and the clinical chemistry literature has become saturated with "kit" comparisons. In a classical review of the performance of steroid assays, isotope-dilution MS was used as the "gold standard" method [2]. The inherent variability of results between kits from different manufacturers and even the same manufacturer was highlighted and the conclusion was drawn that it was important for individual laboratories to understand the limitations of the particular kit that they were using. The more scientific conclusion was that isotope-dilution MS should be used for clinical steroid measurement. The controversy has not diminished and in a recent editorial the question was posed, "How can assays that are grossly inaccurate gain approval for use in the diagnosis and treatment of endocrine abnormalities?" [3].

The areas of clinical diagnostics where MS has made the most impact are toxicology and the diagnosis of inherited metabolic disease. In both areas the initial diagnostic requirement is unequivocal compound identification. In our area of interest, the diagnosis of inherited metabolic disease, the analysis of urine organic acids is now almost exclusively performed by capillary gas chromatography-MS [4] and as a model serves to illustrate the value of MS. The drawback of capillary gas chromatography alone is that compound identification relies on retention time. Originally, in the best laboratories, samples would be run on two columns with different separation characteristics to improve specificity. The introduction of MS detectors enabled unequivocal compound identification using the fragmentation pattern and laboratory derived spectral libraries: as a consequence the analysis was simplified and the interpretation made more specific.

A major problem with the urine organic acid method is the range of potentially diagnostic metabolites that can be detected: simple acids, fatty acids, lactones, sugars, substituted amino acids, and pyrimidines. The problem is exacerbated by a method that uses oxime formation, liquid–liquid extraction, drying, and derivatisation (trimethylsilyl). Controlling each step and the interaction of the samples with the injection liner for such a wide range of compounds is not a trivial process, but in many laboratories the presence of a metabolite in the final chromatogram is the sole criterion. Questions about efficiency of extraction and derivatisation and the patency of the liner to specific compounds are conveniently ignored. Using MS detection allows the inclusion of stable isotope internal standards that provide control of the whole process. This is particularly pertinent for difficult but critically important metabolites like orotic acid where false negative results are a recognised problem. The chromatograms obtained from urine samples containing diagnostically increased concentrations of orotic acid can have no detectable orotic acid, even when the rest of the chromatogram appears unaffected. Addition of stable isotope orotic acid, as internal standard, to the original urine before analysis ensures control of the whole process: if the stable isotope cannot be detected in the final chromatogram then clearly there is a problem with the orotic acid analysis and the analysis can be repeated. Organic acid analysis is considered by many to be a subjective test, where the presence of a particular compound is diagnostic of a specific condition. As the sensitivity of detectors has improved it is clear that in most instances the diagnosis should rely on a quantitative result. With MS, specific and accurate quantitation using stable isotope-dilution methods is routinely applicable.

MS undoubtedly provides the clinical analyst with unequivocal compound identification and "gold standard" isotope-dilution quantitation. However, to be accepted into routine clinical diagnostics, sample preparation and any chromatography must be minimised. Electrospray MSMS has the potential for "dilute and shoot" sample preparation coupled to mass chromatography and collision induced fragmentation and this has led to an on-going revolution in clinical mass spectrometry. The early application of the technology to neonatal screening for inherited amino acid and fat oxidation disorders demonstrated the unique capabilities of MSMS for rapid, simultaneous quantitation of classes of diagnostic metabolites [5,6].

The initial method required a significant sample preparation stage, involving butylation, to improve both specificity and sensitivity but, using neutral loss and precursor ion scanning, a large number of specific inherited disorders could be diagnosed. We suggested that direct assay of the primary screening metabolites using multiple reaction monitoring mode (MRM) acquisition could also be a valid approach and introduced neonatal screening for phenylketonuria, an inherited disorder of phenylalanine metabolism, based on blood spot measurement of phenylalanine and tyrosine [7,8]. The only sample preparation required is the addition of 150 μL of methanol/water (83:17) containing d_5-phenylalanine and d_4-tyrosine as internal standards and mixing for 20 min at 300 cycles/min. Only 2 μL of the supernatant is injected, loop injection, into a mobile phase (acetonitrile/water (1:1) with 0.025% formic acid) flowing at 50 μL/min. Four MRMs, phenylalanine (166/120), d_5-phenylalanine (171/125), tyrosine (182/136), and d_4-tyrosine (186/140) are acquired for 60 s and the inject-to-inject time is less than 90 s. For neonatal screening this approach has the advantage of technical simplicity and improved sensitivity for more accurate and precise measurements of the primary screening metabolites. The latter significantly reduces the potential for false positive and, more importantly, false negative results.

We have continued this simple approach, wherever possible, towards the diagnosis of inherited disorders of metabolism. For instance, plasma amino acids [9] and acylcarnitines [10] are measured after mixing 5 μL of plasma with 150 μL of methanol containing appropriate stable isotope internal standards, and centrifugation. The availability of various scanning modes is very seductive and there is no doubt that, visually, they can be very valuable for diagnosis but we would always recommend quantitation using MRMs and isotope-dilution.

The importance of isotope-dilution in clinical mass spectrometry cannot be overemphasised. In complicated assay systems, like the urine organic acid example above, a stable isotope internal standard corrects for losses at any stage of the process. In simple loop-injection electrospray MSMS of biological samples, stable isotope internal standards are necessary to correct for ion suppression. Some compounds, e.g., creatinine, appear to ionise easily from any biological matrix whereas others, e.g., paracetamol, suffer significant ion suppression. Blood and urine samples can present a very variable matrix, particularly in critically ill patients, neonates, patients with renal failure, hepatic failure, crush injuries, and/or those receiving polypharmacy. The simplest way to manage the problem of ion suppression is to use an appropriate stable isotope internal standard. Consequently stable isotope internal standards are becoming the life-blood of clinical mass spectrometry.

The role of chromatography in clinical MS deserves consideration. It is absolutely essential in single quadrupole and ion-trap MS but, as the original neonatal screening application suggests, there are going to be many applications where the real power of MSMS is in eliminating the need for chromatography. Initial mass filtering coupled to production of a specific fragment ion is entirely specific for many clinically diagnostic compounds. In establishing a loop-injection MSMS method it is important that results are initially compared with those from an MSMS method incorporating an appropriate chromatographic step. If the method is only going to be applied in the investigation of patients with, for instance, renal disease then the comparison should be, predominantly, with samples from this group of patients. If a more general clinical application is envisaged then the comparison should be applied to samples from as many clinical conditions as possible.

Chromatography can also be applied to compounds that suffer significant ion suppression in order to increase sensitivity. If the suppression is due to salts then often a rapid guard column separation will suffice to remove the problem and the analysis time will be relatively unaffected. If the suppression is due to co-eluting compounds, then the required separation could introduce significant assay development and analytical time overheads. The dilemma of isobaric compounds is frequently resolved using chromatography but the first consideration using MSMS must be to utilise specific fragment ions. An interesting example demonstrating both the problem of isobaric compounds and an MSMS solution to a specific clinical problem, is the measurement of leucine and isoleucine. Increased blood concentrations are seen in branched chain 2-oxo acid dehydrogenase deficiency (BCOADD), an inherited disorder of branched chain amino acid metabolism. The compounds are isobaric and the situation is further complicated by increased allo-isoleucine concentrations in this condition. The most sensitive MRM for the three amino acids is 132/86. At higher fragmentation energies the 132/43 transition is specific for leucine and can be used to accurately quantify leucine in the presence of isoleucine and allo-isoleucine. Clinically this is extremely valuable. Patients with BCOADD can become acutely unwell and blood leucine increases rapidly. Dialysis is instituted to bring down the leucine concentrations that can now be monitored hourly to determine the effectiveness of the treatment.

The ability to be specific and measure simultaneously a series of diagnostic metabolites is an important concept that clinical laboratories are only starting to come to terms with. We are now in a position in any given clinical situation to pose the question, "What do we really want to measure?" In the differential diagnosis of hyperammonaemia, for example, it is possible to define all the potential diagnostic metabolites and measure them simultaneously. This is convenient not only at the point of clinical requesting, the "one-stop" shop concept, but also, when reporting, it is absolutely clear what the clinical relevance and consequences of the test are.

The open nature of the technology has seen an explosion in clinical applications across virtually every area of diagnostics, therapeutics, and research. Providing a complete reference list is beyond the scope of this communication but it is pertinent to make the point that MSMS is not just a method for measuring the concentrations of small molecules. Enzyme activities, now described as functional proteomics, can be easily measured by monitoring the rate of product formation [11]. Proteomics will play an increasingly important part of clinical diagnostics and, at present, the principal application is the characterisation of haemoglobin variants [12]. Cellular profiling, where cells are incubated with a stable isotope labelled substrate and metabolic variants determined by tracking the fate of the label, has already become an important confirmatory tool for the diagnosis of fat oxidation disorders [13]. The obvious next stage, to move to stable isotope tracking in vivo, has always been an important research tool [14]. The analytical simplicity of MSMS means that the advent of routine in vivo metabolic profiling is not so far away and will be essential in assessing gene transfer therapeutics.

Despite all this potential, in the routine clinical laboratory there are, at the moment, only a handful of applications for which MSMS is cost effective. Once a machine is installed, then other, more esoteric, tests can be included at marginal cost. In UK clinical laboratories the first electrospray MSMS machines were pioneered for neonatal screening and were largely charity funded. In the more routine clinical diagnostic setting virtually every electrospray MSMS instrument has relied on a business plan that substitutes MSMS immunosuppressive assays, particularly cyclosporin, for current immunoassay kits. In the clinical market, because of the number of assays performed and the cost of the immunoassays, the measurement of immunosuppressive drugs by MSMS must be considered the prime motivator for installation of MSMS. The published method [15] performs significantly better than any of the immunoassays but, in our opinion, lacks a very important ingredient for clinical mass spectrometry, a stable isotope internal standard. Provision of a cyclosporin kit for MSMS is clearly an objective but it must come at a commercial cost that does not totally remove the current cost benefit associated with MSMS.

Clinical MS can be uniquely specific, highly sensitive and importantly, from a modern environmental perspective, a relatively reagent free technique. The challenges for the widespread introduction of the technology into routine clinical diagnostics include automated sample preparation, "real-time" calculation and presentation of results, insuring continuing supplies of appropriate stable isotopes, and defining more clearly the clinical questions that we are trying to answer. However, just as the potential for a significant leap forward in routine clinical diagnostics using MS appears tangible, the real test will be to develop simultaneous multi-analyte isotope-dilution quantitation in a format that does not erode the cost benefits yet fulfils all the regulatory and compliance requirements of the new In Vitro Diagnostic (IVD) directive and certification legislation.

References

1. K. Spencer, Review, *Ann. Clin. Biochem.* **23** (1986) 1.

2. J. Middle, *Ann. Clin. Biochem.* **35** (1998) 354.

3. D.A. Herold and R.L. Fitzgerald, *Clin. Chem.* **49** (2003) 1250.

4. L. Sweetman, *Techniques in Diagnostic Human Biochemical Genetics: A Laboratory Manual*, Wiley–Liss, 1991, p. 143.

5. D.S. Millington, N. Kodo, D.L. Norwood, and C.R. Roe, *J. Inherit. Metab. Dis.* **13** (1990) 321.

6. M.S. Rashed, P.T. Ozand, M.P. Bucknall, and D. Little, *Pediatr. Res.* **38** (1995) 324.

7. S. Bird, C. Turner, and R.N. Dalton, In: *Proceedings 4th Meeting of the International Society for Neonatal Screening*, 1999, P3, p. 48, Abstract.

8. C. Turner, S. Bird, and N. Dalton, Abstract, *J. Inherit. Metab. Dis.* **22** A75 (1999) 146.

9. R.N. Dalton, S. Bird, and C. Turner, In: *Proceedings of Pathology* 2000, p. 116, Abstract.

10. R.N. Dalton and C. Turner, In: *Proceedings of Pathology* 2000, p. 116, Abstract.

11. A. Sankaralingam, C. Turner, and N. Dalton, Abstract, *J. Inherit. Metab. Dis.* **22** A76 (1999) 147.

12. B.J. Wild, B.N. Green, E.K. Cooper, M.R.A. Lalloz, S. Erten, A.D. Stephens, and D.M. Layton, *Blood Cells Mol. Dis.* **27** (2001) 691.

13. C.R. Roe and D.S. Roe, *Mol. Gen. Metab.* **68** (1999) 243.

14. M. Yudkoff, Y. Daikhin, I. Nissim, A. Jawad, J. Wilson, and M. Batshaw, *J. Clin. Invest.* **98** (1996) 2167.

15. B.G. Keevil, D.P. Tierney, D.P. Cooper, and M.R. Morris, *Clin. Chem.* **48** (2002) 69.

Advances in Mass Spectrometry, Volume 16
A.E. Ashcroft, G. Brenton and J.J. Monaghan (Editors)

CHAPTER 18

Mass Spectrometry: A Chemical-Biology Tool for Probing Enzymatic Systems

Julie A. Leary, Na Pi, and Yonghao Yu

University of California, Department of Chemistry, Berkeley, CA 94720, USA

Two novel ESI mass spectrometric (ESI-MS) assays have been developed to:
(1) screen combinatorial libraries of small molecule enzyme substrates and inhibitors, and (2) determine enzyme kinetics and the associated mechanisms of action for sulfotransferase and phosphotransferase enzymes. These assays can be
used to analyze the reaction using natural substrates without the need of a chro-
mophore or radiolabels. Multiplex screening is accomplished through the use of
immobilized enzymes in which spectra are taken of the library compounds before
and after incubation with the enzyme. The K_m and k_{cat} values of the various substrates and enzymes are generated and, similarly, K_i's of individual inhibitors detected in the screening assay are calculated. The data clearly indicate whether the
inhibitors are competitive or noncompetitive. In addition to determination of kinetic parameters, this technique is applied in order to determine the specific mechanism of action of the enzyme and its complexes. Noncovalent complexes of the
enzyme with both substrate and inhibitor have been generated using FTICR mass
spectrometry. Tryptic digests of the enzyme–substrate complex using ^{34}S labeling
provide crucial information on the structure and binding domain of the enzyme
intermediate.

18.1 Introduction

Mass spectrometry (MS) is an extremely useful tool to measure molecular weight
to a mass accuracy of 1 ppm [1]. It is particularly suitable for the study of a covalent
modification to an enzyme in a reaction that results in a mass shift on the original
protein. For example, it has been shown that when 3-ketobutylidene-β-2-chloro-
4-nitrophenyl is used as the substrate and incubated with *Thermococcus litoralis*

(a)   *(b)*

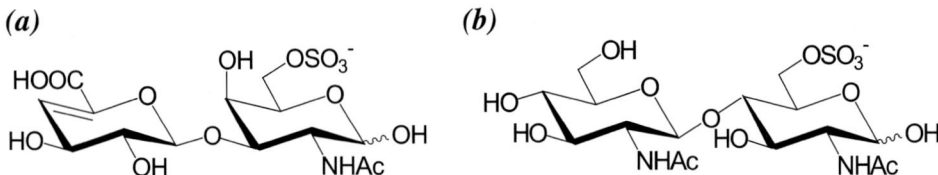

Structure of internal standard and monitored product. (a) Internal standard, α-ΔUA-[1 → 3]-GalNAc-6S (ΔD$_i$-6S), with $m/z = 458$. (b) Monitored product, chitobicose-6-OSO$_3^-$, with $m/z = 503$.

4-α-glucanotransferase (TLGT), the glycosylated enzyme intermediate is formed via a Ping-Pong Bi-Bi mechanism. The intermediate can be readily trapped and detected using MALDI time-of-flight mass spectrometry and the catalytic residue of this enzyme was identified using postsource decay [2]. The use of internal standards can also greatly facilitate the quantification of substrates and products when subjected to MS analysis [3,4]. Since the enzymatic reaction products are separated from each other based on their masses, radiolabels, and chromophogenic substrates are not needed. Thus enzymatic kinetic studies can be easily performed using mass spectrometry.

Recently, soft ionization methods such as electrospray ionization (ESI) and matrix-assisted laser desorption/ionization (MALDI) mass spectrometry have been demonstrated to be complementary techniques to conventional spectrophotometric methods for enzyme kinetic studies [5–9]. The assay described herein is a facile and broadly applicable kinetics assay for any sulfotransferase or phosphotransferase using ion trap electrospray ionization mass spectrometry (ESI-MS) [3,4]. Enzyme kinetic parameters such as the *Michaelis–Menten* constant K_M, inhibition constant K_i, and the reaction turn over number k_{cat} for some enzymes determined by the ESI-MS assay are in excellent agreement with those of the traditional spectrometric assay thus showing proof of principle. This assay was subsequently used to determine the K_i of inhibitors identified by the multiplex assay [10–12].

Herein, we also apply this ESI-MS assay for determining the catalytic mechanism of the NodST catalyzed sulfuryl group transfer reaction, in which the sulfate group is transferred from PAPS to chitobiose, yielding PAP and chitobiose-6-OSO$_3^-$ as products. In particular, we have been able to differentiate the various catalytic mechanisms and unambiguously assign a specific mechanism to NodST. Utilizing a chondroitin disaccharide, alpha-ΔUA-[1, 3]-GalNAc-6S (ΔD$_i$-6S), as an internal standard (a), a single-point normalization factor between the product and the internal standard was obtained and used for product quantification. The catalytic mechanism of NodST was subsequently determined by initial rate kinetic analysis and product inhibition studies using the ESI-MS assay, and identification by MS of a covalent intermediate was confirmed. This is the first mechanistic data reported for GlcNAc-6-O-carbohydrate sulfotransferase NodST and the data are highly suggestive of a hybrid double-displacement Ping-Pong mechanism. To our knowledge, this is also the first time that mass spectrometry has been shown to unambiguously allow for enzyme mechanistic studies on a class of sulfotransferases.

18.2 Experimental

The multiplex assay for identifying potential inhibitors of various enzymes has been published previously [10]. All parameters and experimental conditions used for the kinetics aspects of this work have been described in detail [3,4]. Briefly, each sample containing quenched enzyme plus substrate A at different concentrations and excess of substrate B was analyzed using single ion monitoring (SIM) on an ion trap instrument. A Finnigan LCQ ion trap mass spectrometer equipped with ESI source and HPLC pump (Thermo-Finnigan, San Jose, CA, USA) was used. The capillary temperature and the spray voltage were kept at 200 °C and 3.2 kV, respectively. Approximately 40 μL of each sample solution was delivered via a LC pump at a flow rate of 20 μL/min. The product ion (m/z 503) and the internal standard ion (m/z 458) were monitored in the negative ion mode using SIM. The signals for the ions of interest were optimized by using the automatic tuning option on the instrument. The optimized conditions were then applied in subsequent experiments. When the signal intensity for one sample decreased from approximately 1×10^6 detector counts per scan to 1×10^4 detector counts per scan, indicating the consumption of the former injection, the next sample was injected. The sums of the intensities within 0.8 mass units around the center of the product ion peak and internal standard ion peak were used to determine their intensity ratio (I_p/I_{is}). This ratio was subsequently used to obtain a single-point normalization factor, quantify the product amount and calculate the initial velocity in the enzymatic reaction according to the three equations shown below (v_0 is the initial velocity)

$$R = (I_p)([\text{internal std}])/(I_{is})([\text{product}]), \tag{18.1}$$

$$[\text{product}] = (I_p)([\text{internal std}])/(I_{is})(R), \tag{18.2}$$

$$v_0 = [\text{product}]/\text{quench time}, \tag{18.3}$$

K_m, k_{cat}, V_{max}, and K_i were calculated from typical Linweaver–Burke plots generated from the velocity and product concentration parameters derived from the above equations. For plots used to determine the mechanism of action of the NodST enzyme, PAPS, and Chitobiose concentrations were each varied. These plots clearly indicated the presence of a sulfated intermediate. For the competition experiments, each reaction was initiated by addition of 25 µL of NodST stock solution, such that the NodST concentration in each reaction was 25 nM. A 20-µL aliquot of each reaction solution was quenched in 80 µL methanol with 6.25 µM ΔDi-6S (internal standard) at an optimized reaction time. The 25 quenched samples were analyzed by ESI-MS and the amount of product in each sample was quantified. The mode of inhibition was evaluated by analyzing the pattern of the double reciprocal plots using GraFit program (Version 4.0.12, from Erithacus Software Ltd., Horley, Surrey, UK). At the same time, the K_i value of PAP in regards to the substrate chitobiose was obtained, as was the K_i for PAP with respect to PAPS.

For the experiments showing the presence of a sulfated intermediate, we used an FTICR mass spectrometer. A Bruker FT-ICR mass spectrometer equipped with an actively shielded 7 T superconducting magnet was used for analyzing tryptic digests of NodST and the NodST-PAPS complex. Solutions were infused into an Analytica electrospray source (Bradford, CT, USA) at a rate of 1 µL/min. The nitrogen nebulizing and drying gas pressures were maintained at 50 and 30 psi, respectively. All protein samples were analyzed using 20 mM ammonium acetate (pH = 7.5). The bias on the glass capillary was kept at 4600 V and 102 °C drying gas was used to assist the desolvation process. A throttle valve was installed at the nozzle-skimmer region and the pressure was adjusted to $\sim 1 \times 10^{-5}$ mbar. Ions were externally accumulated in a radio frequency-only hexapole for 1–2 s before transfer into the ICR cell for mass analysis through a series of electrostatic ion optics. All samples were collected using gated trapping. Ions with high kinetic energy were cooled down and trapped by colliding with argon pulsed into the cell to a pressure of 10^{-7} mbar. Eight ion injection loops were used per scan. A pump down time of 0.02 s was used between each ion injection loop and a further pump down of 1 s was applied when all ion packages were injected. The trapping voltage was then lowered to ~ 0.5 V and a final stage 2 s pump down was applied, allowing for high resolution detection. Each spectrum was an average of 16–80 transients composed of 1024 k data points acquired.

The sulfated intermediate was detected from a solution of 33 µM NodST, and 140 µM of PAPS that were added to 100 µL of 20 mM ammonium acetate buffer (pH = 7.5); the mixture was incubated on ice for 0.5 h. The enzyme was denatured and 0.4 µg of sequencing grade trypsin was added to digest NodST and the NodST-PAPS complex. The solution was then analyzed using FT-ICR MS and negative ion detection without further desalting or separation of the resultant tryptic peptides. The theoretical tryptic digest pattern of NodST was generated using Masslynx 3.3 Build 004 (Manchester, UK). The enzyme itself was also digested with trypsin and the tryptic fragments compared with those of the enzyme digested in the presence of PAPS.

18.3 Results and discussion

The IEMS assay (immobilized enzyme mass spectrometry) compares the difference in relative abundance from the original ESI-MS spectrum taken of the compound library, with that detected after the library has been incubated with the immobilized enzyme [10]. Potential ligand binders show up in the supernatant after the immobilized enzyme and the library are centrifuged at high speed. Those compounds that bind will decrease in relative abundance when comparing the spectra after incubation with that taken of the original library before incubation. However, the IEMS assay is meant to be a very qualitative and crude first pass method to identify compounds that bind to the enzyme. It does not identify specific or non-specific binding and thus does not distinguish competitive from noncompetitive inhibitors. Figure 18.1 is a graphical representation of the results obtained from the IEMS assay and the previously used TLC-phosphorimaging assay [11,12]. The original intent of the IEMS assay was to identify potential binders that show 40% or more decrease in relative abundance between the before and after spectra as this is believed to identify binders in the medium to strong binding category. As can be seen from Fig. 18.1, the assays track reasonably well, although there are clearly more potential binders that show up in the TLC assay. We have evidence that the reason for this discrepancy lies in the fact that the compounds tested in the TLC assay are directly from the well from which the synthesis was performed. Thus, although a binder may well be detected, there is no evidence that the well component is a pure compound. In fact, our assay has shown that in many instances the wells are not representative of only one compound and it is the impurity, or synthetic by-product, which exhibits inhibition. Since our assay is based on mass, it is straightforward to determine which component is actually responsible for inhibition. Once a compound is identified as a strong or moderate inhibitor, it is then subjected to our kinetics assay to measure K_i. One such compound, PAP, was identified and it was this inhibitor that was used to help determine the remaining kinetic constants and the mechanism of catalysis by NodST.

18.3.1 Kinetics

In order to obtain accurate and comprehensive mechanistic information for NodST using our kinetic method, the appropriate substrate concentration ranges were chosen to be $0.2 \sim 5.0$ K_M. This was based on previously published Michaelis–Menten constants. The concentration of the donor substrate PAPS was varied between 1.25 and 25 µM according to its known apparent K_M value of 6.7 µM at a chitobiose concentration of 1 mM, while the concentration of the sulfate acceptor chitobiose extended from 0.05 to 1 mM. This was based on its apparent K_M value of 0.28 mM at a PAPS concentration of 25 µM. Preliminary experiments were performed to make sure that inhibition of both substrates was negligible within the

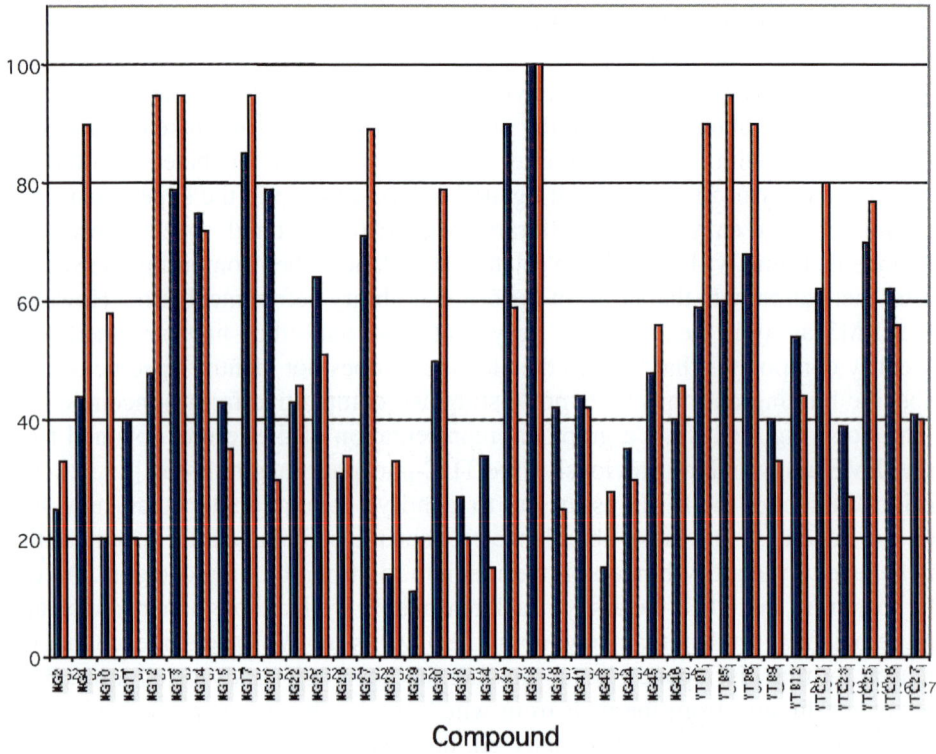

Figure 18.1 Comparison of IEMS assay and TLC assay results for a series of potential inhibitors of NodST. Blue bars (left side bars) represent the % decrease in relative abundance that the compounds show in the IEMS assay. Red bars (right side bars) represent the % inhibition found from the TLC-phosphorimaging assay. The *x* axis represents the sample number for the compound tested.

chosen concentration ranges. For each substrate, five different concentrations were used in the ESI-MS assay. The initial reaction rate was determined as a function of PAPS concentration at different fixed chitobiose concentrations and as a function of chitobiose at different PAPS concentrations. The resulting kinetic data was fitted to two mechanistic models (sequential mechanism and Ping-Pong mechanism) of bisubstrate reaction in the SAS program, and the best fit was obtained in the case of a Ping-Pong mechanism model. The two double reciprocal plots shown in Fig. 18.2 (A and B) are an average of four replicate experiments and both of them resulted in an array of five parallel lines. This is clearly suggestive of a Ping-Pong type mechanism and thus a sulfated intermediate must be present.

In order to obtain more information about the catalytic mechanism of NodST, we evaluated the product inhibition patterns of PAP with respect to both substrates.

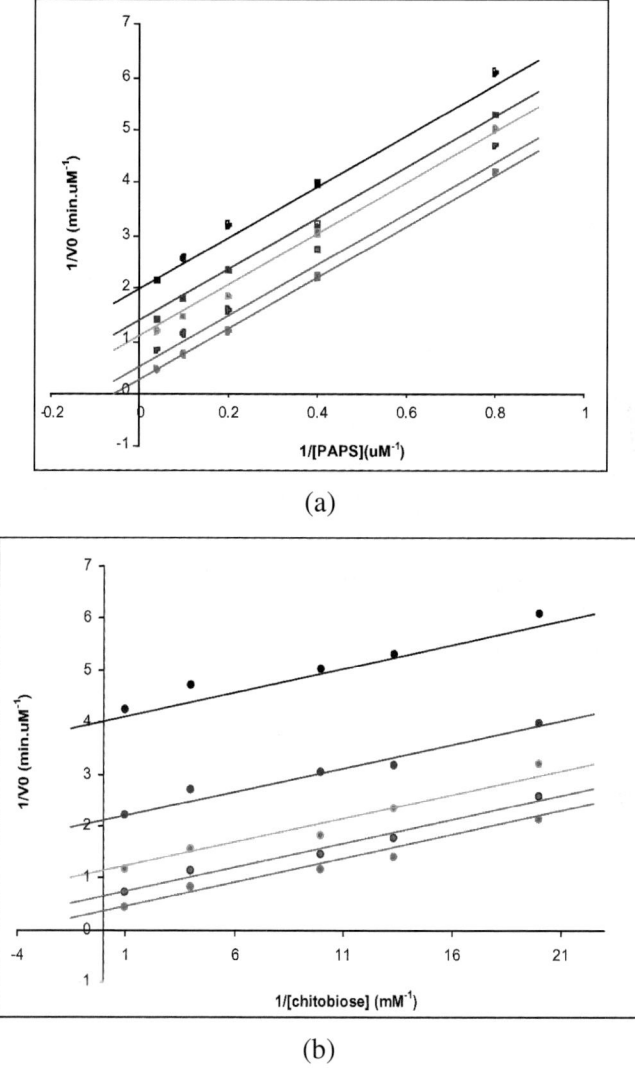

(a)

(b)

Figure 18.2 Double reciprocal plots for NodST-catalyzed sulfate group transfer from PAPS to chitobiose. (a) $1/V_0$ vs. $1/[PAPS]$ at different [chitobiose] (▲ [chitobiose] = 0.05 mM, ◣ [chitobiose] = 0.075 mM, ● [chitobiose] — 0.1 mM, μ [chitobiose] = 0.25 mM, ν [chitobiose] = 1 mM). (b) $1/V_0$ vs. $1/$[chitobiose] at different [PAPS] (▲ [PAPS] = 1.25 μM, ◣ [PAPS] = 2.5 μM, ● [PAPS] = 5 μM, μ [PAPS] = 10 μM, ν [PAPS] = 25 μM). Reproduced with permission (Pi et al., Proteine Science, 2004, in press).

In our earlier studies, PAP was shown to inhibit PAPS binding competitively with a K_i value of 1.80 μM. Herein, the inhibition study of PAP to the sulfate acceptor

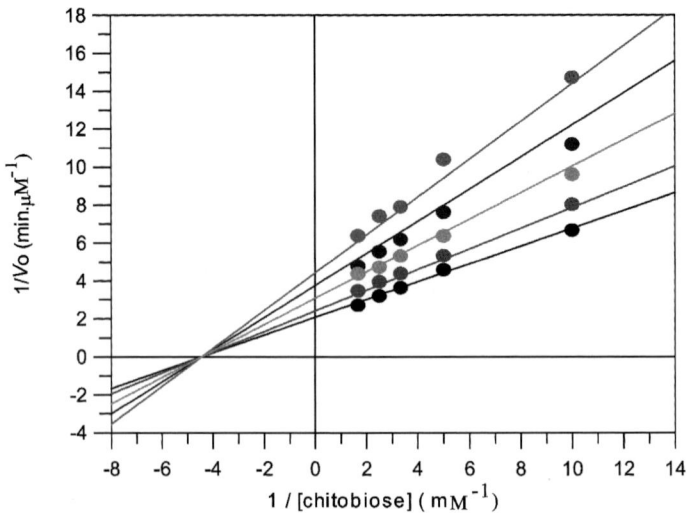

Figure 18.3 Double reciprocal plots for inhibition of NodST by PAP at varied chitobiose concentrations. $1/V_0$ vs. $1/$[chitobiose] at different [PAP]. [PAPS] was fixed at 25 μM (μ [PAP] $= 0$ μM, ● [PAP] $= 1$ μM, □ [PAP] $= 3$ μM, ν [PAP] $= 5$ μM, ◪ [PAP] $= 7$ μM). Reproduced with permission (Pi et al., Proteine Science, 2004, in press).

substrate, chitobiose, was performed. As depicted in Fig. 18.3, the best data fit was obtained with a noncompetitive inhibition model.

This is in contrast to the inhibition pattern of PAP with respect to PAPS, which clearly indicated competitive inhibition.

18.3.2 Detection of a sulfated intermediate

Incubation of an enzyme that follows a Ping-Pong mechanism with the donor substrate results in the formation of a covalent bound enzyme intermediate if the acceptor substrate is absent. Satishchandran et al. reported the characterization of a phosphorylated adenosine 5'-phosphosulfate (APS) kinase intermediate by incubating the enzyme with $[\gamma-^{32}P]$ adenosine 5'-triphosphate (ATP) followed by gel filtration removal of the nucleotides [13]. They showed that phosphoryl linkage in the phosphorylated enzyme intermediate was quite stable and that the covalently bound intermediate can survive the gel filtration process. The sulfuryl group transfer reaction catalyzed by the nucleotide diphosphate (NDP) kinase proceeds via a Ping-Pong mechanism and the sulfated NDP intermediate was also isolated based on the premise that the intermediate is stable without the addition of the second substrate [14].

In order to analyze for the sulfated intermediate, a series of trypsin experiments was performed. The enzyme was incubated with PAPS, denatured at room temperature and digested with trypsin to yield the soluble peptides. Negative ion mode

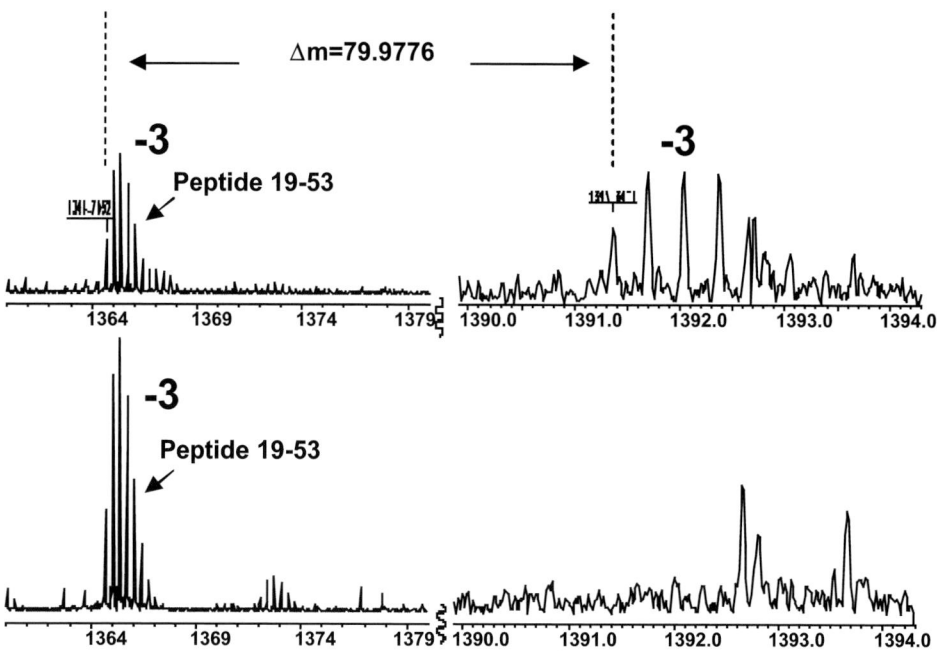

Figure 18.4 Identification of sulfated NodST intermediate. (Upper trace) ESI-MS spectra of 33 μM NodST with 165 μM PAPS after trypsin digestion. (Lower trace) ESI-MS spectra of 33 μM NodST after trypsin digestion. (The mass range around m/z 1391 has been expanded to show isotopic resolution and magnified by 13 fold.) Reproduced with permission (Pi et al., Proteine Science, 2004, in press).

detection was used in the ESI-FT-ICR mass spectrometric analysis [15]. The digestion product was then infused into the mass spectrometer in an NH_4OAc buffer (pH = 7.5) to avoid any potential hydrolysis of the sulfated peptide. As seen in Fig. 18.4a, when NodST was digested as the PAPS complex, an ion at m/z 1364.7 was identified to be the −3 charge state ion corresponding to $[T_{2-3}-3H]^{3-}$ of the tryptic peptide T_{2-3} with a mass error of less than 0.2 Da.

The T_{2-3} peptide corresponds to the amino acid sequence 19–53 (TGTH-YLEELVNEHPNVLSNGELLNTYDTNWPDKER) with a missed cleavage site at Lys. This Lys residue is included in the amino acid sequence DKE, which has been shown to exhibit missed cleavages when treated with Trypsin [16]. Since PAPS was prepared as the lithium salt, a small amount of lithium adduct was also observed as $[T_{2-3}+Li-4H]^{3-}$. In the spectrum of Fig. 18.4a, another −3 charge state ion that is 79.98 Da higher than $[T_{2-3}-3H]^{3-}$ was clearly observed in the spectrum, but missing from the spectrum resulting from digestion of the enzyme alone (Fig. 18.4b). The mass increment of 79.98 is expected for a covalently bound sulfate, and is attributed to the sulfated T_{2-3} peptide resulting from the tryptic di-

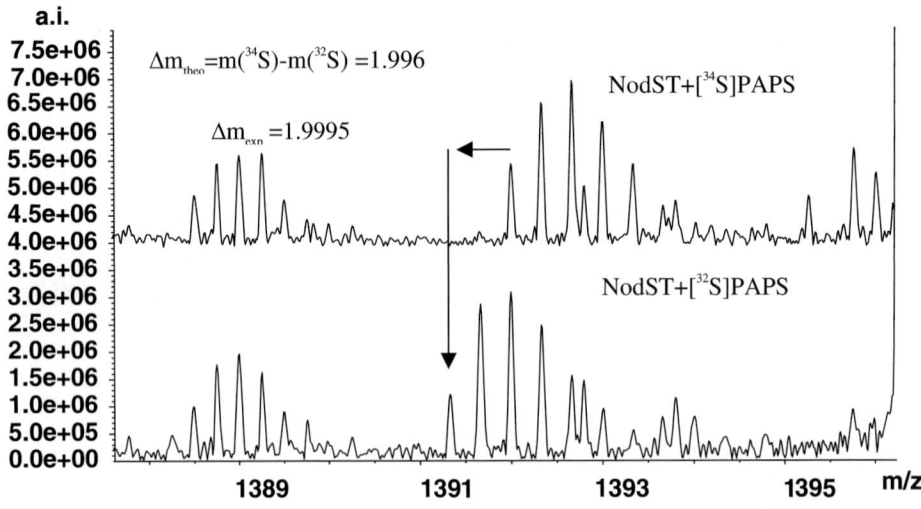

Figure 18.5 ESI-FTICR MS spectra of $[T_{2-3}–SO_3H]^{3-}$ when NodST was incubated with (upper trace) $[^{34}S]$PAPS and (lower trace) $[^{32}S]$PAPS followed by trypsin digestion.

gest of the sulfated NodST intermediate. As a control experiment, the enzyme was also incubated with ammonium sulphate. However, no sulfated intermediate was detected in this experiment thus ruling out the possibility of a gas phase artifact.

Figure 18.5 is the comparison of the sulfated T_{2-3} tryptic peptide when NodST was labeled using $[^{34}S]$PAPS and $[^{32}S]$PAPS. The monoisotopic peak of the $[T_{2-3}–SO_3H]^{3-}$ was shifted 2000 Da when NodST was incubated with $[^{34}S]$PAPS (Fig. 18.5a) compared with $[^{32}S]$PAPS (Fig. 18.5b) followed by trypsin digestion. This is in good agreement with the mass difference (1996 Da) between ^{34}S and ^{32}S. The observation of the mass shift of the sulfated T_{2-3} peptide when using the isotopically labeled substrate ruled out the possibility that this ion originates from the isobaric phosphorylated peptide.

An enzyme-catalyzed sulfuryl group transfer reaction can proceed via two basic mechanisms. One is a direct transfer between the two bound substrates at the enzyme active site in a ternary complex. This is called a sequential Bi-Bi mechanism. The other one, a Ping-Pong Bi-Bi mechanism, involves a modified enzyme intermediate in a double-displacement pathway. Our results in the initial rate kinetic study are a family of parallel lines, which supports a Ping-Pong (substituted enzyme) catalytic mechanism for NodST reaction. Thus, the kinetic pathway observed for NodST is different from that reported for cytosolic sulfotransferases which were suggested to follow a sequential mechanism reflected in the initial velocity pattern as a family of intersecting lines. This is not observed in our experiments. Although contradictory to the mechanisms observed for most sulfotransferases, the data generated herein are very compelling and quite supportive of a

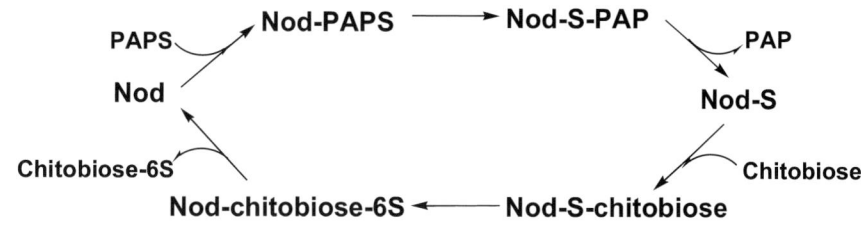

Scheme 18.1 Classical Ping-Pong mechanism for NodST.

Ping-Pong mechanism. The Ping-Pong Bi-Bi mechanism indicated by our initial velocity patterns is further substantiated by the identification of a sulfated enzyme intermediate when NodST was incubated with PAPS in the absence of chitobiose.

Although, all the evidences discussed above are consistent with a classical Ping-Pong Bi-Bi mechanism, the product inhibition patterns are *not* obtained. The classical Ping-Pong mechanism of NodST (Scheme 18.1) requires the initial sulfation of NodST by the donor substrate PAPS. A sulfated NodST intermediate is generated after PAP leaves the enzyme active site. Chitobiose, the acceptor substrate, binds to the sulfated-NodST intermediate and the enzyme-bound sulfate group is transferred to chitobiose to produce chitobiose-6-OSO_3^-. Thus, PAPS and chitobiose-6-OSO_3^- bind to a common site on the free enzyme form, while PAP and chitobiose also bind to the same common site but with the modified enzyme form, the sulfated NodST. Consequently, the product inhibition pattern of PAP with respect to PAPS will be noncompetitive and that of PAP in regards to chitobiose will be competitive. However, the *opposite* product inhibition pattern was observed in our product inhibition studies, in which PAP was determined to be a competitive inhibitor with respect to PAPS and a noncompetitive inhibitor in regards to chitobiose. This result is consistent with a hybrid Ping-Pong rapid equilibrium random two-site mechanism [17,18].

The trypsin digestion experiment further supports our hypothesis of a Ping-Pong mechanism. The MS data show that the T_{2-3} tryptic peptide, which corresponds to amino acid sequence 19–53 of NodST, is sulfated when NodST is incubated with PAPS. Since the 5'-phosphosulfate binding (5'PSB) motif of NodST was identified to be the amino acid sequence 17–24 in the putative PAPS binding pocket, the flexible domain used to transfer the sulfate group in NodST's hybrid two-site Ping-Pong mechanism might reside in the PAPS/PAP binding site, but not the chitobiose/chitobiose-6-OSO_3^- binding site. The possible sulfation site of NodST might be located in the overlapping part of NodST's 5'PSB-loop sequence and the T_{2-3} tryptic peptide sequence, which starts at Trp[19] residue and ends at Leu[24] residue. Thus, the sulfate-modified nucleophilic amino acids of NodST include Thr[19], Thr[21], His[22], and Tyr[23]. Ongoing investigations are focusing on the characterization of the sulfated NodST intermediate and identification of the sulfated residue at the active site of the enzyme.

References

1. A.G. Marshall, C.L. Hendrickson, and G.S. Jackson, *Mass Spectrom. Rev.* **17** (1998) 1.

2. H. Imamura, S. Fushinobu, B.S. Jeon, T. Wakagi, and H. Matsuzawa, *Biochemistry* **40** (2001) 12400.

3. X. Ge, T.L. Sirich, M.K. Beyer, H. Desaire, and J.A. Leary, *Anal. Chem.* **73** (2001) 5078.

4. N. Pi, J.I. Armstrong, C.R. Bertozzi, and J.A. Leary, *Biochemistry* **41** (2002) 13283.

5. B. Bothner, R. Chavez, J. Wei, C. Strupp, Q. Phung, A. Schneemann, and G. Siuzdak, *J. Biol. Chem.* **275** (2000) 13455.

6. D.L. Zechel, L. Konermann, S.G. Withers, and D.J. Douglas, *J. Biochem.* **37** (1998) 7664.

7. P.R. Newton, N. Croot, J. Geyschem, P.E. Diffley, T.J. Walton, M.A. Bayliss, F.M. Harris, D.E. Games, and A.G. Brenton, *Rapid Commun. Mass Spectrom.* **11** (1997) 189.

8. P.R. Newton, M.A. Bayliss, J.A. Khan, A. Bastani, A.C.R. Wilkins, D.E. Games, T.J. Walton, A.G. Brenton, and F.M. Harris, *Rapid Commun. Mass Spectrom.* **13** (1999) 574.

9. H. Gao and J.A. Leary, *J. Amer. Soc. Mass Spectrom.* **14** (2003) 173.

10. M.T. Cancilla, M.D. Leavell J. Chow, and J.A. Leary, *Proc. Natl. Acad. Sci. USA* **97** (2000) 12008.

11. D.E. Verdugo, M.T. Cancilla, X. Ge, N.S. Gray, Y.T. Chang, P.G. Schultz, J.A. Leary, and C.R. Bertozzi, *J. Med. Chem.* **44** (17) (2001) 2683.

12. J.I. Armstrong, X. Ge, D.E. Verdugo, K.A. Winans, J.A. Leary, and C.R. Bertozzi, *Org. Lett.* **3** (17) (2001) 2657.

13. C. Satishchandran, Y.N. Hickman, and G.D. Markham, *Biochemistry* **31** (1992) 11684.

14. J.A. Peliska and M.H. O'Leary, *Biochemistry* **30** (1991) 1049.

15. C. Seibert, M. Cadene, A. Sanfiz, B.T. Chait, and T.P. Sakmar, *Proc. Natl. Acad. Sci. USA* **99** (2002) 11031.

16. B. Thiede, S. Lamer, J. Mattow, F. Siejak, C. Dimmler, T. Rudel, and P.R. Jungblut, *Rapid Commun. Mass Spectrom.* **14** (2000) 496.

17. D.B. Northrop, *J. Biol. Chem.* **244** (1969) 5808.

18. L.J. Song and S.S. Wong, *Biochemistry* **22** (1983) 4637.

Advances in Mass Spectrometry, Volume 16
A.E. Ashcroft, G. Brenton and J.J. Monaghan (Editors)

CHAPTER 19

Development of Multi-turn Time-of-Flight Mass Spectrometers, 'MULTUM Linear plus' and 'MULTUM II'

Michisato Toyoda*, Daisuke Okumura, Morio Ishihara, and Itsuo Katakuse

Department of Physics, Graduate School of Science, Osaka University, 1-16 Machikaneyama,
Toyonaka, Osaka 560-0043, Japan

Abstract

A new type of multi-turn time-of-flight mass spectrometer, the 'MULTUM Linear plus,' has been developed. It consists of four cylindrical electric sectors and 28 electric quadrupole lenses. The size of the vacuum chamber is $60 \times 70 \times 20$ cm. It was demonstrated that the mass resolution increased according to the number of cycles of the ions take through the ion optical system. A mass resolution of 350,000 ($m/z = 28$, FWHM) was achieved after 500.5 cycles. An improved multi turn time-of-flight mass spectrometer, the 'MULTUM II,' has also been developed to eliminate the use of the quadrupole lenses and consists of only four toroidal electric sectors.

19.1 Introduction

The time-of-flight (TOF) mass spectrometer became an especially powerful and attractive instrument in the field of mass spectrometry after the introduction of pulsed laser ionization methods and the developments of very fast electronics. Especially, the introduction of matrix-assisted laser desorption/ionization (MALDI) extended the application area into the biochemical field [1]. The merits of a TOF

---

*E-mail address: toyodam@phys.sci.osaka-u.ac.jp (M. Toyoda).

mass spectrometer are well recognized: (1) essentially unlimited mass range; (2) very high transmission; and (3) the ability to record the whole mass spectrum simultaneously. However, a major limitation of TOF mass spectrometers has often been their relatively poor mass resolution. In order to improve the mass resolution, the following methods were successfully introduced:

(1) In order to minimize the effects of spatial and velocity dispersion on time resolution, two-stage acceleration [2], time-lag focusing [2], and orthogonal acceleration ion sources [3] were introduced.
(2) In order to achieve energy focusing, the ion mirror was invented [4].
(3) In order to achieve isochronous focusing and space focusing, a novel combination of electric sectors was proposed by Poschenrieder [5]. As an example of this method, a TOF mass spectrometer consisting of four toroidal electric sectors was constructed and examined by Sakurai and co-workers [6].
(4) In order to obtain a long flight path, multi-turn ion optical geometries using electric sectors were proposed by Poschenrieder [5], Matsuo et al. [7], and Sakurai et al. [8]. For the same purpose, Wollnik and Przewloka [9] proposed electrostatic multi-pass mirror systems. The multi-pass mirror instrument was constructed by Casares et al. [10].

We studied the ion optics of a multi-turn TOF mass spectrometer consisting of electric sector fields [7,11]. By introducing symmetry in the arrangement of sectors, multiple time and space focusing are easily satisfied [11,12]. Using this principle, we have found the ion optical systems for a multi-turn TOF mass spectrometer. One of these proposed TOF systems, the 'MULTUM Linear plus,' has been constructed as a laboratory model of the COSAC project of the ROSETTA mission [13,14]. It consists of four cylindrical electrostatic sectors and eight electric quadrupole lenses. In order to inject and eject ions, a linear TOF mass spectrometer was combined into its construction. Four electric quadrupole triplets were used to achieve perfect space focusing of the linear type TOF mass spectrometer. It was experimentally demonstrated that the mass resolution increased according to the number of cycles the ions execute in the ion optical system. Mass resolution > 350,000 was achieved after 500 cycles.

The analyzer of the 'MULTUM Linear plus,' however, was not a simple design; 28 electric quadrupole lenses were used. In order to eliminate the use of quadrupole lenses, another multi-turn TOF mass spectrometer, the 'MULTUM II,' has also been developed [15]. The instrument consists of only four toroidal sectors. We also demonstrated that the mass resolution increased according to the number of cycles.

In this article, the theory of ion optics for a multi-turn TOF mass spectrometer and the results of experiments of the multi-turn TOF mass spectrometers, the 'MULTUM Linear plus' and 'MULTUM II,' are reported.

19.2 Ion optics for multi-turn TOF mass spectrometers

Here we simply explain the future of the ion optical system of the new multi-turn TOF mass spectrometers. There are two conditions required for multi-turn systems. The first one is the geometrical conditions for the multi-turn systems, namely, the criterion necessary to close the ion optical orbit. Multi-turn TOF mass spectrometer geometries incorporating this principle have been proposed previously; however, they did not satisfy a second condition, namely the 'perfect focusing condition' [11]. Therefore, ion beam divergence and consequential decrease in mass resolution arises with increasing number of cycles around a system. To avoid this problem, ions should return to the point of origin in the system. In other words, the absolute value of the position and angle at the detector plane should be exactly the same as the initial conditions, both in the horizontal and the vertical planes. Such conditions can be expressed using the transfer matrix as

$$
\begin{pmatrix} x_f \\ \alpha_f \\ y_f \\ \beta_f \\ \gamma \\ \delta \\ l_f \end{pmatrix} = \begin{pmatrix} \pm 1 & \underline{0} & 0 & 0 & 0 & \underline{0} & 0 \\ \underline{0} & \pm 1 & 0 & 0 & 0 & \underline{0} & 0 \\ 0 & 0 & \pm 1 & \underline{0} & 0 & 0 & 0 \\ 0 & 0 & \underline{0} & \pm 1 & 0 & 0 & 0 \\ 0 & 0 & 0 & 0 & 1 & 0 & 0 \\ 0 & 0 & 0 & 0 & 0 & 1 & 0 \\ \underline{0} & \underline{0} & 0 & 0 & R(l|\gamma) & \underline{0} & 1 \end{pmatrix} \begin{pmatrix} x_i \\ \alpha_i \\ y_i \\ \beta_i \\ \gamma \\ \delta \\ l_i \end{pmatrix}. \quad (19.1)
$$

It should be noted here that the character $\underline{0}$ (zero with an underline) refers to the matrix element which should be forced to be zero, while 0 (zero without an underline) means always zero by definition. Angular focusing ($R(x|\alpha) = 0$), energy focusing ($R(x|\delta) = 0$), and the condition $R(x|x) = \pm 1$ for lateral magnification are required to conserve the absolute value of the lateral deviation ($|x| = |x_0|$) in the horizontal direction. In the same way, angular focusing ($R(y|\beta) = 0$) and the condition $R(y|y) = \pm 1$ for lateral magnification are required to conserve the absolute value of the lateral deviation ($|y| = |y_0|$) in the vertical direction. Moreover, $R(\alpha|x) = R(\alpha|\delta) = 0$ and $R(\alpha|\alpha) = \pm 1$ are required to conserve the absolute value of the angle ($|\alpha| = |\alpha_0|$) in the horizontal direction and $R(\beta|y) = 0$ and $R(\beta|\beta) = \pm 1$ are required to conserve the absolute value of the angle ($|\beta| = |\beta_0|$) in the vertical direction. In the case of the TOF mass spectrometer, triple time focusing $R(l|x) = R(l|\alpha) = R(l|\delta) = 0$ is also required. Accordingly, we require "nine-fold focusing," i.e., the nine $\underline{0}$ elements should be zero to satisfy perfect space and time focusing. The final goal is to find the ion optical systems whose overall transfer matrix satisfies the above. In our experience, it is easy to find the solution of $R(x|x) = \pm 1$ and $R(x|\alpha) = 0$; however, it is very difficult to satisfy these parameters with $R(\alpha|x) = 0$ simultaneously. In order to overcome this difficulty, symmetrical geometries were introduced. By introducing symmetry in the arrangement of the ion optical components, multiple focusing conditions are easily achieved [11,12]. Several ion optical systems for a multi-turn TOF mass spectrometer, which satisfies the perfect focusing have been found by using the principle described in Ref. [12]. In this article, the characteristics of two perfect focusing systems, 'MULTUM' and 'MULTUM II' geometry, are discussed.

19.2.1 MULTUM geometry

The 'MULTUM' geometry consists of four cylindrical electrostatic sectors and eight electric quadrupole lenses. The ion trajectories simulated by TRIO-DRAW [16] are shown in Fig. 19.1. The basic unit comprises four drift spaces, two electric quadrupole lenses and a cylindrical electrostatic sector. In the horizontal direction, perfect focusing is satisfied and the image type is the inverse after a half cycle (two units). Therefore, the image type at one cycle (four units) is normal. In the vertical direction, perfect focusing is satisfied at only one cycle and the image type is inverted.

19.2.2 MULTUM II geometry

The 'MULTUM II' geometry consists of only four toroidal electrostatic sectors. The ion trajectories simulated by TRIO-DRAW are shown in Fig. 19.2. The basic unit comprises of two drift spaces and a toroidal electrostatic sector (the c-value [17] of a toroidal electric sector field is 0.033). Perfect focusing is satisfied after one cycle (four units) and the image type is the inverse in both the horizontal and the vertical directions.

19.3 The multi-turn TOF mass spectrometer 'MULTUM Linear plus'

19.3.1 MULTUM Linear plus

We designed and constructed a multi-turn TOF mass spectrometer using the 'MULTUM' geometry as a laboratory model for cometary exploration. The system consists of four discrete units, each of which comprises an electric quadrupole lens, a cylindrical electrostatic sector and an electric quadrupole lens. The deflection radius of the cylindrical electric sector is 50 mm, the deflection angle is 156.87°, the gap between the electrodes is 7.5 mm, and the applied voltage is ±225 V when

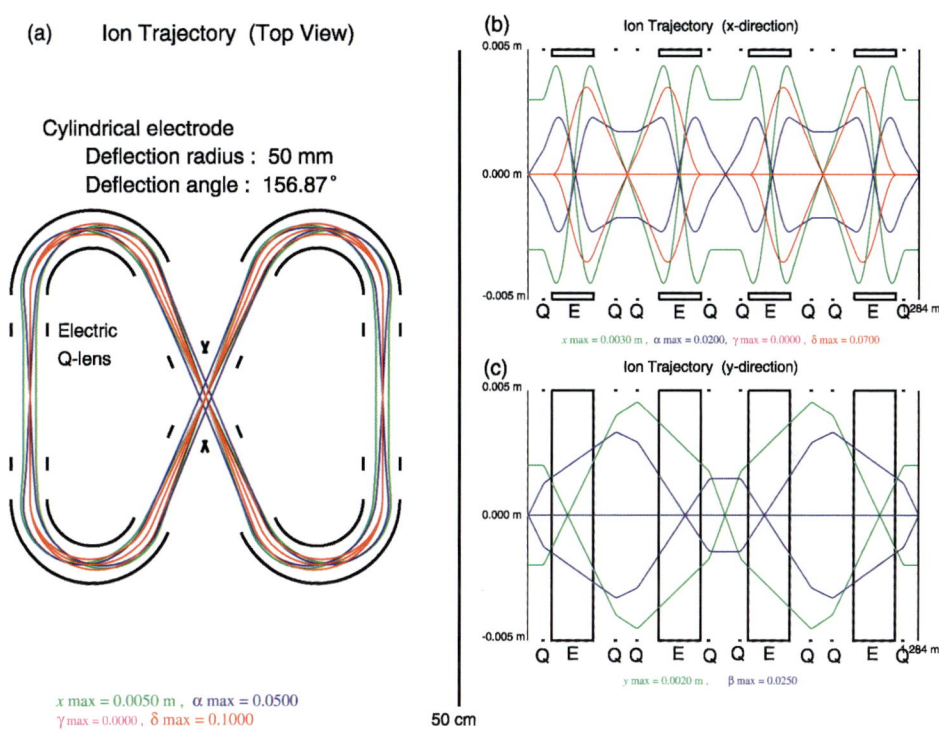

(a) Ion Trajectory (Top View)

Cylindrical electrode
Deflection radius : 50 mm
Deflection angle : 156.87°

Electric
Q-lens

x max = 0.0050 m , α max = 0.0500
γ max = 0.0000 , δ max = 0.1000

50 cm

(b)   Ion Trajectory (x-direction)

x max = 0.0030 m , α max = 0.0200, γ max = 0.0000 , δ max = 0.0700

(c)   Ion Trajectory (y-direction)

y max = 0.0020 m . β max = 0.0250

Figure 19.1 Ion trajectories of the individual beam mode for 'MULTUM' geometry multi-turn TOF mass spectrometer simulated by 'TRIO-DRAW' [16]: (a) top view, (b) x-direction, and (c) y-direction. (Reproduced from Toyoda M., Okumura D., Ishihara M., Katakuse I., Multi-turn time-of-flight mass spectrometers with electrostatic sectors, J. Mass Spectrom. 2003, 38:1129. © John Wiley & Sons Limited, 2003.)

the ion accelerating voltage of ions is 1500 V. The length of the inside quadrupole lens is 10 mm, the radius of the inscribed circle of the electrodes is 5 mm and the applied voltage is ±16.57 V. The length of the other quadrupole lens is 10 mm, the radius of the inscribed circle is 5 mm and the applied voltage is ±49.04 V. The total path length of one cycle is 1.284 m.

In order to inject and eject ions, the linear-type TOF mass spectrometer was combined with the 'MULTUM' and designated the 'MULTUM Linear plus.' The total path length of the linear system is 0.428 m. Four quadrupole triplets were used to achieve perfect space focusing. Ion trajectories simulated by 'TRIO-DRAW' are shown in Fig. 19.3. A schematic drawing and a photograph of 'MULTUM Linear plus' are shown in Figs. 19.4 and 19.5, respectively.

The whole system was fixed on the base plate whose size is 40 × 40 cm. Screws in the vacuum chamber fixed the base plate. The size of the vacuum chamber is

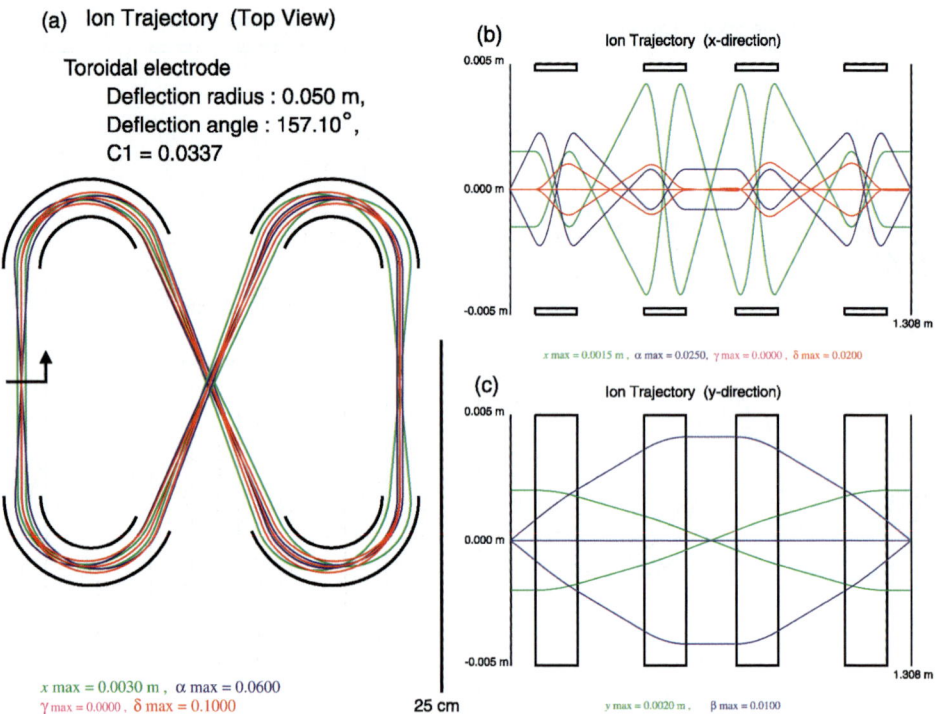

(a) Ion Trajectory (Top View)

Toroidal electrode
Deflection radius : 0.050 m,
Deflection angle : 157.10°,
C1 = 0.0337

x max = 0.0030 m , α max = 0.0600
γ max = 0.0000 , δ max = 0.1000

25 cm

(b) Ion Trajectory (x-direction)

0.005 m

0.000 m

-0.005 m

1.308 m

x max = 0.0015 m , α max = 0.0250, γ max = 0.0000 , δ max = 0.0200

(c) Ion Trajectory (y-direction)

0.005 m

0.000 m

-0.005 m

1.308 m

y max = 0.0020 m , β max = 0.0100

Figure 19.2 Ion trajectories of 'MULTUM II' geometry multi-turn TOF mass spectrometer: (a) top view, (b) x-direction, and (c) y-direction. (Reproduced from Toyoda M., Okumura D., Ishihara M., Katakuse I., Multi-turn time-of-flight mass spectrometers with electrostatic sectors, J. Mass Spectrom. 2003, 38:1131. © John Wiley & Sons Limited, 2003.)

$60 \times 70 \times 20$ cm. The vacuum was maintained at about 2.5×10^{-5} Pa. Two main slits, two collector slits, two α-slits, and four β-slits were installed.

19.3.2 Ion source

A two-stage acceleration electron ionization (EI) ion source [2] was used. This type of ion source can compensate for the flight time deviation caused by the distribution of the initial position of ions. This focusing is satisfied in the linear part of the 'MULTUM Linear plus.'

19.3.3 Detector and data acquisition system

A micro-channel plate (MCP) of 14.5 mm diameter (F4655-10, Hamamatsu Photonics K.K., Shizuoka, Japan) was attached at the position of Detector 2

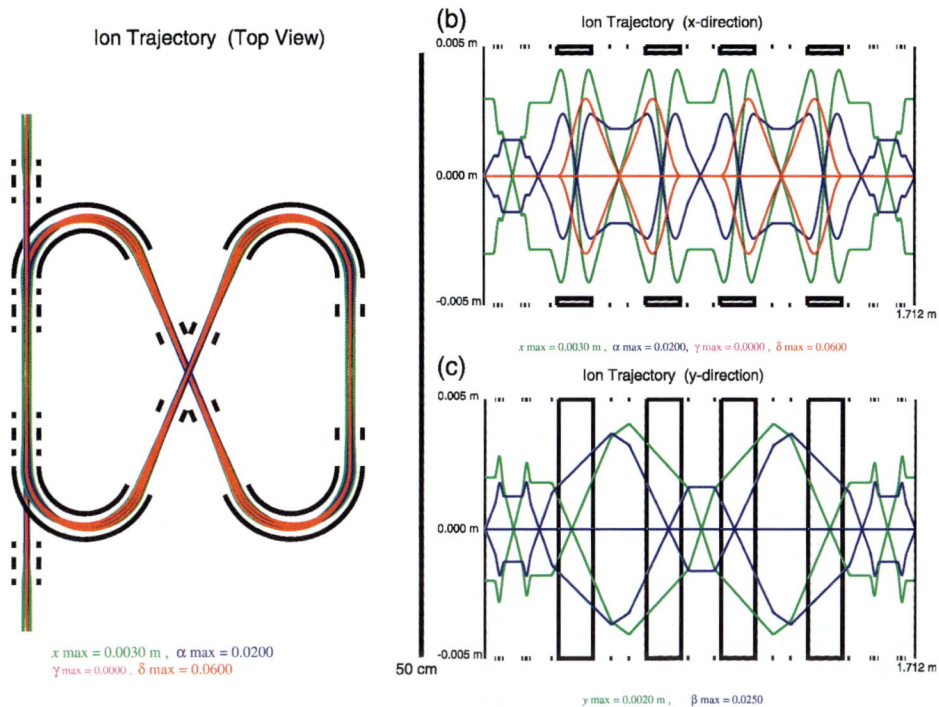

Figure 19.3 Ion trajectories of 'MULTUM Linear plus': (a) top-view, (b) x-direction, and (c) y-direction. (Reproduced from Toyoda M., Okumura D., Ishihara M., Katakuse I., Multi-turn time-of-flight mass spectrometers with electrostatic sectors, J. Mass Spectrom. 2003, 38:1135. © John Wiley & Sons Limited, 2003.)

in Fig. 19.4. The output signals were accumulated with a digital oscilloscope (LC564A, LeCroy Japan Corp., Osaka, Japan).

19.3.4 Operation system

The block diagram of the pulse control circuits is shown in Fig. 19.6. The oscillator provides a trigger pulse for a digital pattern generator (CompuGen T30, Gage Applied Sciences Inc., Montreal, Canada). The digital pattern generator provides the timing signals for the ion source, the ion gate, the sector III and IV electrodes, the quadrupole triplets at the linear TOF mass spectrometer and the digital oscilloscope. As ions were injected into the multi-turn sections, the voltage of sector IV was off and the voltages of quadrupole lenses Q14, Q15, and Q16 were those employed for the linear TOF mode. Before the ions returned, the voltage of sector IV was turned on, the voltages of quadrupole lenses Q14 and Q16 were turned off and the voltage of quadrupole lens Q15 was switched to the voltage for circulation (multi-pass). After the ions had undergone the desired cycles, the voltage

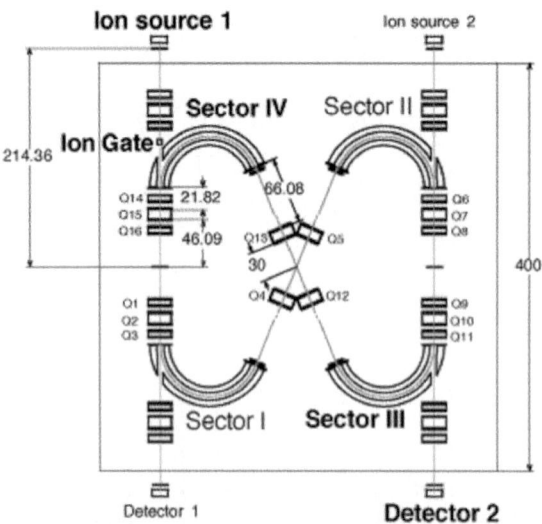

Figure 19.4 Schematic drawing of the 'MULTUM Linear plus.' (Reproduced from Toyoda M., Okumura D., Ishihara M., Katakuse I., Multi-turn time-of-flight mass spectrometers with electrostatic sectors, J. Mass Spectrom. 2003, 38:1136. © John Wiley & Sons Limited, 2003.)

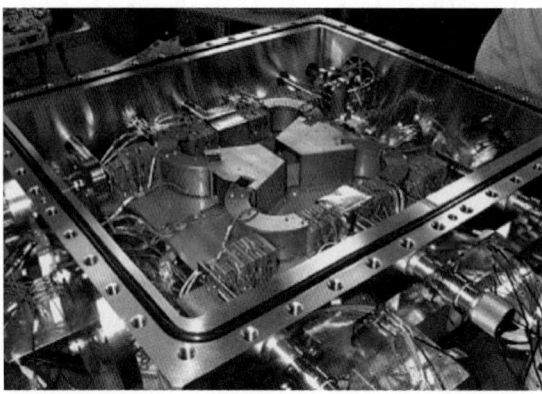

Figure 19.5 Photograph of the 'MULTUM Linear plus' in the vacuum chamber. (Reproduced from Toyoda M., Okumura D., Ishihara M., Katakuse I., Multi-turn time-of-flight mass spectrometers with electrostatic sectors, J. Mass Spectrom. 2003, 38:1136. © John Wiley & Sons Limited, 2003.)

of sector III was turned off and the voltages of quadrupole lenses Q9, Q10, Q11 were switched to those employed for linear TOF mode to eject the ions from the multi-turn part.

Generally, light ions catch up and pass the heavy ions in the multi-turn system. In order to avoid such complicating phenomena, an ion gate was introduced to eject undesired ions.

19.3.5 Experimental

The performance of the 'MULTUM Linear plus' was evaluated using the CO and N_2 doublet. Generally, the mass resolution of a TOF mass spectrometer is directly proportional to its total flight path length. However, if the aberrations are too large,

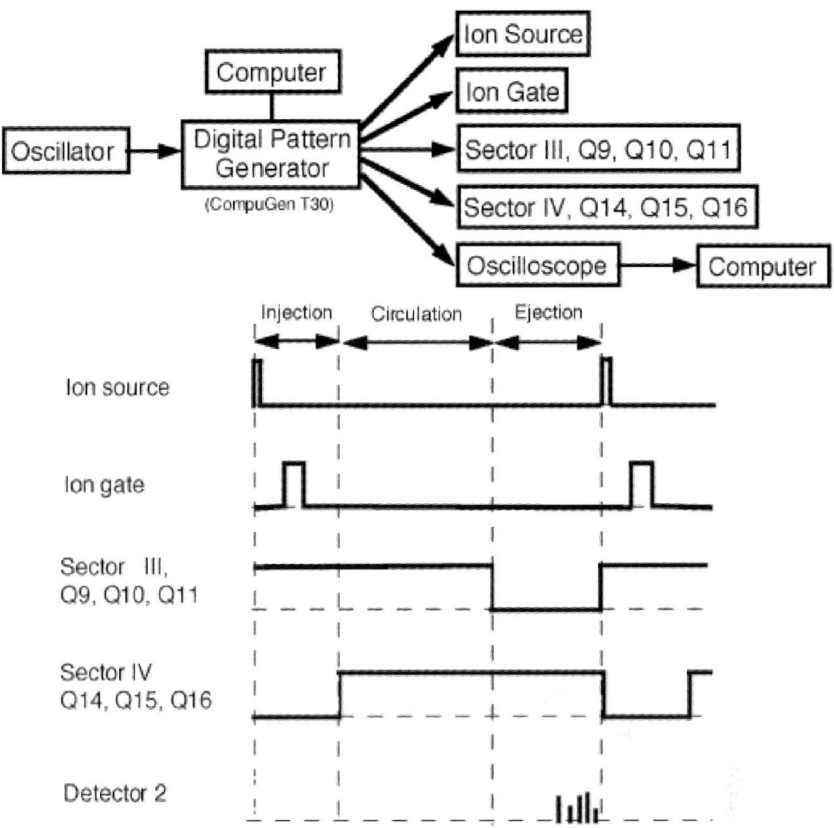

Figure 19.6 Block diagram of the pulse control circuits of the 'MULTUM Linear plus.'

the mass resolution may decrease after many cycles. Hence, we should prove the increase in the mass resolution with the number of cycles experimentally.

The experimental conditions were as follows:

(1) The electron energy was 70 eV.
(2) The electron current was 200 μA.
(3) The pulse voltage applied to the first stage electrode of the ion source was 420 V and the total acceleration voltage of the ions was 1.5 kV.
(4) The background pressure was 6.2×10^{-5} Pa. The pressure increased to 6.7×10^{-5} Pa when CO gas was introduced.
(5) The voltage supplied to the MCP was −2.25 kV.
(6) The sampling rate of the digital oscilloscope was 2 GS/s and the TOF spectra were obtained by summing 5000 spectra on the digital oscilloscope.
(7) The repetition frequency was 100 Hz.

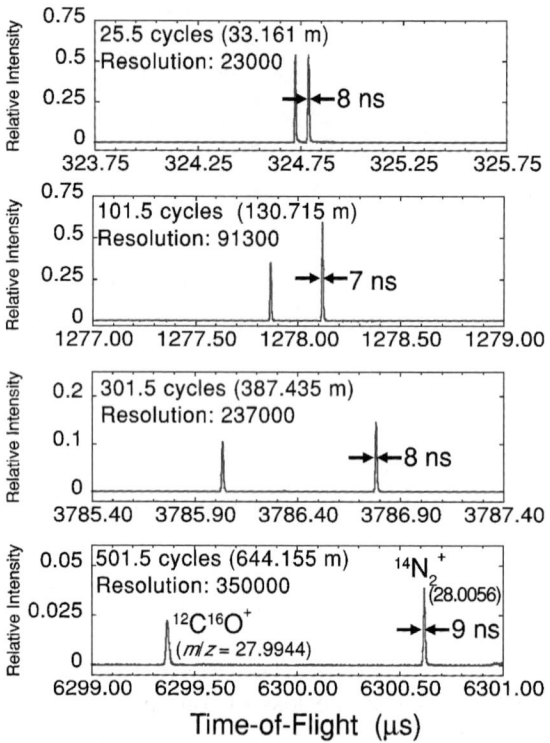

Figure 19.7 TOF spectra of $CO^+-N_2^+$ doublet for 25, 100.5, 300.5, 500.5 cycles. (Reproduced from Toyoda M., Okumura D., Ishihara M., Katakuse I., Multi-turn time-of-flight mass spectrometers with electrostatic sectors, J. Mass Spectrom. 2003, 38:1137. © John Wiley & Sons Limited, 2003.)

  (8) The ion gate was used to inject only ions whose mass-to-charge ratios were in the mass range m/z 25 to 30 into the multi-turn section.

    The TOF spectra of the $CO^+-N_2^+$ doublet for 25.5, 101.5, 301.5, 501.5 cycles are shown in Fig. 19.7. Mass resolution of about 350,000 (at $m/z = 28$, FWHM) was achieved after 500 cycles. The flight times and also the differences in the arrival times between CO^+ and N_2^+ increased linearly with the number of cycles. The peak widths were, however, almost the same. The relationship between the number of cycles and the mass resolution is shown in Fig. 19.8. The mass resolution is defined as the peak width (FWHM) of N_2^+. One can conclude that the mass resolution increases linearly with the number of cycles.

    The transmission through the 'MULTUM Linear plus' was recorded for the number of cycles. This dependence is illustrated in Fig. 19.9. The ordinate is the intensity, which is the sum of the area of both the CO^+ and N_2^+ peaks and which

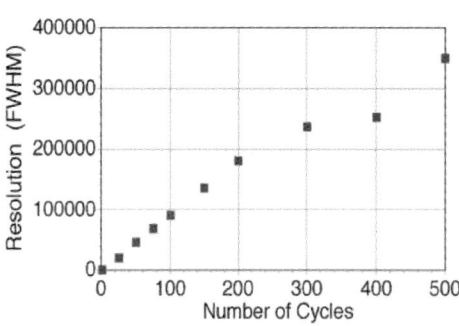
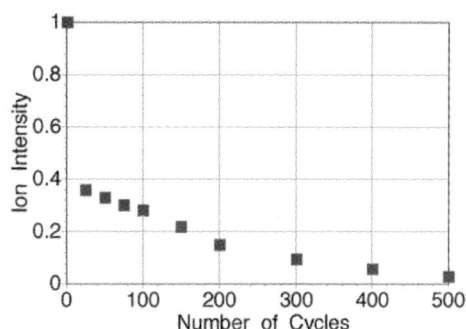

Figure 19.8 The relationship between the number of cycles and the mass resolution.

Figure 19.9 Variation of ion transmission with the numbers of cycles. The ordinate is the intensity which is the area of both the CO^+ and the N_2^+ peaks and is normalized so that the intensity at half cycle is 1.

is normalized so that the intensity at half cycles is 1. The ion intensity after 100 cycles was 20–30% of that after a half cycle. The ion intensity of the first 10 cycles decreased rapidly and then gradually. The major reason for the ion intensity decrease in the first 10 cycles was the loss of ions having large initial deviations caused by collisions with electrodes. After 10 cycles, small angle scattering with neutral particles becomes the dominant ion loss process after which the ion transmission was > 99% per cycle. Consequently, we can conclude that this new TOF mass spectrometer has stable multi-turn orbits.

19.4 The multi-turn TOF mass spectrometer 'MULTUM II'

We have already demonstrated that the multi-turn time-of-flight (TOF) mass spectrometer, the 'MULTUM Linear plus,' achieved mass resolution > 350,000 after 500 cycles (flight length: 642 m) at $m/z = 28$. The analyzer of the 'MULTUM Linear plus' was not, however, a simple design where 28 electric quadrupole lenses were employed. Therefore, it was necessary to reduce the number of ion optical parts. To reduce the number of ion optical parts, we designed and constructed a new multi-turn TOF mass spectrometer using the 'MULTUM II' geometry, consisting of only four toroidal electric sectors and no quadrupole lenses [15]. The schematic drawing and the photograph of the 'MULTUM II' are shown in Figs. 19.10 and 19.11, respectively. It consisted of three interconnected housings; the ion source housing, the analyzer housing and the detector housing.

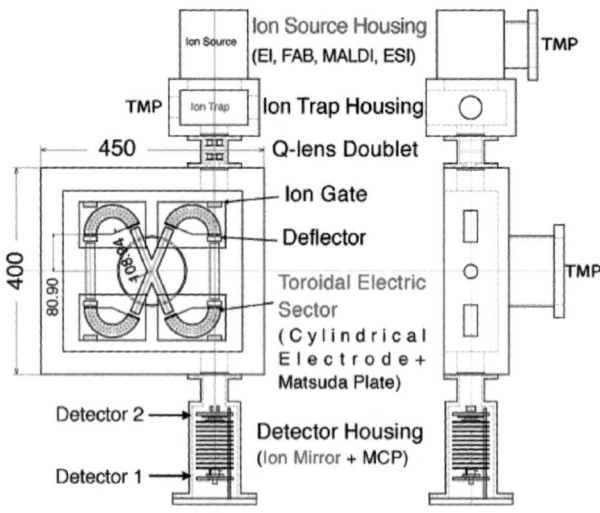

Figure 19.10 Schematic drawing of the 'MULTUM II.'

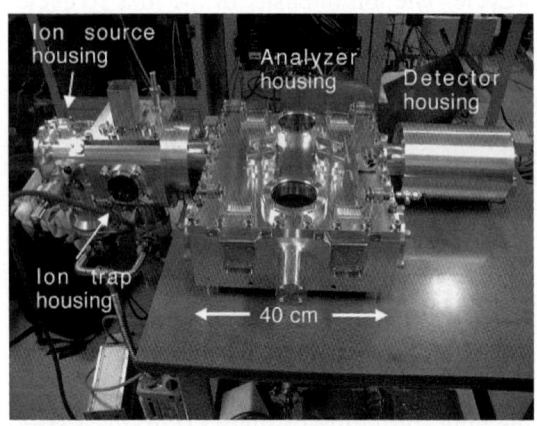

Figure 19.11 Photograph of the 'MULTUM II.' (Reproduced from Toyoda M., Okumura D., Ishihara M., Katakuse I., Multi-turn time-of-flight mass spectrometers with electrostatic sectors, J. Mass Spectrom. 2003, 38:1138. © John Wiley & Sons Limited, 2003.)

19.4.1 Ion source

The ion source is the most important component of any type of mass spectrometer, and the development of mass spectrometers can be seen as a function of ion source development. In the 'MULTUM Linear plus,' only an EI ion source could be attached. Since the ion source was loaded in the same vacuum chamber as the analyzer, the background pressure in the whole chamber rose when the gaseous sample was introduced through a needle valve. However, high vacuum is required to obtain good ion transmittance, because the collision probability of ions with residual gas increases under poorer vacuum pressure. Therefore, it is necessary that a differential ion source vacuum pumping system should be introduced.

We designed the ion source housing to be separate from the analyzer housing and a differential vacuum pumping system was introduced in the 'MULTUM II.' Various types of ion sources (EI, FAB, MALDI, ESI, ICP) can also be attached to the ion source housing. We have already tested EI, FAB, and MALDI sources.

Between the ion source housing and the analyzer housing, an ion trap housing in which an ion trap is installed can also be attached. This configuration will allow us to carry out a wide variety of experiments, for example, ion/molecule reactions, MS/MS applications in the ion trap. The ion trap housing can be removed and the ion source housing can be attached directly to the analyzer housing when the ion trap is not used.

A Q-lens doublet was introduced between the ion source housing and the analyzer housing to improve ion transmission.

19.4.2 Analyzer

The multi-turn part consists of only four toroidal electric sector fields. It is difficult to make toroidal electric sector electrodes because of the small c-value [17]; 0.033. Therefore, a toroidal electric sector field was produced by cylindrical electric sectors and MATSUDA plates [18]. The mean radius of the cylindrical electric sector was 50 mm, the deflection angle was $157.1°$, the gap between the electrodes was 10 mm, and the height of the electrodes was 40 mm. The total pass length of one cycle was 1.308 m. All elements were hung from the top plate of the multi-turn housing.

Ions were injected and ejected through a hole in the outer sector electrodes, similarly to the 'MULTUM Linear plus.' While the ions injected into the multi-turn parts, the voltage of sector IV was off, and it was turned on before the ions returned. After the ions had undergone the desired number of cycles, the voltage of sector I was turned off to eject the ions from the multi-turn section.

19.4.3 Detector

In the detector housing, there were two detectors and an ion mirror. When the ion mirror was off, the ions were ejected from the multi-turn housing flight toward a MCP of 14.5 mm diameter (F4655-13, Hamamatsu Photonics K.K., Shizuoka, Japan) attached at the position of detector 1 in Fig. 19.10. When the ion mirror was used, ions were detected by using a MCP of 27 mm diameter with a center hole (F4294-09, Hamamatsu Photonics K.K., Shizuoka, Japan) attached at the position of detector 2 in Fig. 19.10.

The ion mirror was used to satisfy energy focusing in the linear part. It consisted of 11 ring electrodes. Transmission meshes (85%) were attached to the first, fourth, and last electrodes. The output signals from the MCPs were accumulated with a digital oscilloscope (LC564DL, LeCroy Japan, Osaka, Japan).

19.4.4 Operation system

The operation method is similar to that used in the 'MULTUM Linear plus.' A digital pulse/delay generator (MODEL 555-3, Berkeley Nucleonics Corporation, CA, USA) was used as a substitute for the oscillator and digital pattern generator (CompuGen T30), because the number of necessary timing signals were decreased compared with the 'MULTUM Linear plus.' This generator provides eight-channel 1 ns resolution timing, delaying, gating, pulsing, and synchronising functions.

19.4.5 Experimental

The two-stage acceleration EI ion source, used in the 'MULTUM Linear plus' was attached. Some preliminary test experiments were conducted using the N_2–CO doublet. The experimental conditions were as follows:

(1) The electron energy was 74 eV.
(2) The electron current was 80 μA.
(3) The pulse voltage applied to the first stage electrode of the ion source was 150 V and the total acceleration voltage of ions was 1.5 kV.
(4) The background pressure was 1.5×10^{-5} Pa in the ion source housing and 2.1×10^{-5} Pa in the analyzer housing. The pressure in the ion source housing increased to 1.6×10^{-5} Pa when CO gas was introduced.
(5) The voltage supplied to the MCP was -1.7 kV.
(6) The sampling rate of the digital oscilloscope was 4 GS/s and the following TOF spectra were obtained by summing 25,000 spectra on the digital oscilloscope.
(7) The repetition frequency of the digital pulse/delay generator was 100 Hz.
(8) The ion gate was used to inject only ions whose mass-to-charge ratio were 25 to 30 into the multi-turn parts.
(9) The ion mirror was not used in this experiment.

The TOF spectra of the N_2–CO doublet acquired using different flight path lengths (20, 60, 80, 100, and 150 cycles) are shown in Fig. 19.12. It is clear that the mass resolution increases according to the number of cycles. The relation between the number of cycles and the mass resolution of N_2^+ from 0 to 150 cycles is shown in Fig. 19.13(a). It is clearly shown that the mass resolution increased in proportion to the number of cycles. Mass resolution of 33,000 was achieved after 150 cycles. The peak width was almost the same, 26 to 31 ns. The width was broader than in the case of the 'MULTUM Linear plus,' because the longer the focus length of two-stage acceleration ion source is, the larger the effect of turn around time of ions in ion source becomes. The path length of the linear part of the 'MULTUM II' is 80 cm, whereas that of the 'MULTUM Linear plus' is 40 cm. The focus length

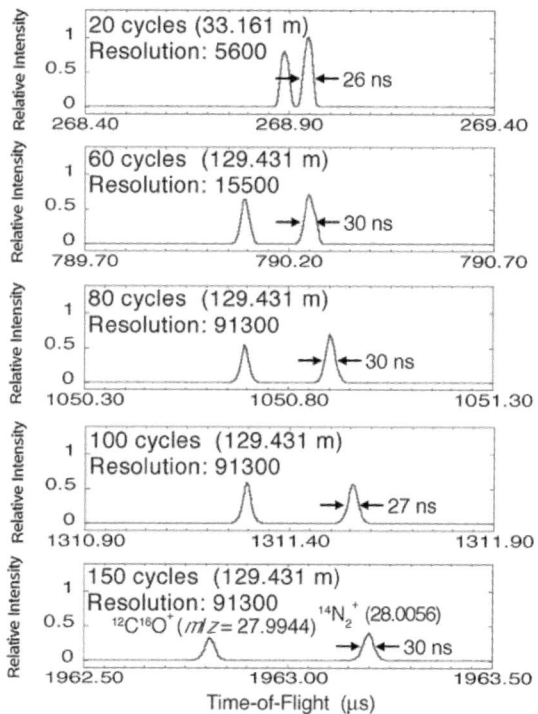

Figure 19.12 The TOF spectra of the N_2–CO doublet acquired using different flight path lengths (20, 60, 80, 100, and 150 cycles).

of an ion source can be short when a two-stage ion mirror was combined with two-stage acceleration ion source. The relation between the number of cycles and the mass resolution of N_2^+ in the case of mirror mode from 0 to 100 cycles is shown in Fig. 19.13(b). The peak width was almost the same, 11 to 14 ns, and mass resolution of 47,000 was achieved after 100 cycles. The experimental conditions were as follows:

(1) The pulse voltage applied to the first stage electrode of the ion source was 220 V and the total acceleration voltage of the ions was 1.5 kV. The focus length of an ion source was 40 cm in this condition.
(2) The voltage supplied to the micro channel plate was −1.7 kV.
(3) The voltage supplied to the fourth electrode of the ion mirror was 588 V and that of last electrode was 1560 V.

Other conditions were the same as in the non-mirror mode.

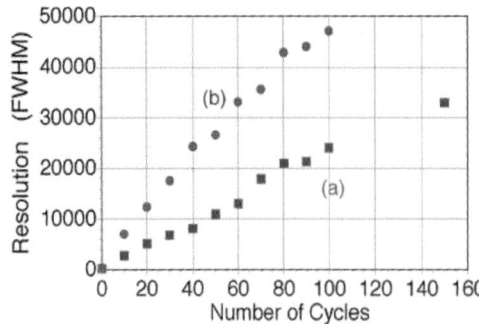

Figure 19.13 The relation between the number of cycles and the mass resolution of N_2^+: (a) without ion mirror and (b) with ion mirror.

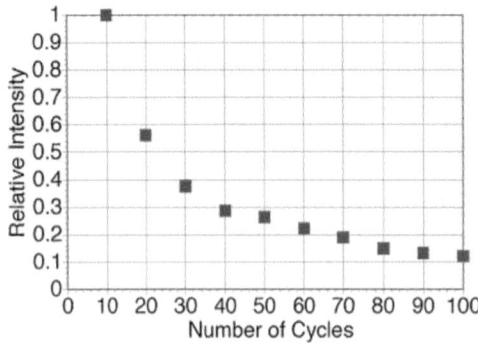

Figure 19.14 The variation of intensity (the sum of the N_2–CO doublet) with number of cycles in the mirror mode. The ordinate was normalized so that the intensity of 10 cycles was 1 unit.

The variation of intensity (the sum of the N_2–CO doublet) with the number of cycles is shown in Fig. 19.14. The ordinate was normalized so that the intensity of 10 cycles was 1 unit because the first 10 cycles represented the escape of ions having large initial deviations, caused by hitting against the electrodes. After which the ion transmission is $> 99\%$ per cycle. This shows that the 'perfect focusing' condition was achieved, similar to that in the 'MULTUM Linear plus.'

19.5 Conclusion

We have constructed a new multi-turn TOF mass spectrometers, the 'MULTUM Linear plus' and the 'MULTUM II.' According to ion optical theory, this TOF mass spectrometer satisfies perfect space and time focusing. We have experimentally demonstrated that the mass resolution can increase according to the number of cycles and that the ion transmission through the system is $> 99\%$ per cycle. Mass resolution of 350,000 was achieved after 500 cycles.

Acknowledgments

The authors express their sincere appreciation and gratitude to the late Dr Takekiyo Matsuo of Osaka University for his encouragement and suggestions. They also express their thanks to the late Dr Toru Sakurai of the Japan Advanced Institute of Science and Technology for using calculation code 'unify' and for the helpful discussion on the ion optics. They are greatly indebted to Mr Shin-ichi Yamaguchi of Shimadzu Corporation for the development of the ion sources and for helpful discussions. They thank to Mr Toshio Ichihara of Osaka University for his technical supports. They express their thanks to Dr Helmut Rosenbauer, Dr Reinhard Roll of Max Planck Institute for Aeronomy, and Dr Arthur Ghielmetti of the Lockheed Martin Advanced Research Center for the construction of the 'MULTUM Linear plus' as a laboratory model of ROSETTA mission. They also thank Professor Masahiro Kimura of Kochi University of Technology, Professor Ryuichi Arakawa of Kansai University, and Dr Yoshinao Wada of Osaka Medical Center and Research Institute for Maternal and Child Health for their helpful discussion. This work was supported by a Grant-in-Aid for Scientific Research (B) (09559012, 10044085, 11559012, 13559005) from the Ministry of Education, Science, Sports and Culture. The Institute of Space and Astronautical Science and the Yamada Science Foundation are thanked for financial support.

References

1. F. Hillenkamp, F. Karas, R.C. Beavis, and R. Nitsche, *Anal. Chem.* **63** (1991) 1193.

2. W.C. Wiley and I.H. McLaren, *Rev. Sci. Instrum.* **26** (1955) 1150.

3. J.H.J. Dawson and M. Guilhaus, *Rapid Commun. Mass Spectrom.* **3** (1989) 155.

4. B.A. Mamyrin, V.I. Karataev, D.V. Shmikk, and V.A. Zagulin, *Sov. Phys. JETP* **37** (1973) 45.

5. W.P. Poschenrieder, *Int. J. Mass Spectrom. Ion Phys.* **9** (1972) 357.

6. T. Sakurai, Y. Fujita, T. Matsuo, H. Matsuda, I. Katakuse, and K. Miseki, *Int. J. Mass Spectrom. Ion Proc.* **66** (1985) 283.

7. T. Matsuo, M. Toyoda, T. Sakurai, and M. Ishihara, *J. Mass Spectrom.* **32** (1997) 1179.

8. T. Sakurai, H. Nakabishi, T. Hiasa, and K. Okanishi, *Nucl. Instrum. Meth. A* **427** (1999) 182.

9. H. Wollnik and M. Przewloka, *Int. J. Mass Spectrom. Ion Proc.* **96** (1990) 267.

10. A. Casares, A. Kholomeev, and H. Wollnik, *Int. J. Mass Spectrom.* **206** (2001) 267.

11. M. Ishihara, M. Toyoda, and T. Matsuo, *Int. J. Mass Spectrom.* **197** (2000) 179.

12. T. Sakurai, T. Matsuo, and H. Matsuda, *Int. J. Mass Spectrom. Ion Proc.* **63** (1985) 273.

13. T. Matsuo, M. Ishihara, M. Toyoda, H. Ito, S. Yamaguchi, R. Roll, and H. Rosenbauer, *Adv. Space Res.* **23** (1999) 341–348.

14. M. Toyoda, M. Ishihara, S. Yamaguchi, H. Ito, T. Matsuo, R. Roll, and H. Rosenbauer, *J. Mass Spectrom.* **35** (2000) 163.

15. D. Okumura, M. Toyoda, M. Ishihara, and I. Katakuse, *J. Mass Spectrom. Soc. Jpn.* **51** (2003) 349.

16. M. Toyoda and T. Matsuo, *Nucl. Instrum. Meth. A* **427** (1999) 375.

17. H. Wollnik, T. Matsuo, and H. Matsuda, *Nucl. Instrum. Meth.* **102** (1972) 13.

18. H. Matsuda and Y. Fujita, *Int. J. Mass Spec. Ion Phys.* **16** (1975) 395.

Author Index

Abbreviations

2D-PAGE	2 dimensional polyacrylamide gel electrophoresis
AE	appearance energy
AMS	accelerator mass spectrometry
APCI	atmospheric pressure chemical ionization
APPI	atmospheric pressure photoionization
BIRD	blackbody infrared radiative dissociation
BSA	bovine serum albumin
CA	collisional activation
CAD	collisionally activated dissociation
CD	circular dichroism
CE	capillary electrophoresis
CI	chemical ionization
CID	collision induced dissociation
CNL	constant neutral loss
Da	Dalton
DNA	deoxyribonucleic acid
DNA ds	deoxyribonucleic acid double strand
DNA ss	deoxyribonucleic acid single strand
DTT	dithiothreitol
EDTA	ethylenediaminetetraacetic acid
EI	electron ionization
ESI	electrospray ionization
FAB	fast-atom bombardment
FDA	US Food & Drug Administration
FRET	fluorescence resonance energy transfer
FTICR	Fourier transform ion cyclotron resonance
FTMS	Fourier transform mass spectrometry
GC/MS	gas chromatography – mass spectrometry
GSH	glutathione
HPLC	high pressure/performance liquid chromatography
ICP	inductively coupled plasma
ICP-MS	inductively coupled plasma mass spectrometry

IE	ionization energy
IEF	isoelectrofocusing
IEMS	immobilised enzyme mass spectrometry
IMS	ion mobility spectrometry
LA-ICP-MS	laser ablation combined with inductively coupled plasma mass spectrometry
LC/MS	liquid chromatography-mass spectrometry
LC-MS/MS	liquid chromatography linked to tandem mass spectrometry
m/z	the mass-to-charge ratio of an ion
MALDI	matrix-assisted laser desorption/ionisation
MS	mass spectrometry
MS/MS	mass spectrometry/mass spectrometry, MS in tandem with multiple mass analysers
MS^n	MS/MS where n stages of analysis occur, e.g., MS^2, MS^3
MudPIT	multi-dimensional protein identification technology
NMR	nuclear magnetic resonance spectroscopy
NodST	sulfotransferase
PA	proton affinity
PAP	$3'$-phosphoadenosine $5'$-phosphate
PAPS	$3'$-phosphoadenosine $5'$-phosphosulfate
PTM	post-translational modification
Q-TOF	quadrupole mass analyzer linked to a time-of-flight mass analyzer
RNA	ribonucleic acid
SELDI	surface enhanced laser desorption ionization
SIM	single ion monitoring
SNP	single nucleotide polymorphism
SORI	sustained off-resonance irradiation
SPR	surface plasmon resonance
TLC	thin layer chromatography
TOF	time-of-flight
TOF-MS	time-of-flight mass spectrometry
TRIO-DRAW	third-order ion optics draw computer program

Subject Index

358